Classical Finite Transformation Semigroups

Algebra and Applications

Volume 9

Managing Editor:

Alain Verschoren
University of Antwerp, Belgium

Series Editors:

Alice Fialowski
Eötvös Loránd University, Hungary

Eric Friedlander
Northwestern University, USA

John Greenlees
Sheffield University, UK

Gerhard Hiss
Aachen University, Germany

Ieke Moerdijk
Utrecht University, The Netherlands

Idun Reiten
Norwegian University of Science and Technology, Norway

Christoph Schweigert
Hamburg University, Germany

Mina Teicher
Bar-llan University, Israel

Algebra and Applications aims to publish well written and carefully refereed monographs with up-to-date information about progress in all fields of algebra, its classical impact on commutative and noncommutative algebraic and differential geometry, K-theory and algebraic topology, as well as applications in related domains, such as number theory, homotopy and (co)homology theory, physics and discrete mathematics.

Particular emphasis will be put on state-of-the-art topics such as rings of differential operators, Lie algebras and super-algebras, group rings and algebras, C*algebras, Kac-Moody theory, arithmetic algebraic geometry, Hopf algebras and quantum groups, as well as their applications. In addition, Algebra and Applications will also publish monographs dedicated to computational aspects of these topics as well as algebraic and geometric methods in computer science.

Olexandr Ganyushkin · Volodymyr Mazorchuk

Classical Finite Transformation Semigroups

An Introduction

 Springer

Olexandr Ganyushkin
Kyiv Taras Shevchenko University
Volodymyrska Street, 64
Kyiv
Ukraine
ganiyshk@univ.kiev.ua

Volodymyr Mazorchuk
Uppsala University
Dept. Mathematics
SE-751 06 Uppsala
Sweden
mazor@math.uu.se

ISBN: 978-1-84996-768-6   e-ISBN: 978-1-84800-281-4
DOI: 10.1007/978-1-84800-281-4

British Library Cataloguing in Publication Data
A catalogue record for this book is available from the British Library

Printed on acid-free paper

Springer Science+Business Media
springer.com

Preface

Semigroup theory is a relatively young part of mathematics. As a separate direction of algebra with its own objects, formulations of problems, and methods of investigations, semigroup theory was formed about 60 years ago.

One of the main motivations for the existence of some mathematical theories are interesting and natural examples. For semigroup theory the obvious candidates for such examples are transformation semigroups. Various transformations of different sets appear everywhere in mathematics all the time. As the usual composition of transformations is associative, each set of transformations, closed with respect to the composition, forms a semigroup.

Among all transformation semigroups one can distinguish three classical series of semigroups: the *full symmetric semigroup* $T(M)$ of all transformations of the set M; the *symmetric inverse semigroup* $IS(M)$ of all partial (that is, not necessarily everywhere defined) injective transformations of M; and, finally, the semigroup $PT(M)$ of all partial transformations of M. If $M = \{1, 2, \ldots, n\}$, then the above semigroups are usually denoted by T_n, IS_n and PT_n, respectively. One of the main evidences for the importance of these semigroups is their universality property: every (finite) semigroup is a subsemigroup of some $T(M)$ (resp. T_n); and every (finite) inverse semigroup is a subsemigroup of some $IS(M)$ (resp. IS_n). Inverse semigroups form a class of semigroups which are closest (in some sense) to groups.

An analogous universal object in group theory is the *symmetric group* $S(M)$ of all bijective transformations of M. Many books are dedicated to the study of $S(M)$ or to the study of transformation groups in general. Transformation semigroups had much less luck. The "naive" search in MathSciNet for books with the keywords "transformation semigroups" in the title results in two titles, one being a conference proceedings, and another one being old 50-page-long lecture notes in Russian ([Sc4]). Just for comparison, an analogous search for "transformation groups" results in 75 titles. And this is in spite of the fact that the semigroup $T(M)$ was studied by Suschkewitsch already in the 1930s. The semigroup $IS(M)$ was introduced by Wagner in 1952, but the first relatively small monograph about it appeared only in 1996 ([Li]). The latter monograph considers some basic questions about $IS(M)$: how one writes down the elements of $IS(M)$, when two elements of $IS(M)$ commute, what is the presentation of $IS(M)$, what are the congruences on $\mathcal{IS}(M)$. For example, such basic semigroup-theoretical notions as ideals and Green's relations are mentioned only in the Appendix without any direct relation to $\mathcal{IS}(M)$.

Much more information about the semigroup $\mathcal{T}(M)$ can be found in the last chapter of [Hi1]; however, it is mostly concentrated around the combinatorial aspects. Otherwise one is left with the options to search through examples in the parts of the abstract theory of semigroups using the classical books [CP1, Gri, Ho3, Ho7, Hi1, Law, Pe] or to look at original research papers.

The aim of the present book is to partially fill the gaps in the literature. In the book we introduce three classical series of semigroups, and for them we describe generating systems, ideals, Green's relations, various classes of subsemigroups, congruences, conjugations, endomorphisms, presentations, actions on sets, linear representations and cross-sections. Some of the results are very old and classical, some are quite young. In order not to overload the reader with too technical and specialized results, we decided to restrict the area of the present book to the above-mentioned parts of the theory of transformation semigroups.

The book was thought to be an elementary introduction to the theory of transformation semigroups, with a strong emphasis on the concrete examples in the form of three classical series of finite transformation semigroups, namely, $\mathcal{T}_n$, $\mathcal{IS}_n$ and $\mathcal{PT}_n$. The book is primarily directed to students, who would like to make their first steps in semigroup theory. The choice of the semigroups $\mathcal{T}_n$, $\mathcal{IS}_n$ and $\mathcal{PT}_n$ is motivated not only by their role in semigroup theory, but also by our strong belief that a good understanding of a couple of interesting and pithy examples is more important for the first acquaintance with some theory than a formal learning of dozens of theorems.

Another good motivation to consider the semigroups $\mathcal{T}_n$, $\mathcal{IS}_n$ and $\mathcal{PT}_n$ at the same time is the observation that many results about these semigroups, which for each of them were obtained independently by different people and in different times, in reality can be obtained in a unified (or almost unified) way.

Several results which will be presented extend in one or another way to the cases of infinite transformation semigroups. However, we restrict ourselves to the case of finite semigroups to make the exposition as elementary and accessible for a wide audience as possible. We are not after the biggest possible generality. Another argument is that we also try to attract the reader's attention to numerous combinatorial aspects and applications of the semigroups we consider.

With our three principal examples of semigroups on the background we also would like to introduce the reader to the basics of the abstract theory of semigroups. So, along the discussion of these examples, we tried to present many important basic notions and prove (or at least mention) as many classical abstract results as possible.

The requirements for the reader's mathematical background are very low. To understand most of the content, it is enough to have a minimal mathematical experience on the level of common sense. Perhaps some familiarity with basic university courses in algebra and combinatorics would be a substantial help. We have tried to define all the notions we use in the book. We have also tried to make all proofs very detailed and to avoid complicated constructions whenever possible.

The penultimate section of each chapter is called "Addenda and Comments." A part of it consists of historical comments (which are by no means complete). Another part consists of some remarks, facts, and statements, which we did not include in the main text of the book. The reason is usually the much less elementary level of these statements or the more complicated character of the proofs. However, we include them in the Addenda as from our point of view they deserve attention in spite of the fact that they do not really fit into the main text. Some statements here are also given with proofs, but these proofs are less detailed than those in the main text. For this part of the book, our requirements for the reader's mathematical background are different and are closer to the standard mathematical university curriculum. In the Addenda, we sometimes also mention some open problems and try to describe possible directions for further investigations.

The division of the book into the main text and the Addenda is not very strict as sometimes the notions and facts mentioned in the Addenda are used in the main text.

The last section of each chapter contains problems. Some problems (not many) are also included in the main text. The latter ones are mostly simple and directed to the reader. Sometimes they also ask to repeat a proof given before for a different situation. These problems are in some sense compulsory for the successful understanding of the main text (i.e., one should at least read them). The additional problems of the last section of each chapter can be quite different. Some of them are easy exercises, while others are much more complicated problems, which form an essential supplement to the material of the chapter. Hints for solutions of the latter ones can be found at the end of the book.

The book was essentially written during the visit of the first author to Uppsala University, which was supported by The Royal Swedish Academy of Sciences and The Swedish Foundation for International Cooperation in Research and Higher Education (STINT). The financial support of The Academy and STINT, and the hospitality of Uppsala University are gratefully acknowledged. We thank Ganna Kudryavtseva, Victor Maltcev, and Abdullahi Umar for their comments on the preliminary version of the book.

Kyiv, Ukraine

Uppsala, Sweden

Olexandr Ganyushkin

Volodymyr Mazorchuk

Contents

Chapter 1

Ordinary and Partial Transformations

1.1 Basic Definitions

The principal objects of interest in the present volume are finite sets and transformations of finite sets. Let M be a finite set, say $M = \{m_1, m_2, \ldots, m_n\}$, where n is a nonnegative integer. *Transformation* of M is an array of the following form:

$$\alpha = \begin{pmatrix} m_1 & m_2 & \cdots & m_n \\ k_1 & k_2 & \cdots & k_n \end{pmatrix}, \tag{1.1}$$

where all $k_i \in M$. If $x \in M$, say $x = m_i$, the element k_i will be called the *value* of the transformation α at the element x and will be denoted by $\alpha(x)$. The fact that α is a transformation of M is usually written as $\alpha : M \to M$. As the nature of elements of M is not important for us, instead of M we shall usually consider the set $\mathbf{N} = \mathbf{N}_n = \{1, 2, \ldots, n\}$.

Apart from the transformations of M we shall also consider the so-called *partial transformations* of M, that is, transformations of the form $\alpha : A \to M$, where $A = \{l_1, l_2, \ldots, l_k\}$ is a subset of M. Note that the set A can be empty. Again, the element α can be written in the following *tabular form*:

$$\alpha = \begin{pmatrix} l_1 & l_2 & \cdots & l_k \\ \alpha(l_1) & \alpha(l_2) & \cdots & \alpha(l_k) \end{pmatrix}. \tag{1.2}$$

Abusing notation, we may also write $\alpha : M \to M$ for a partial transformation, having in mind that such α is only defined on some elements from M. Note that the order of elements in the first row of arrays (1.1) and (1.2) is not important.

With each (partial) transformation α as above we associate the following standard notions:

- The *domain* of α: $\mathrm{dom}(\alpha) = A$

O. Ganyushkin, V. Mazorchuk, *Classical Finite Transformation Semigroups*, Algebra and Applications 9, DOI: 10.1007/978-1-84800-281-4_1, © Springer-Verlag London Limited 2009

- The *codomain* of α: $\overline{\mathrm{dom}}(\alpha) = M \backslash A$

- The *image* of α: $\mathrm{im}(\alpha) = \{\alpha(x) : x \in A\}$

The word *range*, which is also frequently used in the literature, is a synonym of the word *image*. If $\mathrm{dom}(\alpha) = M$, the transformation α is called *full* or *total* .

The set of all total transformations of M is denoted by $\mathcal{T}(M)$, and the set of all partial transformations of M is denoted by $\mathcal{PT}(M)$. Obviously, $\mathcal{T}(M) \subset \mathcal{PT}(M)$. To simplify our notation we set $\mathcal{T}_n = \mathcal{T}(\mathbf{N})$ and $\mathcal{PT}_n = \mathcal{PT}(\mathbf{N})$.

Sometimes it is convenient to use a slightly modified version of (1.2) for some $\alpha \in \mathcal{PT}_n$. In the case of $\mathcal{PT}_n$ it is natural to form the first row of the array for α by simply listing all the elements from $\mathbf{N}$ in their natural order. Then, to define α completely, one needs a special symbol to indicate that some element x belongs to $\overline{\mathrm{dom}}(\alpha)$. We shall use the symbol $\varnothing$. In other words, $\alpha(x) = \varnothing$ means that $x \in \overline{\mathrm{dom}}(\alpha)$. Thus the element α can be written in the following form:

$$\alpha = \begin{pmatrix} 1 & 2 & \dots & n \\ k_1 & k_2 & \dots & k_n \end{pmatrix}, \tag{1.3}$$

where $k_i = \alpha(i)$ if $i \in \mathrm{dom}(\alpha)$ and $k_i = \varnothing$ if $i \in \overline{\mathrm{dom}}(\alpha)$.

Example 1.1.1 Here is the list of all elements of $\mathcal{PT}_2$:

$$\begin{pmatrix} 1 & 2 \\ 1 & 1 \end{pmatrix}, \begin{pmatrix} 1 & 2 \\ 1 & 2 \end{pmatrix}, \begin{pmatrix} 1 & 2 \\ 2 & 1 \end{pmatrix}, \begin{pmatrix} 1 & 2 \\ 2 & 2 \end{pmatrix},$$

$$\begin{pmatrix} 1 & 2 \\ 1 & \varnothing \end{pmatrix}, \begin{pmatrix} 1 & 2 \\ 2 & \varnothing \end{pmatrix}, \begin{pmatrix} 1 & 2 \\ \varnothing & 1 \end{pmatrix}, \begin{pmatrix} 1 & 2 \\ \varnothing & 2 \end{pmatrix}, \begin{pmatrix} 1 & 2 \\ \varnothing & \varnothing \end{pmatrix}.$$

The first row of this list consists of total transformations and hence lists all elements in $\mathcal{T}_2$.

Proposition 1.1.2 *The set $\mathcal{T}_n$ contains n^n elements and the set $\mathcal{PT}_n$ contains $(n+1)^n$ elements.*

Proof. Each element $\alpha \in \mathcal{T}_n$ is uniquely defined by array (1.3), where each $k_i \in \mathbf{N}$. Since the choices of k_is are independent, we have $|\mathcal{T}_n| = n^n$ by the product rule. In the case of $\mathcal{PT}_n$, the elements k_i can be independently chosen from the set $\mathbf{N} \cup \{\varnothing\}$. Hence the product rule implies $|\mathcal{PT}_n|=(n+1)^n$. $\qquad\square$

The cardinality $|\mathrm{im}(\alpha)|$ of the image of a partial transformation $\alpha \in \mathcal{PT}_n$ is called the *rank* of this partial transformation and is denoted by $\mathrm{rank}(\alpha)$. Thus $\mathrm{rank}(\alpha)$ equals the number of different elements in the second row of array (1.2). The number $\mathrm{def}(\alpha) = n - \mathrm{rank}(\alpha)$ is called the *defect* of the partial transformation α.

A partial transformation $\alpha \in \mathcal{PT}_n$ is called

- *Surjective* if $\mathrm{im}(\alpha) = \mathbf{N}$

- *Injective* if $x \neq y$ implies $\alpha(x) \neq \alpha(y)$ for all $x, y \in \mathrm{dom}(\alpha)$

- *Bijective* if α is both surjective and injective

If α is given by (1.2), then surjectivity means that the second row of array (1.2) contains all elements of $\mathbf{N}$; and injectivity means that all elements in the second row of array (1.2) are different. Bijective transformations on $\mathbf{N}$ are also called *permutations* of $\mathbf{N}$.

Proposition 1.1.3 *Let $\alpha \in \mathcal{T}_n$. Then the following conditions are equivalent:*

(a) α is surjective

(b) α is injective

(c) α is bijective

Proof. By the definition of a bijective transformation, it is enough to show that the conditions (a) and (b) are equivalent. We start by proving that injectivity implies surjectivity. Let $\alpha \in \mathcal{T}_n$ be injective and given by (1.3). Because of the injectivity of α, the second row of (1.3) gives n different elements of the set $\mathbf{N}$, namely, $\alpha(1), \alpha(2), \ldots, \alpha(n)$. But $\mathbf{N}$ contains exactly n elements. Hence $\mathbf{N} = \{\alpha(1), \alpha(2), \ldots, \alpha(n)\}$, and thus α is surjective.

Conversely, let $\alpha \in \mathcal{T}_n$ be surjective and given by (1.3). Then the second row of (1.3) contains all n elements of the set $\mathbf{N}$. But this row contains exactly n elements. Hence they all must be different. This implies that α is injective. □

1.2 Graph of a (Partial) Transformation

With each partial transformation α on $\mathbf{N}$ one naturally associates a directed graph Γ_α. A *directed graph* (or a *digraph*) is a pair, $\Gamma = (V, E)$, where V is a set and $E \subset V \times V$. The elements of V are called *vertices* of Γ and the elements of E are called *directed edges* or *arrows* of Γ. If $(a, b) \in E$, then the vertex a is called the *tail* of (a, b) and the vertex b is called the *head* of (a, b).

The graph $\Gamma_\alpha = (V_\alpha, E_\alpha)$ is called the *graph of the transformation α* and is constructed in the following way: The set V_α of vertices coincides with $\mathbf{N}$; for $x, y \in \mathbf{N}$ the element (x, y) belongs to E_α if and only if $x \in \mathrm{dom}(\alpha)$ and $\alpha(x) = y$.

Example 1.2.1 For the transformation

$$\alpha = \begin{pmatrix} 1 & 2 & 3 & 4 & 5 & 6 & 7 & 8 & 9 & 10 & 11 & 12 & 13 & 14 & 15 & 16 \\ 8 & 9 & 7 & 13 & 7 & 16 & \varnothing & 1 & 13 & 11 & 16 & 4 & 12 & 9 & 14 & 13 \end{pmatrix}$$

the graph Γ_α has the following form:

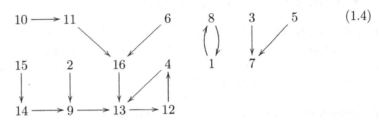

(1.4)

It is obvious that a directed graph $\Gamma = (\mathbf{N}, E)$ will be the graph of some total (partial) transformation of $\mathbf{N}$ if and only if each vertex is the tail of exactly one (at most one) arrow. The graph Γ_α decomposes into a disjoint union of connected components. Intuitively, this is an obvious notion, for example, graph (1.4) has three connected components. The rigorous definition is as follows.

First we define a subgraph. If $\Gamma = (V, E)$ is a directed graph, a *subgraph* of Γ is a directed graph $\Gamma' = (V', E')$ such that $V' \subset V$ and $E' \subset E$. A directed graph $\Gamma = (V, E)$ is called *connected* if for each partition of V into a disjoint union of nonempty subsets V_1 and V_2 there exists $a \in V_1$ and $b \in V_2$ such that either (a, b), or (b, a) is an arrow. If Γ is a directed graph, then *the connected components* of Γ are simply the *maximal connected* subgraphs of Γ, that is, those connected subgraphs of Γ which are not proper subgraphs of any other connected subgraph of Γ.

Exercise 1.2.2 Prove that two different connected components of a directed graph Γ do not have common arrows.

To understand the structure of Γ_α it is of course enough to understand the structure of its connected components. For this we shall need some more graph-theoretical notions. Let $\Gamma(V, E)$ be a directed graph and $a, b \in V$. An *oriented path* from a to b in Γ is a sequence $x_0 = a, x_1, \ldots, x_k = b$ of vertices such that $(x_i, x_{i+1}) \in E$ for each $i = 0, 1, \ldots, k - 1$. Vertex a is called the *tail* of the path and vertex b is called the *head* of the path. If we have an oriented path such that $a = b$ and $x_i \neq x_j$ for all $0 \leq i < j < k$, then such path is called an *(oriented) cycle* and is denoted by $(x_0, x_1, \ldots, x_{k-1})$. If Γ does not contain any arrow with tail b we will say that our path *breaks* at b. Analogously one defines infinite paths. Such paths may be without tails, without heads, or without both tails and heads.

Let Γ be a directed graph and v be a vertex of Γ. We define a *trajectory* of v as any longest possible path with the tail v. There are two possibilities:

either such trajectory is finite and hence breaks at some point, or this trajectory is infinite. For example, 5, 7 is a trajectory of vertex 5 in graph (1.4), which breaks at vertex 7; and 2, 9, 13, 12, 4, 13, 12, 4, 13, ... is a trajectory of point 2. In general, a point can have many different trajectories. However, we have the following obvious statement.

Lemma 1.2.3 *Let Γ be a directed graph. Then the following conditions are equivalent:*

(a) Each vertex of Γ has a unique trajectory

(b) Each vertex of Γ is the tail of at most one arrow

We will say that the infinite trajectory $x_0 = v$, $x_1, \ldots$ *terminates at* the cycle $(x_k, x_{k+1}, \ldots, x_{k+m-1})$ if the path $x_k, x_{k+1}, \ldots, x_{k+m-1}$ is a cycle, $x_i = x_{i+m}$ for all $i \geq k$ and $x_{k-1} \neq x_{k+m-1}$. Thus, the trajectory of vertex 2 in graph (1.4) terminates at the cycle $(13, 12, 4)$; and the trajectory of vertex 4 terminates at the cycle $(4, 13, 12)$.

Proposition 1.2.4 *Let $\alpha \in \mathcal{PT}_n$.*

(i) Every vertex of Γ_α has a unique trajectory.

(ii) The trajectory of each $x \in \mathbf{N}$ in Γ_α either breaks at some vertex or terminates at some cycle.

(iii) α is total if and only if the trajectory of each $x \in \mathbf{N}$ in Γ_α terminates at some cycle.

(iv) Let $x, y \in \mathbf{N}$. If y occurs in the trajectory of x in Γ_α, then the trajectory of y is a subsequence of the trajectory of x.

Proof. Since each vertex of Γ_α is a tail of at most one arrow, the statement (i) follows immediately from Lemma 1.2.3. The statement (iv) follows immediately from (i).

Assume that the trajectory $x = x_0, x_1, \ldots$ of x does not break. Since Γ_α is finite, this trajectory must contain repetitions of some vertices. Let k be minimal for which there exists a repetition of x_k and let x_{k+m} be the first repetition of x_k. Since each vertex of Γ_α is a tail of at most one arrow, the condition $x_k = x_{k+m}$ implies $x_{k+1} = x_{k+m+1}$, which, in turn, implies $x_{k+2} = x_{k+m+2}$, and so on. Hence our trajectory terminates at the cycle $(x_k, x_{k+1}, \ldots, x_{k+m-1})$. This proves (ii).

If α is total, the trajectory of each vertex cannot break. Hence (iii) follows from (ii). □

For $\alpha \in \mathcal{PT}_n$ and $x \in \mathbf{N}$ we denote by $\mathrm{tr}_\alpha(x)$ the trajectory of x in Γ_α. This is well-defined because of Proposition 1.2.4(i). Define now the binary relation ω_α on $\mathbf{N}$ in the following way: For $x, y \in \mathbf{N}$ set $x\,\omega_\alpha\,y$ if $\mathrm{tr}_\alpha(x)$ and $\mathrm{tr}_\alpha(y)$ have at least one common vertex.

Lemma 1.2.5 *The relation ω_α is an equivalence relation.*

Proof. That ω_α is reflexive and symmetric is obvious. To prove the transitivity of ω_α consider $x, y, z \in \mathbf{N}$ such that $x\,\omega_\alpha\,y$ and $y\,\omega_\alpha\,z$. Let a be a common vertex of $\mathrm{tr}_\alpha(x)$ and $\mathrm{tr}_\alpha(y)$ and b be a common vertex of $\mathrm{tr}_\alpha(y)$ and $\mathrm{tr}_\alpha(z)$. Without loss of generality we can assume that the first occurrence of a in $\mathrm{tr}_\alpha(y)$ is not later than the first occurrence of b in $\mathrm{tr}_\alpha(y)$. But this means that b occurs in $\mathrm{tr}_\alpha(a)$ by Proposition 1.2.4(iv). Another application of Proposition 1.2.4(iv) implies that b occurs in $\mathrm{tr}_\alpha(x)$. Hence $x\,\omega_\alpha\,z$, completing the proof. ☐

The equivalence classes of ω_α are called the *orbits* of α. For $x \in \mathbf{N}$ the orbit of x in Γ_α will be denoted by $\mathfrak{o}_\alpha(x)$. From the definition of ω_α it follows that for any $x \in \mathrm{dom}(\alpha)$ we have $x\,\omega_\alpha\,\alpha(x)$. Hence all vertices which occur in $\mathrm{tr}_\alpha(x)$ belong to $\mathfrak{o}_\alpha(x)$. Furthermore, for each $x \in \mathbf{N}$ we can restrict the partial transformation α to the orbit $K = \mathfrak{o}_\alpha(x)$, obtaining a new partial transformation, $\alpha^{(K)} \in \mathcal{PT}(\mathfrak{o}_\alpha(x))$. Certainly, $\alpha^{(K)}$ does not depend on the choice of the vertex in K.

Proposition 1.2.6 *For each $x \in \mathbf{N}$ the graph $\Gamma_{\alpha^{(K)}}$ is a connected component of Γ_α.*

Proof. From the definition of ω_α it follows that the graph $\Gamma_{\alpha^{(K)}}$ is connected and contains all those arrows of Γ_α, for which both the heads and the tails belong to K. Assume now that (x, y) is an arrow of Γ_α such that $x \in K$. Then $y \in \mathrm{tr}_\alpha(x)$ and hence $x\,\omega_\alpha\,y$, that is, $y \in K$. If (x, y) is an arrow of Γ_α such that $y \in K$, then again $y \in \mathrm{tr}_\alpha(x)$ and hence $x\,\omega_\alpha\,y$, that is, $x \in K$. This means that $\Gamma_{\alpha^{(K)}}$ is not properly contained in any connected subgraph of Γ_α, which proves our statement. ☐

As an immediate corollary of Proposition 1.2.6 we have:

Corollary 1.2.7 *The mapping $K \mapsto \Gamma_{\alpha^{(K)}}$ is a bijection between the orbits of α and the connected components of Γ_α.*

A directed graph $\Gamma = \Gamma(V, E)$ is called a *tree* with the *sink* $a \in V$ provided that for each $x \in V$ the trajectory of x in Γ is unique and breaks at a. For instance, in the example (1.4) if $K = \mathfrak{o}_\alpha(3)$, the connected component $\Gamma_{\alpha^{(K)}}$ is a tree with the sink 7. A (nonempty) disjoint union of several trees with sinks is called a *forest* of trees with sinks.

Exercise 1.2.8 Let Γ be a tree with the sink a. Show that Γ is connected; that each $b \neq a$ is a tail of exactly one arrow; and that Γ contains neither oriented nor unoriented cycles.

A directed graph $\Gamma = (V, E)$ is called a *cycle* provided that we can enumerate $V = \{a_1, \ldots, a_k\}$ such that $E = \{(a_1, a_2), (a_2, a_3), \ldots, (a_{k-1}, a_k),$

$(a_k, a_1)\}$. If $\Gamma_i = (V_i, E_i)$, $i \in I$, are directed graphs, then their *union* $\Gamma = \cup_{i \in I} \Gamma_i$ is defined as follows $\Gamma = (V, E)$, where $V = \cup_{i \in I} V_i$ and $E = \cup_{i \in I} E_i$. For example, each directed graph is a union of its connected components.

Theorem 1.2.9 *Each connected component of Γ_α, $\alpha \in \mathcal{PT}_n$, is either*

  (i) a tree with a sink, or

  (ii) a union of a forest of trees with sinks with a cycle on the set of all their sinks.

We note that the forest in Theorem 1.2.9(ii) may contain only one tree with a sink. The union of this tree with a sink with the cycle on its sink is not a tree with a sink anymore. We also note that the forest in Theorem 1.2.9(ii) may also have some *trivial* trees with sinks, that is, trees consisting only of sinks. If all trees in this forest are trivial, Theorem 1.2.9(ii) simply describes a cycle.

Proof. Let $\alpha \in \mathcal{PT}_n$ and $\Gamma_{\alpha(K)} = (K, E_K)$ be a connected component of Γ_α, and $x \in K$. Then Lemma 1.2.3 and Proposition 1.2.4(i) imply that x has a unique trajectory in both Γ_α and $\Gamma_{\alpha(K)}$. Moreover, since K is a connected component of Γ_α we also have that the trajectories of x in Γ_α and $\Gamma_{\alpha(K)}$ coincide (and hence they both are equal to $\mathrm{tr}_\alpha(x)$).

Assume first that $\mathrm{tr}_\alpha(x)$ breaks at some vertex, say a. Let $y \in K$ be arbitrary. Then, by definition, $\mathrm{tr}_\alpha(x)$ and $\mathrm{tr}_\alpha(y)$ have a common vertex, say z. Hence $\mathrm{tr}_\alpha(z)$ is a subsequence of both $\mathrm{tr}_\alpha(x)$ and $\mathrm{tr}_\alpha(y)$. But $\mathrm{tr}_\alpha(x)$ breaks at a, which means that $\mathrm{tr}_\alpha(z)$ must break at a as well. This implies that $\mathrm{tr}_\alpha(y)$ breaks at a. This means, by definition, that $\Gamma_{\alpha(K)}$ is a tree with the sink a, that is, of the type Theorem 1.2.9(i).

Now we assume that $\mathrm{tr}_\alpha(x)$ terminates at some cycle, say $(a_1, a_2, \ldots, a_k)$. For each a_i, $i = 1, \ldots, k$, the trajectory of a_i is unique by Proposition 1.2.4(i) and hence is $a_i, a_{i+1}, \ldots, a_k, a_1, \ldots$. For every $y \in K$, the trajectory $\mathrm{tr}_\alpha(y)$ has a common subsequence with $\mathrm{tr}_\alpha(x)$ and thus must contain some a_i. For $i = 1, \ldots, k$ we denote by K_i the set of all those vertices y from K such that the first vertex from the cycle $(a_1, a_2, \ldots, a_k)$ in $\mathrm{tr}_\alpha(y)$ is a_i. Note that $K_i \cap K_j = \varnothing$ for $i \neq j$ by definition.

Assume that $i \neq j$ and let (x, y) be an arrow in $\Gamma_{\alpha(K)}$ such that $x \in K_i$ and $y \in K_j$. Then $\mathrm{tr}_\alpha(x)$ has the form $x, y, y_1, \ldots$, where $y, y_1, \ldots$ is just $\mathrm{tr}_\alpha(y)$. However, the first element from the cycle $(a_1, a_2, \ldots, a_k)$ in $\mathrm{tr}_\alpha(x)$ is a_i, whereas in $\mathrm{tr}_\alpha(y)$ it is $a_j \neq a_i$. This is possible only in the case of $x = a_i$ and $y = a_j$. This means that every arrow in $\Gamma_{\alpha(K)}$ from some element in K_i to some element in K_j in fact belongs to the cycle $(a_1, a_2, \ldots, a_k)$.

For each $i = 1, \ldots, k$, consider the graph $\Gamma_i = (K_i, E_i)$, where

$$E_i = ((K_i \times K_i) \cap E) \backslash \{(a_i, a_i)\},$$

(this means that E_i consists of all arrows from E, which has both tails and heads in K_i, with the exception of the arrow (a_i, a_i)). Let Γ_0 be the cycle $(a_1, a_2, \ldots, a_k)$. From the previous paragraph, we have that $\Gamma_{\alpha(K)} = \cup_{i=0}^{k} \Gamma_i$. Furthermore, K_is are disjoint for $i = 1, \ldots, k$. To show that $\Gamma_{\alpha(K)}$ is of the form described in Theorem 1.2.9(ii) it remains to show that each Γ_i, $i = 1, \ldots, k$, is a tree with the sink a_i.

By definition, Γ_i does not contain any arrow with the tail a_i. Let $y \in K_i$. Then $\mathrm{tr}_\alpha(y)$ has the form $y = y_0, y_1, \ldots, y_m = a_i, a_{i+1}, \ldots$, where $y_m = a_i$ is the first occurrence of a_i in $\mathrm{tr}_\alpha(y)$. By the definition of K_i, all vertices $y_1, \ldots, y_{m-1}$ do not belong to Γ_0. Hence the definition of E_i implies that $y = y_0, y_1, \ldots, y_m = a_i$ is the trajectory of y in Γ_i, and it breaks at a_i. In other words, the trajectory of each vertex in Γ_i breaks at a_i and thus Γ_i is a tree with the sink a_i. This completes the proof.   □

Example 1.2.10 The graph Γ_α from Example 1.2.1 is given by (1.4) and has three connected components. The third component is a tree with the sink 7. The second component is a cycle (that is, the union of the cycle $(1, 8)$ with the corresponding forest of trivial trees with sinks). The first component is the union of the cycle $(4, 13, 12)$ with the following forest of trees with sinks:

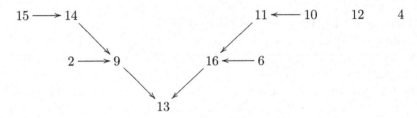

An immediate corollary of Theorem 1.2.9 is the following:

Corollary 1.2.11 *Different cycles of Γ_α belong to different connected components of Γ_α.*

1.3   Linear Notation for Partial Transformations

The graphical presentation of a transformation $\alpha \in \mathcal{PT}_n$ via Γ_α is very transparent, but also rather space consuming. For a plain mathematical text, it would be very useful to have some space-saving alternative. For permutations this is known as the *cyclic* notation and can be easily described by the following example:

Example 1.3.1 For the permutation

$$\alpha = \begin{pmatrix} 1 & 2 & 3 & 4 & 5 & 6 & 7 & 8 & 9 & 10 & 11 & 12 & 13 & 14 & 15 \\ 9 & 8 & 15 & 2 & 10 & 1 & 14 & 4 & 7 & 5 & 6 & 11 & 13 & 3 & 12 \end{pmatrix}$$

the graph Γ_α has the following form:

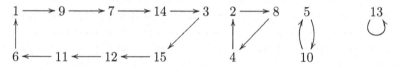

Using the notation for cycles, introduced on page 4 in the paragraph after Exercise 1.2.2, we may write

$$\alpha = (1, 9, 7, 14, 3, 15, 12, 11, 6)(2, 8, 4)(5, 10)(13).$$

Clearly the above notation is not uniquely defined. Writing a cycle we can start from each of its vertices. Moreover, the order of cycles in the cyclic notation can also be chosen in an arbitrary way. In this subsection, we would like to generalize this notation to be able to use it for all elements of $\mathcal{PT}_n$. A very good hint how to do this is given by Theorem 1.2.9, which roughly says that we only have to find a nice notation for trees with sinks. We call our notation *linear* and will define it recursively.

Assume for the moment that the graph Γ_α, where $\alpha \in \mathcal{PT}_n$, is a tree with the sink a. If a is the only vertex of Γ, we shall write $\Gamma_\alpha = [a]$ (or simply $\alpha = [a]$). If Γ_α contains some other vertices, then it has to have the following form:

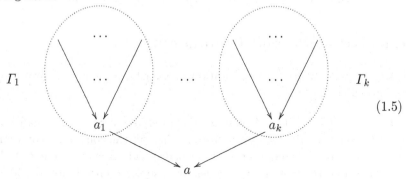

(1.5)

For $i = 1, \ldots, k$, the subgraph Γ_i of the graph (1.5) above is a tree with the sink a_i and has strictly less vertices than Γ_α. Assume that we already have the linear notation $\tilde{\Gamma}_i$ for Γ_i, $i = 1, \ldots, k$. Then the *linear notation* for Γ_α (and α) is defined recursively as follows

$$\Gamma_\alpha = [\tilde{\Gamma}_1, \tilde{\Gamma}_2, \ldots, \tilde{\Gamma}_k; a].$$

This defines the notation for the elements given by Theorem 1.2.9(i).

Assume now that Γ_α is connected and given by Theorem 1.2.9(ii). Then Γ_α is the union of some cycle, say $(a_1, \ldots, a_k)$, with certain disjoint trees Γ_i with sinks a_i, $i = 1, \ldots, k$. Let $\tilde{\Gamma}_i$, $i = 1, \ldots, k$, be the corresponding linear notation. In this case, we define the linear notation for Γ_α (and α) as follows

$$\Gamma_\alpha = (\tilde{\Gamma}_1, \tilde{\Gamma}_2, \ldots, \tilde{\Gamma}_k).$$

Finally, for any $\alpha \in \mathcal{PT}_n$, we define the linear notation for Γ_α (and α) to be the product of linear notation over all connected components of Γ_α, written in an arbitrary order. In the same way as the classical cycle notation for permutations, the linear notation for (partial) transformartions is unique only up to permutation of certain components of the notation. Namely, the connected components can be written in an arbitrary order, and on each step of the recursive procedure the order of the components $\tilde{\Gamma}_1, \ldots, \tilde{\Gamma}_k$ can also be chosen arbitrarily.

Note that, by the above definition, the ordinary cycle $(a_1, a_2, \ldots, a_k)$ is denoted by $([a_1], [a_2], \ldots, [a_k])$. This is of course not very practical, so to avoid this unnecessary complication inside the notation for cycles (but not for trees with sinks!) we shall usually skip the brackets "[ ]" surrounding *trivial* trees with sinks. Sometimes, if n is fixed, one can also skip all loops (i.e., the elements of the form $(x) = ([x])$). This just means that all $x \in \mathbf{N}$, which do not appear in the notation, correspond to loops. It is clear that this does not give rise to any confusion, moreover, it restores the original notation for usual cycles and permutations.

Example 1.3.2 For the transformation α from Example 1.2.1 we have

$$\alpha = ([[[[15]; 14], [2]; 9], [[[10]; 11], [6]; 16]; 13], 12, 4)(1, 8)[[3], [5]; 7].$$

1.4   Addenda and Comments

1.4.1 To use graphs for presentation of transformations was proposed by Suschkewitsch in [Su1].

1.4.2 Let $\alpha \in \mathcal{PT}_n$. The element $x \in \mathbf{N}$ satisfying $\alpha(x) = x$ is usually called a *fixed point* of α. If α is a permutation, then all the connected components of Γ_α are cycles. Fixed points of α correspond to cycles of length 1. In the cyclic notation for α such cycles are usually omitted (for the identity element one thus has to use a special notation, for example ε). If after such simplification the cyclic notation for α contains only one cycle, say of length k, the α is usually called a *cycle of length* k. Cycles of length 2 are called *transpositions*.

1.4.3 An alternative "linear" notation for total transformations was proposed in [AAH]. Although [AAH] works only with total transformations it is fairly straightforward to generalize their notation to cover all partial transformations. Roughly speaking the [AAH]-notation reduces to listing the trajectories of all vertices of the graph Γ_α. If we know the trajectory $x_0 = x, x_1, x_2, \ldots$ of the vertex x, then of course we know the trajectories of each of the vertices $x_1, x_2, \ldots$, so we can omit the latter ones. A trajectory is denoted by $[a_1, a_2, \ldots, a_k | a_j]$, where $1 \le j \le k$. This means the following:

- If $j < k$, the trajectory $a_1, a_2, \ldots$ of the vertex a_1 terminates at the cycle $(a_j, a_{j+1}, \ldots, a_k)$

- If $j = k$ and a_k does not occur previously, then the trajectory $a_1, a_2, \ldots$ of the vertex a_1 terminates at the cycle (a_k)

- If $j = k$ and a_k does occur previously, it means that we already know the trajectory of a_k, in this case the trajectory of the vertex a_1 is obtained by attaching $a_1, a_2, \ldots, a_k$ to the known trajectory of a_k

On each step we choose any vertex, say a, whose trajectory is not yet written down, and we write down the trajectory of a until we either reach a vertex, whose trajectory is already written down, or we terminate the trajectory of a in some cycle. To make the notation as short as possible, on each step one should try to choose a new vertex, which is not a head of any arrow in Γ_α.

For example, for the transformation

$$\alpha = \begin{pmatrix} 1 & 2 & 3 & 4 & 5 & 6 & 7 & 8 & 9 & 10 & 11 & 12 & 13 & 14 & 15 & 16 \\ 8 & 9 & 7 & 13 & 7 & 16 & 7 & 1 & 13 & 11 & 16 & 4 & 12 & 9 & 14 & 13 \end{pmatrix}$$

the [AAH]-notation for α will have the following form:

$$[15, 14, 9, 13, 12, 4|13][10, 11, 16, 13|13][6, 16|16][2, 9|9][8, 1|8][3, 7|7][5, 7|7].$$

For comparison, our notation for α looks as follows:

$$([[[[15]; 14], [2]; 9], [[[10]; 11], [6]; 16]; 13], 12, 4)(1, 8)([[3], [5]; 7]).$$

From our point of view the [AAH]-notation has some disadvantages, namely,

- The cyclic notation for permutations is not a partial case of the [AAH]-notation

- The [AAH]-notation is long, that is, it always contains elements occurring more than one time

- The [AAH]-notation is by far not unique, even up to permutations of certain components of this notation

Another disadvantage of the [AAH]-notation, related to the composition of transformations, will be discussed in 2.9.3. An advantage of the [AAH]-notation in comparison to our notation is that it contains less brackets.

1.4.4 In [Ka] it was shown that the number of those $\alpha \in \mathcal{T}_n$ for which Γ_α is connected equals

$$n! \sum_{k=0}^{n-1} \frac{n^{k-1}}{k!}$$

This can be proved, for example, in the following way:

From Theorem 1.2.9 it follows that Γ_α is connected if and only if it is a union of a forest of trees with sinks with a cycle on the set of all sinks. For each such α and for each $i \geq 0$ we define the set $N_\alpha^{(i)}$ as the set of all $x \in \mathbf{N}$ such that the first element from the cycle occurs in $\mathrm{tr}_\alpha(x)$ on step i. Obviously, $N_\alpha^{(0)}$ consists of the elements of our cycle and $\mathbf{N} = \cup_{i \geq 0} N_\alpha^{(i)}$ is a disjoint union. Furthermore, $\alpha(N_\alpha^{(i)}) \subset N_\alpha^{(i-1)}$ for $i > 0$. The number of those α, for which $|N_\alpha^{(i)}| = n_i$, $0 \leq i \leq t$, and such that $\sum_{i=0}^t n_i = n$ equals

$$\binom{n}{n_0, n_1, \ldots, n_t}(n_0 - 1)! n_0^{n_1} n_1^{n_2} \ldots n_{t-1}^{n_t} \tag{1.6}$$

(here the polynomial coefficient $\binom{n}{n_0, n_1, \ldots, n_t}$ gives the number of ordered partitions of $\mathbf{N}$ into blocks with cardinalities $n_0, \ldots, n_t$, respectively; $(n_0 - 1)!$ is the number of ways to form a cycle out of n_0 elements; and $n_i^{n_{i+1}}$ is the number of maps from the block with n_{i+1} elements to the block with n_i elements). We can rewrite (1.6) as follows

$$\frac{n!}{n_0} \cdot \frac{n_0^{n_1}}{n_1!} \cdot \frac{n_1^{n_2}}{n_2!} \ldots \frac{n_{t-1}^{n_t}}{n_t!}. \tag{1.7}$$

Now to find the number X_{n_0} of those $\alpha \in \mathcal{T}_n$ for which Γ_α is connected and contains a cycle of length n_0 one has to add up all summands of the form (1.7) for all $t \leq n - n_0$ and all decompositions $n_1 + \cdots + n_t = n - n_0$. Let $n - n_0 = k$. Then

$$\frac{n^{k-1}}{k!} = \frac{(n_0 + k)^{k-1}}{k!} = \sum_{i=0}^{k-1} \frac{1}{k!} \cdot \frac{(k-1)!}{i!(k-1-i)!} \cdot n_0^i \cdot k^{k-1-i} =$$

$$= \sum_{n_1=1}^{k} \frac{n_0^{n_1-1}}{(n_1-1)!} \cdot \frac{k^{k-n_1}}{(k-n_1)!},$$

where for the last equality we substituted i by $n_1 - 1$. Continuing in the same way we get

$$\frac{n^{k-1}}{k!} =$$

$$= \sum_{n_1=1}^{k} \frac{n_0^{n_1-1}}{(n_1-1)!} \cdot \sum_{n_2=1}^{k-n_1} \frac{n_1^{n_2-1}}{(n_2-1)!} \cdots \sum_{n_t=1}^{k-n_1-\cdots-n_{t-1}} \frac{n_{t-1}^{n_t-1}}{(n_t-1)!} \cdot n_t^{-1} =$$

$$= \frac{1}{n_0} \sum_{n_1=1}^{k} \frac{n_0^{n_1}}{n_1!} \sum_{n_2=1}^{k-n_1} \frac{n_1^{n_2}}{n_2!} \cdots \sum_{n_t=1}^{k-n_1-\cdots-n_{t-1}} \frac{n_{t-1}^{n_t}}{n_t!} =$$

$$= \sum_{n_1+\cdots+n_t=n-n_0} \frac{1}{n_0} \cdot \frac{n_0^{n_1}}{n_1!} \cdot \frac{n_1^{n_2}}{n_2!} \cdots \frac{n_{t-1}^{n_t}}{n_t!}. \tag{1.8}$$

Hence $X_{n_0} = n! \cdot \frac{n^{k-1}}{k!}$, where $k = n - n_0$. Since $1 \leq n_0 \leq n$, the final answer is now computed as follows

$$\sum_{n_0=1}^{n} X_{n_0} = n! \sum_{k=0}^{n-1} \frac{n^{k-1}}{k!}.$$

1.5 Additional Exercises

1.5.1 Let $\mathbb{N}$ denote the set of all positive integers. Give an example of a transformation $\alpha : \mathbb{N} \to \mathbb{N}$ such that

(a) α is injective but not surjective.

(b) α is surjective but not injective.

1.5.2 Prove that $\lim_{n \to \infty} \frac{|\mathcal{PT}_n|}{|\mathcal{T}_n|} = e$.

1.5.3 Directed graphs $\Gamma_i = (V_i, E_i)$, $i = 1, 2$, are called *isomorphic* provided that there exists a bijection $\varphi : V_1 \to V_2$ which induces a bijection from E_1 to E_2. Compute the number of pairwise nonisomorphic graphs Γ_α, where

(a) $\alpha \in \mathcal{T}_3$.

(b) $\alpha \in \mathcal{T}_4$.

(c) $\alpha \in \mathcal{PT}_2$.

(d) $\alpha \in \mathcal{PT}_3$.

(e) $\alpha \in \mathcal{PT}_4$.

1.5.4 Find the number of those partial transformations $\alpha \in \mathcal{PT}_8$, whose graphs are isomorphic to the following graph:

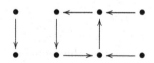

1.5.5 For $\alpha \in \mathcal{PT}_n$ characterize $\mathrm{dom}(\alpha)$, $\mathrm{im}(\alpha)$, $\overline{\mathrm{dom}}(\alpha)$, $\mathrm{rank}(\alpha)$, and $\mathrm{def}(\alpha)$ in terms of Γ_α.

1.5.6 Compute the number of those $\alpha \in \mathcal{T}_n$ (resp. $\alpha \in \mathcal{PT}_n$) for which $\mathrm{im}(\alpha)$

(a) Does not contain given elements $a_1, a_2, \ldots, a_k$.

(b) Contains given elements $a_1, a_2, \ldots, a_k$

(c) Coincides with the given set $\{a_1, a_2, \ldots, a_k\}$

1.5.7 Prove that the number of those $\alpha \in \mathcal{PT}_n$, for which Γ_α is a tree with a sink, equals n^{n-1}.

1.5.8 Prove that the number of those $\alpha \in \mathcal{PT}_n$, for which Γ_α does not contain cycles, equals $\sum_{k=1}^{n} \binom{n-1}{k-1} n^{n-k}$.

1.5.9 (a) Find the number of those $\alpha \in \mathcal{T}_n$ which fix at least one (resp. exactly one) element (that is, $\alpha(x) = x$ for at least one or exactly one element $x \in \mathbf{N}$, respectively).

(b) The same problem for $\mathcal{PT}_n$.

1.5.10 Let $\Gamma = (V, E)$ be a directed graph. Consider the set $\mathcal{X}$, which consists of all possible unordered partitions of V into disjoint unions of nonempty subsets V_is such that for each $i \neq j$ the graph Γ does not contain any arrow from V_i to V_j. The set $\mathcal{X}$ is partially ordered in the natural way with respect to inclusions of components of partitions. Prove that the partition of V, which corresponds to the partition of Γ into connected components, is the minimum of $\mathcal{X}$.

1.5.11 Let $\Gamma = (V, E)$ be a directed graph. Consider the set $\mathcal{Y}$, which consists of all possible unordered partitions of V into disjoint unions of nonempty subsets V_is such that for each i we have that the subgraph $(V_i, (V_i \times V_i) \cap E)$ is connected. The set $\mathcal{Y}$ is partially ordered in the natural way with respect to inclusions of components of partitions. Prove that the partition of V, which corresponds to the partition of Γ into connected components, is the maximum of $\mathcal{Y}$.

1.5.12 For $\alpha \in \mathcal{T}_n$, let $t_k(\alpha)$ denote the number of those $x \in \mathbf{N}$ for which $|\{y \in \mathbf{N} : \alpha(y) = x\}| = k$. Prove that

(a) $\displaystyle\sum_{k=0}^{n} t_k(\alpha) = n,$

(b) $\displaystyle\sum_{k=0}^{n} k t_k(\alpha) = n.$

1.5.13 For $\alpha \in \mathcal{PT}_n$ let $t_k(\alpha)$ denote the number of those $x \in \mathbf{N}$ for which $|\{y \in \mathbf{N} : \alpha(y) = x\}| = k$. Prove that

(a) $\displaystyle\sum_{k=0}^{n} t_k(\alpha) = n,$

(b) $\displaystyle\sum_{k=0}^{n} k t_k(\alpha) \leq n.$

Chapter 2

The Semigroups $\mathcal{T}_n$, $\mathcal{PT}_n$, and $\mathcal{IS}_n$

2.1 Composition of Transformations

Let X and Y be two sets. A *mapping* from X to Y is an array of the form

$$f = \left(\begin{array}{c} x \\ f(x) \end{array} \right)_{x \in X},$$

where all $f(x) \in Y$. This is usually denoted by $f : X \to Y$. The element $f(x)$ is called the *value* of the mapping f at the element x. A transformation, as defined in Sect. 1.1, is just a mapping from a set to itself. Let now X, Y, Y', Z be sets such that $Y \subset Y'$ and let $f : X \to Y$ and $g : Y' \to Z$ be two mappings. In this situation, we can define the *product* or the *composition* gf of f and g by the following rule: The composition gf is the mapping from X to Z such that for all $x \in X$ we have $(gf)(x) = g(f(x))$. In particular, we can always compose two total transformations of the same set and the result will be a total transformation of this set.

The above definition admits a straightforward generalization to partial mappings. A *partial mapping* from X to Y is a mapping $\alpha : X' \to Y$, where $X' \subset X$. In this case, we say that the partial mapping α is *defined* on elements from X'. Again, a partial transformation, as defined in Sect. 1.1, is a partial mapping from a set to itself. One usually abuses notation and writes $\alpha : X \to Y$ just emphasizing that α is a partial mapping. Let $\alpha : X \to Y$ and $\beta : Y \to Z$ be two partial mappings. We define their *product* or *composition* $\beta\alpha$ as the partial mapping, defined on all those $x \in X$ for which α and β are defined on the elements x and $\alpha(x)$, respectively; on such x the value of $\beta\alpha$ is given by $(\beta\alpha)(x) = \beta(\alpha(x))$. In particular, we can always compose two partial transformations of the same set and the result will be another

O. Ganyushkin, V. Mazorchuk, *Classical Finite Transformation Semigroups*, Algebra and Applications 9, DOI: 10.1007/978-1-84800-281-4_2,
© Springer-Verlag London Limited 2009

partial transformation of this set. We also note that the definition of the composition of total transformations is just a special case of that of partial transformations.

Proposition 2.1.1 *The composition of (partial) mappings is associative, that is, if α, β, and γ are partial mappings, then the composition $\gamma(\beta\alpha)$ is defined if and only if the composition $(\gamma\beta)\alpha$ is defined, and if they both are defined, we have $\gamma(\beta\alpha) = (\gamma\beta)\alpha$.*

Proof. Follows immediately from the following picture:

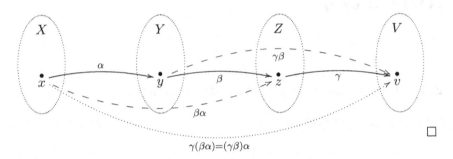

$$\gamma(\beta\alpha)=(\gamma\beta)\alpha$$

Associativity of the composition of partial transformations naturally leads to the notion of a semigroup. Let S be a nonempty set, and let $\cdot :$ $S \times S \rightarrow S$ be a *binary operation* on S. Then $(S, \cdot)$ is called a *semigroup* provided that $\cdot$ is *associative*, that is, $a \cdot (b \cdot c) = (a \cdot b) \cdot c$ for all $a, b, c \in S$. To simplify the notation, in the case when the operation $\cdot$ is clear from the context one usually writes S for $(S, \cdot)$. Furthermore, one usually writes ab instead of $a \cdot b$.

Exercise 2.1.2 Let $(S, \cdot)$ be a semigroup. Show that the value of the product $a_1 a_2 \cdots a_n$, where all $a_i \in S$, does not depend on the way of computing it (that is, of putting brackets into this product).

Let $(S, \cdot)$ be a semigroup. From Exercise 2.1.2 it follows that for every $a \in S$ we have a well-defined element $a^k = \underbrace{a \cdot a \cdots a}_{k\text{times}}$. The number of elements in S is called the *cardinality* of S and is denoted by $|S|$.

By Proposition 2.1.1, in both $\mathcal{T}_n$ and $\mathcal{PT}_n$ the composition of (partial) transformations is an associative operation. Hence we have:

Proposition 2.1.3 *Both $\mathcal{T}_n$ and $\mathcal{PT}_n$ are semigroups with respect to the composition of (partial) transformations.*

The semigroup $\mathcal{T}_n$ is called the *full transformation semigroup* on the set **N** or the *symmetric semigroup* of all transformations of **N**. The semigroup $\mathcal{PT}_n$ is called the *semigroup of all partial transformations* on **N**.

A nonempty subset T of a semigroup $(S, \cdot)$ is called a *subsemigroup* of S provided that T is closed with respect to $\cdot$ (that is, $a \cdot b \in T$ as soon as $a, b \in T$). Obviously, in this case, T itself is a semigroup with respect to the restriction of the operation $\cdot$ to T. The fact that T is a subsemigroup of S is usually denoted by $T < S$.

Exercise 2.1.4 Show that for arbitrary $\alpha, \beta \in \mathcal{PT}_n$ the following is true:

(a) $\mathrm{dom}(\beta\alpha) \subset \mathrm{dom}(\alpha)$

(b) $\mathrm{im}(\beta\alpha) \subset \mathrm{im}(\beta)$

(c) $\mathrm{rank}(\beta\alpha) \leq \min(\mathrm{rank}(\alpha), \mathrm{rank}(\beta))$

2.2 Identity Elements

An element e of a semigroup S is called a *left* or a *right identity* provided that $ea = a$, or $ae = a$, respectively, for all $a \in S$. An element e, which is a left and a right identity at the same time, is called a *two-sided identity* or simply an *identity*.

If S contains some left identity e_l and some right identity e_r we have $e_l = e_l \cdot e_r = e_r$ and hence these two elements coincide. Hence in this case S contains a unique identity element, which is, moreover, a two-sided identity. However, a semigroup may contain many different left identities or many different right identities (see Exercise 2.10.2). It is possible for a semigroup to contain neither left nor right identities. An example of such a semigroup is the semigroup $(\mathbb{N}, +)$. Another example is the semigroup $\{2, 3, 4, \dots\}$ with respect to the ordinary multiplication.

A semigroup which contains a two-sided identity element is called a *monoid*. The absence of an identity element can be easily repaired in the following way.

Proposition 2.2.1 *Each semigroup can be extended to a monoid by adding at most one element.*

Proof. Let $(S, \cdot)$ be a semigroup. If S contains an identity, we have nothing to prove. If S does not contain any identity element, consider the set $S^1 = S \cup \{1\}$, where $1 \notin S$. Define the binary operation $*$ on S^1 as follows: For $a, b \in S^1$ set

$$a * b = \begin{cases} a \cdot b, & a, b \in S; \\ a, & b = 1; \\ b, & a = 1. \end{cases}$$

A direct calculation shows that $*$ is associative, hence S^1 is a semigroup. Furthermore, from the definition of $*$ we have that 1 is the identity element in S^1. Moreover, the restriction of the operation $*$ to S coincides with the original operation $\cdot$. Hence S is a subsemigroup of S^1. $\qquad\square$

Denote by $\varepsilon_n : \mathbf{N} \to \mathbf{N}$ the *identity* transformation

$$\varepsilon_n = \begin{pmatrix} 1 & 2 & \cdots & n \\ 1 & 2 & \cdots & n \end{pmatrix}.$$

If n is clear from the context we shall sometimes write ε instead of ε_n. The following statement is obvious.

Proposition 2.2.2 *The transformation ε_n is the (two-sided) identity element in both $\mathcal{T}_n$ and $\mathcal{PT}_n$. In particular, both, $\mathcal{T}_n$ and $\mathcal{PT}_n$, are monoids.*

Let S be a monoid with the identity element 1. An element $a \in S$ is called *invertible* or a *unit* provided that there exists $b \in S$ such that $ab = ba = 1$. Such an element b, if it exists, is unique. Indeed, assume that b_1 and b_2 are different such elements, then

$$b_1 = b_1 \cdot 1 = b_1(ab_2) = (b_1 a)b_2 = 1 \cdot b_2 = b_2.$$

The element b is called the *inverse* of a and is denoted by a^{-1}. Note that if b is the inverse of a, then a is the inverse of b. In other words, $(a^{-1})^{-1} = a$. The set of all invertible elements of the monoid S is denoted by S^*. Note that $1 \in S^*$ since $1 \cdot 1 = 1$. In particular, S^* is not empty.

The above terminology and notation deserve some explanation. Usually the operation in an abstract semigroup is thought of as a multiplication. Since the element 1 is the identity element in such multiplicative semigroups as $\mathbf{N}$, $\mathbb{Z}$, $\mathbb{R}$, $\mathbb{C}$, it is natural to denote the identity element of an abstract semigroup by the same symbol 1. This also justifies the notions "unit" and "inverse." However, there are many semigroups where the operation is not the multiplication, for example the semigroup $(\mathbb{Z}, +)$. The identity element in this semigroup is the number 0 and not the number 1. And the inverse of the number $n \in \mathbb{Z}$ is the number $-n$ and not the number n^{-1} (note that the latter one is not always defined, and when it is defined, it is not an integer in general).

A monoid in which each element has an inverse is called a *group*.

Proposition 2.2.3 *Let S be a monoid with the identity element 1. Then S^* is a group.*

Proof. Obviously, if $a \in S^*$, then $a^{-1} \in S^*$ as well. If $a, b \in S^*$, then we have

$$ab \cdot b^{-1}a^{-1} = a \cdot bb^{-1} \cdot a^{-1} = a \cdot 1 \cdot a^{-1} = aa^{-1} = 1.$$

Analogously one shows that $b^{-1}a^{-1} \cdot ab = 1$ and hence $b^{-1}a^{-1} = (ab)^{-1}$. In particular, $ab \in S^*$. Thus S^* is a submonoid of S and each element of S^* has an inverse in S^*. The claim follows. □

Proposition 2.2.4 *Let $\alpha \in \mathcal{T}_n$, or $\alpha \in \mathcal{PT}_n$. Then α is invertible if and only if α is a permutation on $\mathbf{N}$.*

Proof. Assume that α is invertible and β is a (partial) transformation such that $\alpha\beta = \beta\alpha = \varepsilon$. Note that $\mathrm{dom}(\varepsilon) = \mathbf{N}$. Hence Exercise 2.1.4(a) implies $\mathrm{dom}(\alpha) = \mathbf{N}$. Further, if $x, y \in \mathbf{N}$ are such that $x \neq y$, then $\varepsilon(x) \neq \varepsilon(y)$. If $\alpha(x) = \alpha(y)$, we would get $\varepsilon(x) = \beta(\alpha(x)) = \beta(\alpha(y)) = \varepsilon(y)$, a contradiction. This means that $\alpha(x) \neq \alpha(y)$. Hence α is everywhere defined and injective and thus is a permutation by Proposition 1.1.3.

Conversely, if

$$\alpha = \begin{pmatrix} 1 & 2 & \cdots & n \\ i_1 & i_2 & \cdots & i_n \end{pmatrix}$$

is a permutation, the element

$$\alpha = \begin{pmatrix} i_1 & i_2 & \cdots & i_n \\ 1 & 2 & \cdots & n \end{pmatrix}$$

is a permutation as well and a direct computation shows that $\alpha\beta = \beta\alpha = \varepsilon$, that is, α is invertible. $\square$

The group $\mathcal{T}_n^* = \mathcal{PT}_n^*$ of all permutations on $\mathbf{N}$ is called the *symmetric group* on $\mathbf{N}$ and is denoted by $\mathcal{S}_n$.

2.3 Zero Elements

An element 0 of a semigroup S is called a *left* or a *right zero* provided that $0a = 0$, or $a0 = 0$, respectively, for all $a \in S$. An element 0 which at the same time is a left and a right zero, is called a *two-sided zero* or simply a *zero*.

If S contains some left zero 0_l and some right zero 0_r, we have $0_l = 0_l \cdot 0_r = 0_r$ and hence these two elements coincide. Hence in this case S contains a unique zero element, which is, moreover, a two-sided zero. The analog of Proposition 2.2.1 is the following statement.

Proposition 2.3.1 *Each semigroup can be extended to a semigroup with zero by adding at most one element.*

Proof. Let $(S, \cdot)$ be a semigroup. If S contains a zero, we have nothing to prove. Otherwise, consider the set $S^0 = S \cup \{0\}$, where $0 \notin S$. Define the binary operation $*$ on S^0 as follows: For $a, b \in S^0$ set

$$a * b = \begin{cases} a \cdot b, & a, b \in S; \\ 0, & b = 0; \\ 0, & a = 0. \end{cases}$$

A direct calculation shows that $*$ is associative, hence S^0 is a semigroup. Furthermore, from the definition of $*$ we have that 0 is the zero element in S^0 and that the restriction of $*$ to S coincides with $\cdot$. Hence S is a subsemigroup of S^0. $\square$

Denote by $\mathbf{0}_n$ the partial transformation $\mathbf{0}_n : \varnothing \to \mathbf{N}$ on $\mathbf{N}$. We have $\mathrm{dom}(\mathbf{0}_n) = \varnothing$. If n is clear from the context, we shall usually write simply $\mathbf{0}$ instead of $\mathbf{0}_n$.

Proposition 2.3.2 $\mathbf{0}_n$ *is the zero element of the semigroup* $\mathcal{PT}_n$.

Proof. The equalities $\alpha \mathbf{0}_n = \mathbf{0}_n \alpha = \mathbf{0}_n$, $\alpha \in \mathcal{PT}_n$, are obvious. $\square$

For $a \in \mathbf{N}$ define the total transformation $0_a : \mathbf{N} \to \mathbf{N}$ via $0_a(x) = a$ for all $x \in \mathbf{N}$. The transformation 0_a is called the *constant* transformation.

Proposition 2.3.3 *For* $n > 1$ *the semigroup* $\mathcal{T}_n$ *does not contain any right zeros.* $\alpha \in \mathcal{T}_n$ *is a left zero if and only if* $\alpha = 0_a$ *for some* $a \in \mathbf{N}$. *In particular, the semigroup* $\mathcal{T}_n$ *contains exactly n left zeros.*

Proof. From the obvious equality $0_a \beta = 0_a$ for any $\beta \in \mathcal{T}_n$ we obtain that each constant transformation on $\mathcal{T}_n$ is a left zero. Hence $\mathcal{T}_n$ contains at least n different left zeros. In particular, for $n > 1$ the semigroup $\mathcal{T}_n$ cannot contain any right zeros.

Note that constant transformations are the only transformations of rank 1. Let $\alpha \in \mathcal{T}_n$ be a transformation of rank at least 2. Then for any 0_a the rank of $\alpha 0_a$ is 1 by Exercise 2.1.4(c) and hence the equality $\alpha 0_a = \alpha$ is not possible. This implies that α is not a left zero of $\mathcal{T}_n$. $\square$

We note that the semigroup $\mathcal{T}_1$ consists of the identity element only. This element is the zero element at the same time.

2.4   Isomorphism of Semigroups

Let $(S, \cdot)$ and $(T, *)$ be two semigroups. The semigroups S and T are said to be *isomorphic* (denoted by $S \cong T$) provided that there exists a bijection $\varphi : S \to T$ such that

$$\varphi(a) * \varphi(b) = \varphi(a \cdot b) \quad \text{for all } a, b \in S. \tag{2.1}$$

The bijection φ is called an *isomorphism* from S to T. Note that if $\varphi : S \to T$ is an isomorphism, then $\varphi^{-1} : T \to S$ is an isomorphism as well.

Exercise 2.4.1 Show that the relation of being isomorphic semigroups is an equivalence relation.

Example 2.4.2 It is rather handy to present small semigroups using their *multiplication tables*, which is also called *Cayley tables*. Such table is a square matrix with $|S|$ rows and $|S|$ columns, which are indexed by the elements of S. At the intersection of the ath row and the bth column, $a, b \in S$, one writes the product ab. For example, the semigroup $T_2 = \{\varepsilon, (12), 0_1, 0_2\}$ has the following Cayley table:

$\cdot$	ε	(12)	0_1	0_2
ε	ε	(12)	0_1	0_2
$(1,2)$	$(1,2)$	ε	0_2	0_1
0_1	0_1	0_1	0_1	0_1
0_2	0_2	0_2	0_2	0_2

(2.2)

The following set of 2×2 matrices is also a semigroup with respect to the usual matrix multiplication

$$S = \left\{ \begin{pmatrix} 1 & 0 \\ 0 & 1 \end{pmatrix}, \begin{pmatrix} 0 & 1 \\ 1 & 0 \end{pmatrix}, \begin{pmatrix} 1 & 1 \\ 0 & 0 \end{pmatrix}, \begin{pmatrix} 0 & 0 \\ 1 & 1 \end{pmatrix} \right\}.$$

To simplify the notation we denote

$$E = \begin{pmatrix} 1 & 0 \\ 0 & 1 \end{pmatrix}, \ A = \begin{pmatrix} 0 & 1 \\ 1 & 0 \end{pmatrix}, \ B = \begin{pmatrix} 1 & 1 \\ 0 & 0 \end{pmatrix}, \ C = \begin{pmatrix} 0 & 0 \\ 1 & 1 \end{pmatrix}.$$

The Cayley table for S has the following form

$\cdot$	E	A	B	C
E	E	A	B	C
A	A	E	C	B
B	B	B	B	B
C	C	C	C	C

(2.3)

It is easy to check that the bijection

$$\varphi = \begin{pmatrix} \varepsilon & (12) & 0_1 & 0_2 \\ E & A & B & C \end{pmatrix}$$

transforms the Cayley table (2.2) to the Cayley table (2.3) and hence is an isomorphism from T_2 to S.

The primary importance of the semigroup T_n is revealed by the following statement.

Theorem 2.4.3 (Cayley's Theorem) *Each finite semigroup S of cardinality n is isomorphic to a subsemigroup of either T_n or T_{n+1}.*

Proof. First we assume that S contains the identity element 1. Let $S = \{a_1 = 1, a_2, \ldots, a_n\}$ and define the mapping $\varphi : S \to T_n$ as follows

$$\varphi(a) = \begin{pmatrix} 1 & 2 & \cdots & n \\ i_1 & i_2 & \cdots & i_n \end{pmatrix},$$

where for $k = 1, \ldots, n$, the number i_k is uniquely determined by the equality $aa_k = a_{i_k}$.

Since $a_k \cdot 1 = a_k$, we have $\varphi(a_i)(1) = i$ for all $i = 1, \ldots, n$. In particular, $\varphi(a_i) \neq \varphi(a_j)$ if $i \neq j$ and hence the mapping φ is injective. The mapping $\varphi : S \to \varphi(S)$ is therefore even bijective.

Further, the equality $(ab)a_k = a(ba_k)$ implies the equality $\varphi(ab)(k) = \varphi(a)(\varphi(b)(k))$ for all $k = 1, \ldots, n$ and $a, b \in S$. This means that $\varphi(ab) = \varphi(a)\varphi(b)$ for all $a, b \in S$ and thus φ is an isomorphism from S to the subsemigroup $\varphi(S)$ of $\mathcal{T}_n$.

If S does not contain any identity element, we can consider S as a subsemigroup of the semigroup S^1 and $|S^1| = n + 1$. By the above, S^1 is isomorphic to a subsemigroup of $\mathcal{T}_{n+1}$, and the claim follows. $\square$

2.5   The Semigroup $\mathcal{IS}_n$

Obviously, the composition $\alpha\beta$ of two partial injections α and β is again a partial injection. This observation leads to the definition of another very important subsemigroup of $\mathcal{PT}_n$, namely, the semigroup $\mathcal{IS}_n$ of all partial injective transformations of $\mathbf{N}$, which is usually called the *symmetric inverse semigroup*.

Theorem 2.5.1

$$|\mathcal{IS}_n| = \sum_{k=0}^{n} \binom{n}{k}^2 \cdot k!.$$

Proof. Each partial injection $\alpha : \mathbf{N} \to \mathbf{N}$ can be considered as a bijection $\alpha : \operatorname{dom}(\alpha) \to \operatorname{im}(\alpha)$. Let us count the number of such bijections of rank k. The set $A = \operatorname{dom}(\alpha)$ can be chosen in $\binom{n}{k}$ different ways. The set $B = \operatorname{im}(\alpha)$ can be independently chosen in $\binom{n}{k}$ different ways. If A and B are fixed, there are exactly $k!$ different bijections from A to B. Hence we have exactly $\binom{n}{k} \cdot \binom{n}{k} \cdot k!$ bijections of rank k. Since k can be an arbitrary integer between 0 and n, the statement of the theorem is obtained by applying the sum rule. $\square$

Partial injections $\alpha : \mathbf{N} \to \mathbf{N}$ are also called *partial bijections* or *partial permutations* on $\mathbf{N}$.

If $\alpha : \mathbf{N} \to \mathbf{N}$ is a partial injection, then each $x \in \mathbf{N}$ has at most one preimage. In particular, the graph Γ_α is much simpler than for general partial transformations. From Theorem 1.2.9 it follows that each connected component of Γ_α is either a cycle or has the form

$$a_1 \longrightarrow a_2 \longrightarrow a_3 \longrightarrow \cdots \longrightarrow a_{k-1} \longrightarrow a_k . \qquad (2.4)$$

In our linear notation such a graph is denoted by

$$[[[\ldots [[[a_1]; a_2]; a_3]; \ldots]; a_{k-1}]; a_k].$$

Since the order of the brackets in this expression is fixed, we can simplify the notation by omitting all inner brackets. So, in what follows the graph (2.4) (and the corresponding element) will be denoted by $[a_1, a_2, a_3, \ldots, a_{k-1}, a_k]$. Such element is called a *chain*. This allows us to use for partial permutations a simplified version of linear notation, which we call the *chain-cycle notation*. For example, the partial permutation

$$\alpha = \begin{pmatrix} 1 & 2 & 3 & 4 & 5 & 6 & 7 & 8 & 9 & 10 & 11 & 12 & 13 & 14 & 15 \\ 14 & 12 & 11 & 13 & 7 & \varnothing & 3 & 15 & \varnothing & \varnothing & 5 & 2 & 1 & 9 & 10 \end{pmatrix},$$

written in the chain-cycle notation has the following form:

$$\alpha = (5, 7, 3, 11)(2, 12)[4, 13, 1, 14, 9][8, 15, 10][6].$$

Obviously the chain-cycle notation for a partial permutation α is unique up to a permutation of components and cyclic permutation of elements in each cycle. Note that no permutations of elements in chains are allowed.

As $\mathcal{IS}_n$ contains the identity element ε and the zero element $\mathbf{0}$ of the bigger semigroup $\mathcal{PT}_n$, these elements will be the identity element and the zero element of $\mathcal{IS}_n$, respectively. Moreover, we have inclusions $\mathcal{S}_n \subset \mathcal{IS}_n^* \subset \mathcal{PT}_n^* = \mathcal{S}_n$, which imply $\mathcal{IS}_n^* = \mathcal{S}_n$. The following diagram characterizes the connection between the principal objects of the present book:

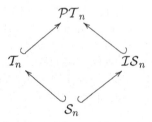

Note that all these inclusions are proper for $n > 1$.

2.6 Regular and Inverse Elements

An element a of a semigroup S is called *regular* provided that there exists $b \in S$ such that $aba = a$. Elements $a, b \in S$ form a *pair of inverse elements* provided that $aba = a$ and $bab = b$. Set

$$V_S(a) = \{b \in S : a \text{ and } b \text{ is a pair of inverse elements}\}.$$

If a and b is a pair of inverse elements, then one says that a is an *inverse* of b and b is an *inverse* of a. This might be slightly confusing, since in such generality the inverse element of a given element is not uniquely defined. Note that this notion extends the notion of an inverse element for invertible elements: if $a \in S$ is invertible, then $aa^{-1}a = a$ and $a^{-1}aa^{-1} = a^{-1}$. Hence a and a^{-1} is a pair of inverse elements.

Exercise 2.6.1 Let $a \in S$ be invertible. Show that $V_S(a) = \{a^{-1}\}$.

If a has at least one inverse (i.e., $V_S(a) \neq \varnothing$), then the element a is obviously regular. The converse is also true.

Proposition 2.6.2 *Let $a \in S$ be regular and $b \in S$ be such that $aba = a$. Then a and $c = bab$ is a pair of inverse elements.*

Proof. Follows from the following computation

$$aca = a \cdot bab \cdot a = aba \cdot ba = aba = a$$
$$cac = bab \cdot a \cdot bab = b \cdot aba \cdot bab = b \cdot aba \cdot b = bab = c.$$

$\square$

The semigroup S is called *regular* provided that every element of S is regular. Obviously, each group is a regular semigroup.

Theorem 2.6.3 *The semigroups $\mathcal{T}_n$, $\mathcal{PT}_n$, and $\mathcal{IS}_n$ are regular.*

Proof. Let $\alpha \in \mathcal{PT}_n$. Let us construct the element $\beta \in \mathcal{T}_n$ as follows: For each $x \in \text{im}(\alpha)$ take some $y \in \mathbf{N}$ such that $\alpha(y) = x$ and set $\beta(x) = y$. For $x \notin \text{im}(\alpha)$ define $\beta(x) = 1$. A direct calculation shows that $\alpha\beta\alpha = \alpha$. It follows that both $\mathcal{T}_n$ and $\mathcal{PT}_n$ are regular semigroups.

Let now $\alpha \in \mathcal{IS}_n$. Define $\beta \in \mathcal{IS}_n$ as follows: First of all we set $\text{dom}(\beta) = \text{im}(\alpha)$, and for each $x \in \text{im}(\alpha)$ we define β in the same way as before. Since α was a partial bijection, that is, a bijection from $\text{dom}(\alpha)$ to $\text{im}(\alpha)$, the element β is simply the inverse bijection from $\text{im}(\alpha)$ to $\text{dom}(\alpha)$. In particular, $\beta \in \mathcal{IS}_n$. As before, we still have the equality $\alpha\beta\alpha = \alpha$ and hence the semigroup $\mathcal{IS}_n$ is regular as well. $\square$

Let us study the structure of inverse elements in the semigroups $\mathcal{T}_n$ and $\mathcal{PT}_n$ in more detail.

Theorem 2.6.4 *Let $\alpha, \beta \in \mathcal{PT}_n$. Then the element β is inverse to the element α if and only if the following conditions are satisfied:*

(a) $\text{dom}(\beta) \supset \text{im}(\alpha)$.

(b) $\beta(a) \in \{x \in \mathbf{N} : \alpha(x) = a\}$ *for all $a \in \text{im}(\alpha)$.*

(c) $\text{im}(\beta) = \beta(\text{im}(\alpha))$.

Proof. Let $\text{im}(\alpha) = \{a_1, a_2, \ldots, a_k\}$. For $i = 1, \ldots, k$ set $B_i = \{x \in \mathbf{N} : \alpha(x) = a_i\}$. Then the equality $\alpha\beta\alpha = \alpha$ is equivalent to the following condition:

$$\beta(a_i) \in B_i \quad \text{for all} i = 1, 2, \ldots, k. \tag{2.5}$$

Assume now that the condition (2.5) is satisfied. Then for all i we have $(\alpha\beta)(a_i) = a_i$ and $(\beta\alpha\beta)(a_i) = \beta(a_i)$. Set $B = \{\beta(y) : y \in \text{im}(\alpha)\}$. From the previous equality we have $(\beta\alpha)(b) = b$ for all $b \in B$. As $\operatorname{im}(\beta\alpha\beta) \subset B$, the equality $\beta\alpha\beta = \beta$ requires $\operatorname{im}(\beta) \subset B$. However, the latter condition is also sufficient. Indeed, as $\beta(y) \in B$ for any $y \in \operatorname{dom}(\beta)$, we have $(\beta\alpha\beta)(y) = (\beta\alpha)(\beta(y)) = \beta(y)$ and thus $\beta\alpha\beta = \beta$. This completes the proof. □

Theorem 2.6.4 is also true for the semigroup $\mathcal{T}_n$, just in this case the condition (a) is superfluous since it is automatically satisfied.

Corollary 2.6.5 *Let* $\alpha \in \mathcal{PT}_n$, $\operatorname{im}(\alpha) = \{a_1, a_2, \ldots, a_k\}$, $B_i = \{x \in \mathbf{N} : \alpha(x) = a_i\}$ *and* $m_i = |B_i|$. *Then*

(i) $|V_{\mathcal{PT}_n}(\alpha)| = m_1 m_2 \cdots m_k \cdot (k+1)^{n-k}$;

(ii) if α *is total, then* $|V_{\mathcal{T}_n}(\alpha)| = m_1 m_2 \cdots m_k \cdot k^{n-k}$.

Proof. For each $a_i \in \operatorname{im}(\alpha)$ we have to choose $\beta(a_i) \in B_i$. As for different i these choices are independent, we have $m_1 m_2 \cdots m_k$ different ways to define β on $\operatorname{im}(\alpha)$.

After this we have to define β on the set $\mathbf{N}\backslash\operatorname{im}(\alpha)$. From Theorem 2.6.4 it follows that the restriction of β to $\mathbf{N}\backslash\operatorname{im}(\alpha)$ is an arbitrary (partial in the case of $\mathcal{PT}_n$) mapping from $\mathbf{N}\backslash\operatorname{im}(\alpha)$ to the set $\{\beta(y) : y \in \operatorname{im}(\alpha)\}$. The latter set contains exactly k elements. Hence the restriction of β to $\mathbf{N}\backslash\operatorname{im}(\alpha)$ can be chosen in k^{n-k} different ways for $\mathcal{T}_n$ and in $(k+1)^{n-k}$ different ways for $\mathcal{PT}_n$. □

Exercise 2.6.6 Show that for any $a \in \mathbf{N}$ we have

(a) $|V_{\mathcal{T}_n}(0_a)| = n$.

(b) $|V_{\mathcal{PT}_n}(0_a)| = n \cdot 2^{n-1}$.

A regular semigroup S is called an *inverse* semigroup provided that each $a \in S$ has a unique inverse element. This inverse element is usually denoted by a^{-1} (this can now be justified by the requirement that it is unique). From Corollary 2.6.5 it follows that in the case of $n > 1$ the semigroups $\mathcal{T}_n$ and $\mathcal{PT}_n$ are not inverse semigroups.

Theorem 2.6.7 *The semigroup* $\mathcal{IS}_n$ *is an inverse semigroup.*

Proof. Let $\alpha \in \mathcal{IS}_n$ and $\beta \in V_{\mathcal{PT}_n}(\alpha)$. Let us try to analyze in which case β is a partial permutation. Since α is injective, the sets B_i from the proof of Theorem 2.6.4 consist of one element each. Hence β is uniquely defined on $\operatorname{im}(\alpha)$ by Theorem 2.6.4(b). At the same time, the injectivity of β and Theorem 2.6.4(c) imply that β must be undefined on $\mathbf{N}\backslash\operatorname{im}(\alpha)$. Thus $V_{\mathcal{PT}_n}(\alpha) \cap \mathcal{IS}_n$ contains a unique element, implying that $\mathcal{IS}_n$ is an inverse semigroup. □

Since the notion of an inverse element is a natural extension of the corresponding notion for invertible elements, inverse semigroups are very natural generalizations of groups. In some sense they form the class of semigroups, which is "closest" to groups.

Exercise 2.6.8 Prove the following identities in any inverse semigroup

(a) $(a^{-1})^{-1} = a$,

(b) $(ab)^{-1} = b^{-1}a^{-1}$.

2.7   Idempotents

Identities and zero elements in semigroups are special cases of a more general notion. An element e of a semigroup S is called an *idempotent* provided that $e^2 = e$. The set of all idempotents of S is denoted by $\mathcal{E}(S)$.

Theorem 2.25 describes idempotents in $\mathcal{PT}_n$. To formulate it we need some notation. If $\alpha \in \mathcal{PT}_n$ and $B \subset \mathbf{N}$, one says that B is *invariant* with respect to α provided that $\alpha(x) \in B$ for all $x \in B \cap \mathrm{dom}(\alpha)$. In case B is invariant with respect to α we can define the *restriction* $\alpha|_B$ of α to B as follows: $\alpha|_B$ is a partial transformation on B, $\mathrm{dom}(\alpha|_B) = B \cap \mathrm{dom}(\alpha)$, and for $x \in \mathrm{dom}(\alpha|_B)$ we have $\alpha|_B(x) = \alpha(x)$.

Exercise 2.7.1 Show that for each $\alpha \in \mathcal{PT}_n$ the sets $\mathbf{N}$, $\varnothing$, and $\mathrm{im}(\alpha)$ are invariant with respect to α.

Theorem 2.7.2 $\alpha \in \mathcal{PT}_n$ *is an idempotent if and only if* $\mathrm{im}(\alpha) \subset \mathrm{dom}(\alpha)$ *and the restriction* $\overline{\alpha} = \alpha|_{\mathrm{im}(\alpha)}$ *is the identity transformation on* $\mathrm{im}(\alpha)$.

Proof. If $\alpha^2 = \alpha$, for all $x \in \mathrm{dom}(\alpha)$ we have $\alpha(x) = \alpha^2(x) = \alpha(\alpha(x))$. Hence $\mathrm{im}(\alpha) \subset \mathrm{dom}(\alpha)$ and for each $y \in \mathrm{im}(\alpha)$ we have $\overline{\alpha}(y) = y$.

Conversely, if α acts as the identity on $\mathrm{im}(\alpha)$, we have $\alpha^2(x) = \alpha(x)$ for any $x \in \mathrm{dom}(\alpha)$. Since $\mathrm{dom}(\alpha^2) \subset \mathrm{dom}(\alpha)$, it follows that $\alpha^2 = \alpha$.   □

Corollary 2.7.3 *The element* $\alpha \in \mathcal{IS}_n$ *is an idempotent if and only if* α *is the identity transformation of some* $A \subset \mathbf{N}$. *In particular,* $\mathcal{IS}_n$ *contains exactly* 2^n *idempotents.*

Proof. For $\alpha \in \mathcal{IS}_n$ we have $|\mathrm{im}(\alpha)| = |\mathrm{dom}(\alpha)|$ by the injectivity of α. Hence $\mathrm{im}(\alpha) \subset \mathrm{dom}(\alpha)$ in this case implies $\mathrm{im}(\alpha) = \mathrm{dom}(\alpha)$. The condition that $\alpha|_{\mathrm{im}(\alpha)}$ is the identity transformation means exactly that α is the identity transformation on $\mathrm{im}(\alpha)$. Since $\mathbf{N}$ has exactly 2^n subsets, the claim follows.   □

For $A \subset \mathbf{N}$ we denote by ε_A the unique idempotent of $\mathcal{IS}_n$ such that $\mathrm{dom}(\varepsilon_A) = A$. We have $\varepsilon = \varepsilon_{\mathbf{N}}$ and $\mathbf{0} = \varepsilon_{\varnothing}$.

Corollary 2.7.4 *The number u_n of idempotents in the semigroup $\mathcal{T}_n$ equals*

$$u_n = \sum_{k=1}^{n} \binom{n}{k} k^{n-k}.$$

Proof. To define an idempotent α of rank k we have to choose a k-element set $\text{im}(\alpha)$ (this can be done in $\binom{n}{k}$ different ways), and then we have to define a mapping from $\mathbf{N}\backslash\text{im}(\alpha)$ to $\text{im}(\alpha)$ in an arbitrary way (this can be done in k^{n-k} different ways). Hence $\mathcal{T}_n$ contains exactly $\binom{n}{k}k^{n-k}$ idempotents of rank k. The statement is now obtained applying the sum rule. $\square$

Corollary 2.7.5 *The number u'_n of idempotents in the semigroup $\mathcal{PT}_n$ equals*

$$u'_n = \sum_{k=0}^{n} \binom{n}{k} (k+1)^{n-k}.$$

Proof. The difference with the proof of Corollary 2.7.4 is just the facts that $\mathcal{PT}_n$ contains idempotents of rank 0, and that the mapping from $\mathbf{N}\backslash\text{im}(\alpha)$ to $\text{im}(\alpha)$ may be partial. $\square$

There is a very close connection between the idempotents and the regular elements. Indeed, each idempotent e is a regular element as $e \cdot e \cdot e = e$. Conversely, if a is a regular element and $aba = a$, then the elements ab and ba are idempotents: $ab \cdot ab = aba \cdot b = ab$ and $ba \cdot ba = b \cdot aba = ba$. In terms of idempotents, the inverse semigroups can be detected in the set of all regular semigroups using the following:

Theorem 2.7.6 *A regular semigroup S is inverse if and only if all idempotents of S commute.*

Proof. Let S be a regular semigroup in which all idempotents commute. Let further $a \in S$ and $b_1, b_2 \in S$ be two inverse elements to a. Then $ab_1a = a$, $b_1ab_1 = b_1$, $ab_2a = a$, $b_2ab_2 = b_2$, and as we mentioned above, the elements ab_1, ab_2, b_1a, and b_2a are idempotents. Hence we have

$$b_1 = b_1ab_1 = b_1ab_2ab_1 = b_1 \cdot ab_2 \cdot ab_1 = b_1 \cdot ab_1 \cdot ab_2 = b_1ab_2 =$$
$$= b_1ab_2ab_2 = b_1a \cdot b_2a \cdot b_2 = b_2a \cdot b_1a \cdot b_2 = b_2 \cdot ab_1a \cdot b_2 = b_2ab_2 = b_2.$$

This means that each $a \in S$ has a unique inverse and thus S is inverse.

Conversely, let S be an inverse semigroup and $e, f \in S$ be two idempotents. Let us first prove that their product ef is an idempotent. Consider $a = (ef)^{-1}$. We have

$$ef \cdot a \cdot ef = ef, \quad a \cdot ef \cdot a = a.$$

Using this we obtain the following equalities:

$$ef \cdot ae \cdot ef = ef \cdot a \cdot e^2 f = ef \cdot a \cdot ef = ef,$$
$$ae \cdot ef \cdot ae = a \cdot e^2 f \cdot ae = a \cdot ef \cdot a \cdot e = ae.$$

Hence ae is also an inverse to ef, which means $ae = a$. Analogously one shows that $fa = a$. Thus

$$a^2 = ae \cdot fa = a \cdot ef \cdot a = a,$$

and hence a is an idempotent. In particular, $a^{-1} = a$, and thus $a = ef$.

Analogously one shows that fe is an idempotent as well. Moreover, we have

$$fe \cdot ef \cdot fe = f \cdot e^2 \cdot f^2 \cdot e = fe \cdot fe = fe,$$
$$ef \cdot fe \cdot ef = e \cdot f^2 \cdot e^2 \cdot f = ef \cdot ef = ef.$$

This means that $(ef)^{-1} = fe$. However, since ef is an idempotent, we also have $(ef)^{-1} = ef$. Thus $fe = ef$ and we are done. $\qquad\qquad\square$

A semigroup S is called *commutative* or *abelian* if $ab = ba$ for all $a, b \in S$. From Theorem 2.7.6 it follows that the set $E(S)$ of idempotents of an inverse semigroup S is a commutative subsemigroup of S. In particular, $E(\mathcal{IS}_n)$ is a commutative semigroup of order 2^n.

Exercise 2.7.7 Prove that $\varepsilon_A \cdot \varepsilon_B = \varepsilon_{A \cap B}$ for all $A, B \subset \mathbf{N}$ and use this to show that the semigroup $E(\mathcal{IS}_n)$ is isomorphic to the semigroup of all subsets of $\mathbf{N}$ with respect to the operation of the intersection of subsets.

Exercise 2.7.8 (a) Show that $E(\mathcal{T}_2)$ is a noncommutative subsemigroup of $\mathcal{T}_2$.

(b) Show that $E(\mathcal{T}_n)$ is not a subsemigroup of $\mathcal{T}_n$ for all $n > 2$.

Exercise 2.7.9 Show that $E(\mathcal{PT}_n)$ is not a subsemigroup of $\mathcal{PT}_n$ for all $n > 1$.

We know already that for $n > 1$ the semigroups $\mathcal{T}_n$ and $\mathcal{PT}_n$ are not inverse. This fact also follows from Theorem 2.7.6 and Exercises 2.7.8 and 2.7.9.

2.8   Nilpotent Elements

An element a of a semigroup S with the zero 0 is called *nilpotent* or a *nil-element* provided that $a^k = 0$ for some $k \in \mathbb{N}$. The minimal k for which $a^k = 0$ is called the *nilpotency degree* or *nilpotency class* of the element a and is denoted by $\mathrm{nd}(a)$.

Proposition 2.8.1 $\alpha \in \mathcal{PT}_n$ *is nilpotent if and only if the graph* Γ_α *does not contain cycles.*

Proof. If Γ_α contains a cycle, say $(a_1, a_2, \ldots, a_m)$, then $a_1 \in \text{dom}(\alpha^k)$ for all $k > 0$ (it is easy to see that $\alpha^k(a_1) = a_{(1+k) \bmod m}$). As the zero element $\mathbf{0}$ of $\mathcal{PT}_n$ satisfies $\text{dom}(\mathbf{0}) = \varnothing$, we have $\alpha^k \neq \mathbf{0}$ for all $k > 0$ and hence the element α cannot be nilpotent.

If Γ_α does not contain any cycle, the trajectory of each vertex a must break. More precisely, this trajectory has the form $a_0 = a, a_1, a_2, \ldots, a_m$, where all a_i are pairwise different and Γ_α does not contain any arrow with the tail a_m (i.e., $a_m \notin \text{dom}(\alpha)$). This means that $\alpha^m(a) = a_m$ and that $a \notin \text{dom}(\alpha^{m+1})$ since otherwise we would have $\alpha^{m+1}(a) = \alpha(a_m)$, which does not make sense since $a_m \notin \text{dom}(\alpha)$. Since the length of any trajectory does not exceed n, we have $\text{dom}(\alpha^n) = \varnothing$ and thus α is nilpotent. $\square$

From the proof of Proposition 2.8.1 we, in fact, derive the following:

Proposition 2.8.2 *The nilpotency degree of a nilpotent element $\alpha \in \mathcal{PT}_n$ equals the length of the longest trajectory in Γ_α.*

Corollary 2.8.3 *The partial permutation*

$$\alpha = (a_1, \ldots, a_k) \cdots (b_1, \ldots, b_l)[c_1, \ldots, c_p] \cdots [d_1, \ldots, d_q] \in \mathcal{IS}_n$$

is nilpotent if and only if it does not contain cycles. If $\alpha \in \mathcal{IS}_n$ is nilpotent, the nilpotency degree of α equals the maximum $\max(p, \ldots, q)$ of the lengths of all chains.

Denote by $\mathbf{S}(n, k)$ the *Stirling number of the second kind*, that is, the number of partitions $\mathbf{N} = A_1 \cup A_2 \cup \cdots \cup A_k$ into an unordered disjoint union of k nonempty blocks.

Theorem 2.8.4 ([LU1]) *The semigroup $\mathcal{PT}_n$ contains $\binom{n}{k} \cdot \mathbf{S}(n, k+1) \cdot k!$ nilpotent elements of rank k.*

Proof. We have the unique nilpotent element of rank 0, namely, the element $\mathbf{0}$ itself. On the other hand, $\binom{n}{0} \cdot \mathbf{S}(n, 1) \cdot 0! = 1$. Hence for the rest of the proof we may assume $k \geq 1$.

Let $\alpha \in \mathcal{PT}_n$ be a nilpotent element of rank k. Assume that $\text{im}(\alpha) = \{a_1, a_2, \ldots, a_k\}$. Then the sets $A_i = \{x \in \mathbf{N} : \alpha(x) = a_i\}$, $i = 1, \ldots, k$, and $A_{k+1} = \overline{\text{dom}}(\alpha)$ form a partition $\mathbf{N} = A_1 \cup A_2 \cup \cdots \cup A_{k+1}$ of the set $\mathbf{N}$ into $k+1$ block (note that $A_{k+1} \neq \varnothing$ since α is nilpotent). This observation suggests the following procedure for constructing all nilpotent elements of rank k (we construct some nilpotent element α): First we choose $\text{im}(\alpha) = \{a_1, \ldots, a_k\}$. This is a k-element subset of $\mathbf{N}$ and hence it can be chosen in $\binom{n}{k}$ different ways. Then we choose a partition, $\mathbf{N} = B_1 \cup B_2 \cup \cdots \cup B_{k+1}$ into an unordered disjoint union of nonempty blocks. This can be done in $\mathbf{S}(n, k+1)$ different ways. Now for each a_i we have to choose its preimage, which is one of the blocks $B_1, \ldots, B_{k+1}$, such that the resulting element is nilpotent. The choice of the preimage for a_i means that we choose some arrows for Γ_α in such a way that the resulting graph does not contain any cycles.

The preimage of a_1 can be any of the blocks $B_1, \ldots, B_{k+1}$, which does not contain a_1 (since otherwise Γ_α would contain a loop at a_1, which is a cycle of length 1). This means that the preimage of a_1 can be chosen in k different ways.

We proceed by induction. Assume that we have already chosen the preimages for $a_1, \ldots, a_m$ (and let these preimages be the blocks $B_{i_1}, \ldots, B_{i_m}$, respectively). Denote $X = B_{i_1} \cup \cdots \cup B_{i_m}$ and consider two different cases.

Case 1: $a_{m+1} \notin X$. This means that there is no arrow with the tail a_{m+1} yet. Hence the preimage of a_{m+1} could be any of the remaining blocks, which does not contain a_{m+1}. This preimage can be chosen in $(k+1-m)-1 = k-m$ different ways.

Case 2: $a_{m+1} \in X$. Consider the trajectory $a^{(0)} = a_{m+1}, a^{(1)}, \ldots, a^{(t)}$ of a_{m+1} in that part of Γ_α, which is already constructed. Since the constructed part of Γ_α does not contain cycles, this trajectory breaks at $a^{(t)} \notin X$. In order to ensure that the choice of the preimage for a_{m+1} does not create a cycle in Γ_α, it is necessary and sufficient to make sure that this preimage does not contain $a^{(t)}$. Hence we again have $(k+1-m)-1 = k-m$ ways to choose the preimage for a_{m+1}.

From the above we have that for fixed $\{a_1, \ldots, a_k\}$ and $\mathbf{N} = B_1 \cup B_2 \cup \cdots \cup B_{k+1}$ the element α can be constructed in $k(k-1) \cdots (k-(k-1)) = k!$ different ways. Hence the total number of nilpotent elements of rank k equals $\binom{n}{k} \cdot S(n, k+1) \cdot k!$. $\square$

Theorem 2.8.5 *The number of nilpotent elements of defect k in the semigroup* $\mathcal{IS}_n$ *equals the signless Lah number* $L'_{n,k} = \frac{n!}{k!}\binom{n-1}{k-1}$.

Proof. By Corollary 2.8.3, a nilpotent partial permutation of defect k has the following form:

$$[i_1, i_2, \ldots, i_{m_1}][i_{m_1+1}, i_{m_1+2}, \ldots, i_{m_2}] \cdots [i_{m_{k-1}+1}, i_{m_{k-1}+2}, \ldots, i_{m_k}].$$

To get the above expression from the permutation $i_1, i_2, \ldots, i_n$ of $1, 2, \ldots, n$, it is enough to choose the ends $i_{m_1}, i_{m_2}, \ldots, i_{m_{k-1}}$ of the first $k-1$ chains (as $m_k = n$ automatically). This can be done in $\binom{n-1}{k-1}$ different ways. Going through all permutations we will get chain-cycle notation for all nilpotent elements of defect k. Since the order of chains in the chain-cycle notation is not important (because all chains in the chain-cycle notation commute), every nilpotent element of defect k will be counted $k!$ times. Indeed, the chains in each chain-cycle notation can be permuted in $k!$ different ways without changing the corresponding partial transformation of $\mathbf{N}$; however, each of these permutations of chains corresponds to a different permutation of **N**. This means that $n!$ permutations of **N** give $k!$ repetitions of each nilpotent element. Hence the number of nilpotent elements of defect k equals $\frac{n!}{k!}\binom{n-1}{k-1}$. □

Corollary 2.8.6 *The semigroup* $\mathcal{IS}_n$ *contains*

$$\sum_{k=1}^{n} \frac{n!}{k!}\binom{n-1}{k-1}$$

nilpotent elements.

2.9 Addenda and Comments

2.9.1 A subset $M \subset \mathbf{N}$ is called invariant with respect to $S \subset \mathcal{PT}_n$ provided that M is invariant with respect to each $\alpha \in S$. Obviously, the union and the intersection of invariant subsets is an invariant subset. Further, for each $\alpha \in \mathcal{PT}_n$ and $x \in \mathbf{N}$ the set of vertices of $\mathrm{tr}_\alpha(x)$ is invariant with respect to α.

Let $\alpha \in \mathcal{PT}_n$. If there exists a partition $\mathbf{N} = \mathbf{N}_1 \cup \mathbf{N}_2$ of $\mathbf{N}$ into a disjoint union of two invariant subsets, then the study of the action of α on $\mathbf{N}$ reduces to the study of the action of α on $\mathbf{N}_1$ and $\mathbf{N}_2$. Continuing this procedure, the study of the action of α on $\mathbf{N}$ reduces to the study of the action of α on all its orbits:

Proposition 2.9.1 *Each orbit of α is invariant with respect to α. If $\mathbf{N} = \mathbf{N}_1 \cup \mathbf{N}_2$ is a partition into invariant subsets, then each of them is a union of orbits.*

Hence the partition of **N** into orbits is the finest possible partition of **N** into a disjoint union of nonempty invariant subsets.

2.9.2 Let $\alpha \in \mathcal{PT}_n$ and let $\mathbf{N} = \mathbf{N}_1 \cup \mathbf{N}_2$ be a partition into invariant subsets. For each $i = 1, 2$ set

$$\alpha^{(i)}(x) = \begin{cases} \alpha(x), & x \in \mathbf{N}_i, \\ x, & \text{otherwise.} \end{cases}$$

In other words, $\alpha^{(i)}$ acts on $\mathbf{N}_i$ in the same way as α, and $\alpha^{(i)}$ acts on the complement $\mathbf{N}_{1-i}$ as the identity transformation. It is easy to see that

$$\alpha = \alpha^{(1)}\alpha^{(2)} = \alpha^{(2)}\alpha^{(1)}.$$

Analogous transformations $\alpha^{(i)}$ and the corresponding decomposition $\alpha = \alpha^{(1)}\alpha^{(2)} \cdots \alpha^{(m)}$ can be defined for any partition $\mathbf{N} = \mathbf{N}_1 \cup \mathbf{N}_2 \cup \cdots \cup \mathbf{N}_m$ of **N** into a disjoint union of arbitrarily many invariant subsets. In particular, if some $\mathbf{N}_i$ is an orbit, the graph of the transformation $\alpha^{(i)}$ has the following

form: one of its components coincides with the corresponding component of
Γ_α (the one which corresponds to the orbit $\mathbf{N}_i$), and all other components
consist of loops. In some sense $\alpha^{(i)}$ can be identified with the corresponding
connected component. Then the decomposition of α described in Sect. 1.3
can be interpreted as a decomposition of α into a product of its connected
components. In particular, for permutation we get the usual decomposition
into a product of independent cycles.

2.9.3 Let us interpret $[a_1, a_2, \ldots, a_k | a_i]$ as the notation for the element from
$\mathcal{T}_n$, which acts as the element

$$\begin{pmatrix} a_1 & a_2 & \cdots & a_{k-1} & a_k \\ a_2 & a_3 & \cdots & a_k & a_i \end{pmatrix}$$

on the set $A = \{a_1, a_2, \ldots, a_k\}$, and as the identity on the set $\mathbf{N} \backslash A$. Then
the linear notation

$$\mu = [a_1, a_2, \ldots, a_k | a_i][b_1, b_2, \ldots, b_l | b_j] \cdots [c_1, c_2, \ldots, c_m | c_h],$$

described in 1.4.3 can be considered as the following decomposition of μ:

$$\mu = [c_1, c_2, \ldots, c_m | c_h] \cdots [b_1, b_2, \ldots, b_l | b_j][a_1, a_2, \ldots, a_k | a_i]$$

(note that the factors should be taken in the reverse order). In the general case, the factors of this decomposition do not correspond to connected
components, the factors themselves (and even the number of these factors)
depend on the choice of certain elements along the way to write down this
decomposition, and these factors do not commute. This is another disadvantage of the notation from [AAH].

2.9.4 Let $\alpha \in \mathcal{PT}_n$. As each orbit of α is invariant with respect to α
and the intersection of invariant sets is invariant, each minimal (with re-
spect to inclusions) nonempty invariant set is contained in some orbit. From
Theorem 1.2.9 it follows that the trajectory of each vertex from a fixed connected component either contains a cycle (if it exists in this component) or
terminates at components sink. On the other hand, the set of all vertices
in a cycle, and the set consisting of the sink are obviously invariant. Hence
only these sets are minimal (with respect to inclusions) nonempty invariant
sets.

Let now K be an orbit of α. If the corresponding connected component
of Γ_α contains a cycle, the set of all vertices of this cycle is called the *kernel*
of the orbit K. If K does not contain any cycle, we say that the kernel of
this orbit is empty.

Proposition 2.9.2 *Let $\alpha \in \mathcal{PT}_n$, R be an orbit of α containing a cycle and
K be the kernel of R. Assume $|K| = k > 0$. Then*

(i) $\alpha^k(x) = x$ for all $x \in K$.

(ii) If $y \in R$ and $m > 0$ are such that $\alpha^m(y) = y$, then $y \in K$, $k|m$ and
    $K = \{y, \alpha(y), \ldots, \alpha^{k-1}(y)\}$.

2.9.5 Let $\alpha \in \mathcal{PT}_n$. The *stable image* of α is the set

$$\mathrm{stim}(\alpha) = \bigcap_{k \in \mathbb{N}} \mathrm{im}(\alpha^k).$$

Proposition 2.9.3 *For each $\alpha \in \mathcal{PT}_n$ the set $\mathrm{stim}(\alpha)$ is invariant with respect to α. The restriction of α to $\mathrm{stim}(\alpha)$ is a permutation, moreover, $\mathrm{stim}(\alpha)$ is the maximum subset of $\mathbf{N}$ (with respect to inclusions) such that the restriction of α to this subset is defined and is a permutation.*

It is easy to see that $\mathrm{stim}(\alpha)$ is the union of kernels of all orbits of α. The restriction of α to $\mathrm{stim}(\alpha)$ is called the *permutational part* of α.

2.9.6 Let $\alpha \in \mathcal{PT}_n$. The *stable rank* of α is defined as the following number: $\mathrm{strank}(\alpha) = \min_{k \in \mathbb{N}} \mathrm{rank}(\alpha^k)$.

Proposition 2.9.4 $\mathrm{strank}(\alpha) = |\mathrm{stim}(\alpha)|$.

2.9.7 The notation for the elements of $\mathcal{IS}_n$, introduced in Sect. 2.5, is close to the notation used in [Li].

2.9.8 The role of $\mathcal{IS}_n$ in the theory of inverse semigroups is analogous to the role of $\mathcal{S}_n$ in group theory, and to the role of $\mathcal{T}_n$ in semigroup theory. Namely, we have the following analog of Cayley's Theorem for inverse semigroups:

Theorem 2.9.5 (Preston–Wagner) *Any inverse semigroup T is isomorphic to a subsemigroup of the semigroup $\mathcal{IS}(T)$ of all partial injective transformations of the set T.*

The idea of the proof is as follows: for each $a \in T$ we consider the transformation $\rho_a : a^{-1}T \to aT$, defined via $\rho_a(x) = ax$. This transformation turns out to be bijective. Hence we can consider ρ_a as a partial permutation on T. For the mapping $\varphi : T \to \mathcal{IS}(T)$, $\varphi(a) = \rho_a$, one easily checks that $\varphi(ab) = \varphi(a)\varphi(b)$ and that φ is injective. For details, see [CP1, Theorem 1.20].

2.9.9 If $(S, \cdot)$ is a semigroup, we can consider the new semigroup $(S, *)$, where $a * b = b \cdot a$. The semigroup $(S, *)$ is called the *opposite semigroup* or the *dual semigroup* or simply the *dual* of S. It is usually denoted by $\overleftarrow{S}$. Obviously, the second dual of S is isomorphic to S. If $\overleftarrow{S} \cong S$, the semigroup S is called *self-dual*.

If S is an inverse semigroup, from Exercise 2.6.8 it follows that the mapping $a \mapsto a^{-1}$ is an isomorphism from S to $\overleftarrow{S}$. Hence all inverse semigroups are self-dual, in particular, the semigroup $\mathcal{IS}_n$ is self-dual.

However, not any semigroup is self-dual. For example, for $n > 1$ the semigroup $\mathcal{T}_n$ is not self-dual. Indeed, we know that $\mathcal{T}_n$ contains left zeros but not right zeros (see Proposition 2.3.3). Hence $\overleftarrow{\mathcal{T}_n}$ contains right zeros but no left zeros. At the same time each isomorphism must map left zeros to left zeros, and right zeros to right zeros.

2.9.10 The number of nilpotent elements in $\mathcal{PT}_n$ is given by the following statement.

Theorem 2.9.6 ([LU1]) $\mathcal{PT}_n$ *contains exactly* $(n+1)^{n-1}$ *nilpotent elements.*

Proof. For $\alpha \in \mathcal{PT}_n$ and $k \geq 0$ let N_k denote the set of all $x \in \mathbf{N}$ for which the trajectory of x breaks at the kth step, that is, has the form $x_0 = x, x_1, \ldots, x_k$. Assume that the nilpotency degree of α equals t. In this case from the proof of Proposition 2.8.1 it follows that the sets $N_0, N_1, \ldots, N_t$ are not empty and form a partition of $\mathbf{N}$. Moreover, $\alpha(N_i) \subset N_{i-1}$ for all i. Hence each nilpotent element of nilpotency degree t can be obtained in the following way: First we choose an ordered partition $\mathbf{N} = N_0 \cup N_2 \cup \cdots \cup N_t$ of $\mathbf{N}$ into $t+1$ nonempty blocks (this can be done in $\binom{n}{n_0,n_1,\ldots,n_t}$ different ways, where $n_i = |N_i|$, $i = 0, \ldots, t$). Then for each $i > 0$ we define some mapping from N_i to N_{i-1} (altogether such mappings can be defined in $n_0^{n_1} n_1^{n_2} \cdots n_{t-1}^{n_t}$ different ways).

The total number of nilpotent elements in $\mathcal{PT}_n$ is denoted by $\mathbf{n}(\mathcal{PT}_n)$. By the above, we have

$$\mathbf{n}(\mathcal{PT}_n) = \sum_{\substack{1 \leq t < n \\ n_0+n_1+\cdots+n_t=n}} \binom{n}{n_0, n_1, \ldots, n_t} n_0^{n_1} n_1^{n_2} \cdots n_{t-1}^{n_t} =$$

$$= \sum_{\substack{1 \leq t < n \\ n_0+n_1+\cdots+n_t=n}} \frac{n!}{n_0!} \cdot \frac{n_0^{n_1}}{n_1!} \cdot \frac{n_1^{n_2}}{n_2!} \cdots \frac{n_{t-1}^{n_t}}{n_t!}.$$

Using (1.8) we have

$$\mathbf{n}(\mathcal{PT}_n) = \sum_{n_0=1}^{n} \frac{n!}{(n_0-1)!} \sum_{\substack{1 \leq t < n-n_0 \\ n_1+\cdots+n_t=n-n_0}} \frac{1}{n_0} \cdot \frac{n_0^{n_1}}{n_1!} \cdot \frac{n_1^{n_2}}{n_2!} \cdots \frac{n_{t-1}^{n_t}}{n_t!} =$$

$$= \sum_{n_0=1}^{n} \frac{n!}{(n_0-1)!} \cdot \frac{n^{k-1}}{k!} = \sum_{k=0}^{n-1} \frac{(n-1)!}{(n-1-k)!k!} \cdot n^k = (n+1)^{n-1}.$$

$\square$

In [LU1] Theorem 2.9.6 is proved in a different way. One more way to prove it is to use [Hi1, Lemma 6.1.4]. Yet another way to prove this theorem is to use Cayley's theorem on the number of labeled trees: If we have a tree, say Γ on the set $\{1, 2, \ldots, n+1\}$ of vertices, we can consider it as a rooted tree with the root $(n+1)$. In particular, we have a well-defined notion of a distance of a vertex to the root. Now we can define the nilpotent element α from $\mathcal{PT}_n$ as follows: If i is a vertex, let j denote the unique vertex of Γ such that (i, j) is an edge and whose distance to the root is smaller than that of i. Define $\alpha(i) = j$ if j is not the root and $\alpha(i) = \varnothing$ otherwise. One verifies that the correspondence $\Gamma \mapsto \alpha$ is a bijection from the set of all labeled trees on $\{1, 2, \ldots, n+1\}$ to the set of all nilpotent elements in $\mathcal{PT}_n$.

From Theorem 2.9.6 it follows that

$$\frac{\mathrm{n}(\mathcal{PT}_n)}{|\mathcal{PT}_n|} = \frac{(n+1)^{n-1}}{(n+1)^n} = \frac{1}{n+1}.$$

From Theorems 2.8.4 and 2.9.6 we get the following identity.

Corollary 2.9.7

$$\sum_{k=0}^{n-1} \binom{n}{k} S(n, k+1) k! = (n+1)^{n-1}.$$

2.10 Additional Exercises

2.10.1 Let S be a nonempty set such that $0 \in S$. For all $a, b \in S$ set $a \cdot b = 0$. Show that $(S, \cdot)$ is a semigroup (it is called a semigroup with *zero multiplication* or a *null semigroup*).

2.10.2 Let S be a nonempty set. For $a, b \in S$ define $a \cdot b = b$. Show that $(S, \cdot)$ is a semigroup, each element of which is at the same time a left identity and a right zero (such S is usually called a *semigroup of left identities* or a *right zero semigroup*; analogously, using $a \cdot b = a$ one defines a *left zero semigroup* or a *semigroup of right identities*).

2.10.3 Show that $\mathcal{T}_{n+1}$ contains a subsemigroup, isomorphic to the semigroup $\mathcal{PT}_n$.

2.10.4 Let X be a nonempty set and $\mathcal{B}(X)$ be the set of all subsets of X (the *Boolean* of X). Show that both $(\mathcal{B}(X), \cup)$ and $(\mathcal{B}(X), \cap)$ are commutative semigroups. Furthermore, show that these two semigroups are isomorphic.

2.10.5 Let K be an orbit of some transformation $\alpha \in \mathcal{PT}_n$. Prove that the kernel of K coincides with the set

$$\cap_{x \in K} \{\alpha^m(x) : m \geq 0\}.$$

2.10.6 Let $\alpha \in \mathcal{T}_n$. Prove that $x \in \mathbf{N}$ belongs to the kernel of an orbit K if and only if $\{y : \alpha^k(y) = x \text{ for some } k > 0\} = K$.

2.10.7 Let $\alpha \in \mathcal{PT}_n$. Characterize $\mathrm{stim}(\alpha)$ and $\mathrm{strank}(\alpha)$ in terms of Γ_α.

2.10.8 Let $\alpha \in \mathcal{PT}_n$.

(a) Prove that $\mathrm{stim}(\alpha)$ is invariant with respect to α^k for each $k > 0$.

(b) Let $\beta \in \mathcal{PT}_n$ be such that $\alpha = \beta^k$ for some $k > 0$. Prove that $\mathrm{stim}(\alpha)$ is invariant with respect to β.

2.10.9 ([HS]) Denote by $\mathbf{u}_n$ the number of idempotents in $\mathcal{T}_n$. For a prime p show that

(a) $\mathbf{u}_{p+1} \equiv 2 \bmod p$;

(b) $\mathbf{u}_{p+2} \equiv 7 \bmod p$.

2.10.10 ([HS]) Let $\mathbf{u}_n$ be as in 2.10.9 and set $\mathbf{u}_0 = 1$. Show that

$$\mathbf{u}_{n+1} = \sum_{k=0}^{n} \binom{n}{k}(k+1)\mathbf{u}_{n-k}.$$

2.10.11 Show that for each $\alpha \in \mathcal{PT}_n$ the set $V_{\mathcal{PT}_n}(\alpha) \cdot \alpha$ is a left zero semigroup and the set $\alpha \cdot V_{\mathcal{PT}_n}(\alpha)$ is a right zero semigroup.

2.10.12 Prove that each regular semigroup with only one idempotent is a group.

2.10.13 Prove that the operations of extending a semigroup with the element 1 (as in Proposition 2.2.1) and with the element 0 (as in Proposition 2.3.1) commute, that is, $(S^1)^0 = (S^0)^1$ (note the equality and not the isomorphism sign).

2.10.14 Prove that the set $A \times B$ with respect to the operation $(a_1, b_1) * (a_2, b_2) = (a_1, b_2)$ is a semigroup in which each pair of elements is a pair of inverse elements (such a semigroup is called a *rectangular band*).

2.10.15 Show that the elements

$$\alpha = (a_1, \ldots, a_k) \cdots (b_1, \ldots, b_l)[c_1, \ldots, c_p] \cdots [d_1, \ldots, d_q],$$
$$\beta = [d_q, \ldots, d_1] \cdots [c_p, \ldots, c_1](b_l, \ldots, b_1) \cdots (a_k, \ldots, a_1)$$

of the semigroup $\mathcal{IS}_n$ form a pair of inverse elements.

2.10.16 Prove that for $n > 1$ the set of left zeros in the semigroup $\mathcal{T}_n$ forms a noncommutative semigroup.

2.10.17 Prove or disprove the following statements:

(a) If the idempotents a and b commute, their product ab is an idempotent.

(b) If the product ab of two idempotents, a and b, is an idempotent, then a and b commute.

2.10.18 ([GH1]) Prove that

(a) $\mathcal{IS}_n$ contains $n!$ nilpotent elements of defect 1,

(b) $\mathcal{PT}_n$ contains $n!$ nilpotent elements of defect 1.

2.10.19 ([LU1]) Let N_n denote the total number of nilpotent elements in the semigroup $\mathcal{IS}_n$. Prove that $\mathsf{N}_n = |\mathcal{IS}_n| - n|\mathcal{IS}_{n-1}|$, $n > 1$.

2.10.20 ([BRR]) Prove the following recursive relation (for $n > 2$):

$$|\mathcal{IS}_n| = 2n|\mathcal{IS}_{n-1}| - (n-1)^2|\mathcal{IS}_{n-2}|.$$

2.10.21 ([GM, Theorem 9]) For N_n as in 2.10.19 show that

$$\lim_{n\to\infty} \frac{\mathsf{N}_n}{|\mathcal{IS}_n|} = 0.$$

2.10.22 For $\alpha \in \mathcal{PT}_n$ denote by $\mathsf{c}(\alpha)$ the number of connected components of the graph Γ_α. Prove the following:

(a)
$$\frac{1}{|\mathcal{S}_n|} \sum_{\alpha\in\mathcal{S}_n} \mathsf{c}(\alpha) = 1 + \frac{1}{2} + \cdots + \frac{1}{n}.$$

(b) ([Hi1, Lemma 6.1.12])

$$\frac{1}{|\mathcal{T}_n|} \sum_{\alpha\in\mathcal{T}_n} \mathsf{c}(\alpha) = \sum_{k=1}^{n} \frac{n!}{k(n-k)!n^k}.$$

(c) ([GM5, Corollary 1])

$$\sum_{\alpha\in\mathcal{IS}_n} \mathsf{c}(\alpha) = \sum_{k=1}^{n} \left(1 + \frac{1}{k}\right) |\mathcal{IS}_{n-k}| n(n-1)\cdots(n-k+1).$$

2.10.23 Prove that the semigroup $\mathcal{PT}_n$ is not self-dual for $n > 1$.

2.10.24 (a) Let $\alpha, \beta \in \mathcal{T}_n$. Show that we either have $\mathcal{S}_n\alpha\mathcal{S}_n = \mathcal{S}_n\beta\mathcal{S}_n$, or $\mathcal{S}_n\alpha\mathcal{S}_n \cap \mathcal{S}_n\beta\mathcal{S}_n = \varnothing$.

(b) For $\alpha \in \mathcal{T}_n$ let $t_k(\alpha)$ be as in Exercise 1.5.12. Set

$$t(\alpha) = (t_0(\alpha), t_1(\alpha), \ldots, t_n(\alpha))$$

and call this vector the *type* of α. Show that $\mathcal{S}_n \alpha \mathcal{S}_n = \mathcal{S}_n \beta \mathcal{S}_n$ if and only if $t(\alpha) = t(\beta)$.

2.10.25 (a) Let $\alpha, \beta \in \mathcal{PT}_n$. Show that we either have $\mathcal{S}_n \alpha \mathcal{S}_n = \mathcal{S}_n \beta \mathcal{S}_n$, or $\mathcal{S}_n \alpha \mathcal{S}_n \cap \mathcal{S}_n \beta \mathcal{S}_n = \varnothing$.

(b) For $\alpha \in \mathcal{PT}_n$ let $t_k(\alpha)$ be as in Exercise 1.5.13. Set

$$t(\alpha) = (t_0(\alpha), t_1(\alpha), \ldots, t_n(\alpha))$$

and call this vector the *type* of α. Show that $\mathcal{S}_n \alpha \mathcal{S}_n = \mathcal{S}_n \beta \mathcal{S}_n$ if and only if $t(\alpha) = t(\beta)$.

Chapter 3

Generating Systems

3.1 Generating Systems in $\mathcal{T}_n$, $\mathcal{PT}_n$, and $\mathcal{IS}_n$

Let S be a semigroup and $A \subset S$ be a set. An element $s \in S$ is said to be *generated* by A provided that s can be written as a finite product of elements from A. The set of all elements from S generated by A is usually denoted by $\langle A \rangle$. If S is a monoid, we will understand the identity element of this monoid as a trivial finite product of elements from A (of length 0). Thus $1 \in \langle A \rangle$ for any $A \subset S$. A subset A of the semigroup S is called a *generating system* or a *generating set* for S provided that $\langle A \rangle = S$. A generating system is called *irreducible* provided that each proper subset of A is no longer a generating system.

One of the first most natural questions in the study of some semigroup is to describe some (irreducible) generating system and to classify all (irreducible) generating systems.

Until the end of this section we let S denote one of the semigroups $\mathcal{T}_n$, $\mathcal{PT}_n$, or $\mathcal{IS}_n$. Let us recall that each of these semigroups contains $\mathcal{S}_n$ as the group of units (see Proposition 2.2.4).

Lemma 3.1.1 *(i) Each generating system of S must contain a generating system of $\mathcal{S}_n$.*

(ii) If A is an irreducible generating system of S, then $A \cap \mathcal{S}_n$ is an irreducible generating system of $\mathcal{S}_n$.

Proof. $\mathcal{S}_n$ coincides with the set of all elements of S of rank n. By Exercise 2.1.4(c), an element of rank n can be a product of elements of rank n only. This implies both (i) and (ii). $\qquad\square$

Lemma 3.1.2 *Each generating system of S must contain at least one element of rank $(n-1)$.*

Proof. The semigroup S contains elements of rank $(n-1)$. The set of elements of rank n coincides with $\mathcal{S}_n$ and is closed under composition. Hence

O. Ganyushkin, V. Mazorchuk, *Classical Finite Transformation Semigroups*, Algebra and Applications 9, DOI: 10.1007/978-1-84800-281-4_3,

from the inequality of Exercise 2.1.4(c) it follows that an element of rank $(n-1)$ can be written as a product of elements of rank $(n-1)$ and n only. Moreover, an element of rank $(n-1)$ cannot be written as a product of elements of rank n. The claim follows. $\qquad\square$

Theorem 3.1.3 *Let A be a generating system of $\mathcal{T}_n$. Then this system is irreducible if and only if $A = A_1 \cup \{\alpha\}$, where A_1 is an irreducible generating system of $\mathcal{S}_n$ and $\mathrm{rank}(\alpha) = n - 1$.*

Proof. Taking Lemmas 3.1.1 and 3.1.2 into account, it is enough to show that any set of the form $A = A_1 \cup \{\alpha\}$, where A_1 is an irreducible generating system of $\mathcal{S}_n$ and $\mathrm{rank}(\alpha) = n - 1$ is a generating system of $\mathcal{T}_n$.

As A_1 is an irreducible generating system of $\mathcal{S}_n$, we have $\mathcal{S}_n = \langle A_1 \rangle \subset \langle A \rangle$. Let us first show that $\langle A \rangle$ contains all elements of $\mathcal{T}_n$ of rank $(n-1)$. Assume that

$$\alpha = \begin{pmatrix} 1 & 2 & \cdots & i-1 & i & i+1 & \cdots & j-1 & j & j+1 & \cdots & n \\ a_1 & a_2 & \cdots & a_{i-1} & a & a_{i+1} & \cdots & a_{j-1} & a & a_{j+1} & \cdots & a_n \end{pmatrix}$$

and let

$$\beta = \begin{pmatrix} 1 & 2 & \cdots & k-1 & k & k+1 & \cdots & l-1 & l & l+1 & \cdots & n \\ b_1 & b_2 & \cdots & b_{k-1} & b & b_{k+1} & \cdots & b_{l-1} & b & b_{l+1} & \cdots & b_n \end{pmatrix}$$

be an arbitrary element from $\mathcal{T}_n$ of rank $(n-1)$. Let $\pi \in \mathcal{S}_n$ be an arbitrary permutation satisfying $\pi(k) = i$ and $\pi(l) = j$. Then

$$\alpha\pi = \begin{pmatrix} 1 & 2 & \cdots & k-1 & k & k+1 & \cdots & l-1 & l & l+1 & \cdots & n \\ c_1 & c_2 & \cdots & c_{k-1} & a & c_{k+1} & \cdots & c_{l-1} & a & c_{l+1} & \cdots & c_n \end{pmatrix}$$

is an element of rank $(n-1)$. Consider now the permutation

$$\tau = \begin{pmatrix} c_1 & c_2 & \cdots & c_{k-1} & c_{k+1} & \cdots & c_{l-1} & c_{l+1} & \cdots & c_n & a & x \\ b_1 & b_2 & \cdots & b_{k-1} & b_{k+1} & \cdots & b_{l-1} & b_{l+1} & \cdots & b_n & b & y \end{pmatrix},$$

where $\{x\} = \mathbf{N}\backslash\mathrm{im}(\alpha)$ and $\{y\} = \mathbf{N}\backslash\mathrm{im}(\beta)$. A direct calculation shows that $\beta = \tau\alpha\pi$ and hence $\beta \in \langle A \rangle$.

So, $\langle A \rangle$ contains all elements of $\mathcal{T}_n$ of defects 0 and 1. Let us prove by induction on k that $\langle A \rangle$ contains all elements of $\mathcal{T}_n$ of defect k. We only now have to prove the induction step. Let $\gamma \in \mathcal{T}_n$ be an element of defect k, $k > 1$. Since $\mathrm{def}(\gamma) > 0$, there exists $a \in \mathrm{im}(\gamma)$ such that its preimage $B = \{x \in \mathbf{N} : \gamma(x) = a\}$ contains more than one element. Let $b_1, b_2 \in B$ be such that $b_1 < b_2$. Finally, let $a' \in \mathbf{N}\backslash\mathrm{im}(\gamma)$. Consider the idempotent

$$\mu = \begin{pmatrix} 1 & 2 & \cdots & b_1-1 & b_1 & b_1+1 & \cdots & b_2-1 & b_2 & b_2+1 & \cdots & n \\ 1 & 2 & \cdots & b_1-1 & b_1 & b_1+1 & \cdots & b_2-1 & b_1 & b_2+1 & \cdots & n \end{pmatrix}$$

of rank $(n-1)$ and the transformation δ, which coincides with γ on $\mathbf{N}\backslash\{b_2\}$, and such that $\delta(b_2) = a'$. Then $\mathrm{def}(\delta) = (k-1)$ and hence $\delta \in \langle A \rangle$ by induction. We also know that $\mu \in \langle A \rangle$ since $\mathrm{def}(\mu) = 1$. At the same time, we have $\gamma = \delta\mu$ and hence $\gamma \in \langle A \rangle$. This completes the induction step and the proof. $\square$

Theorem 3.1.4 *Let A be a generating system of $\mathcal{IS}_n$. Then this system is irreducible if and only if $A = A_1 \cup \{\alpha\}$, where A_1 is an irreducible generating system of $\mathcal{S}_n$ and $\mathrm{rank}(\alpha) = (n-1)$.*

Proof. Repeat Theorem 3.1.3 with the following changes: For the partial permutations

$$\alpha = \begin{pmatrix} 1 & 2 & \cdots & i-1 & i & i+1 & \cdots & n \\ a_1 & a_2 & \cdots & a_{i-1} & \varnothing & a_{i+1} & \cdots & a_n \end{pmatrix}$$

and

$$\beta = \begin{pmatrix} 1 & 2 & \cdots & j-1 & j & i+1 & \cdots & n \\ b_1 & b_2 & \cdots & b_{j-1} & \varnothing & b_{j+1} & \cdots & b_n \end{pmatrix}$$

the permutations π and τ should be chosen as follows: $\pi(j) = i$;

$$\tau = \begin{pmatrix} c_1 & c_2 & \cdots & c_{j-1} & c_{j+1} & \cdots & c_n & x \\ b_1 & b_2 & \cdots & b_{j-1} & b_{j+1} & \cdots & b_n & y \end{pmatrix}.$$

For the partial permutation

$$\gamma = \begin{pmatrix} i_1 & i_2 & \cdots & i_k & i_{k+1} & \cdots & i_n \\ j_1 & j_2 & \cdots & j_k & \varnothing & \cdots & \varnothing \end{pmatrix}$$

the partial permutations δ and μ should be chosen as follows:

$$\delta = \begin{pmatrix} i_1 & i_2 & \cdots & i_k & i_{k+1} & i_{k+2} & \cdots & i_n \\ j_1 & j_2 & \cdots & j_k & j_{k+1} & \varnothing & \cdots & \varnothing \end{pmatrix}$$

$$\mu = \begin{pmatrix} i_1 & i_2 & \cdots & i_k & i_{k+1} & i_{k+2} & \cdots & i_n \\ i_1 & i_2 & \cdots & i_k & \varnothing & i_{k+2} & \cdots & i_n \end{pmatrix}.$$

     $\square$

Theorem 3.1.5 *Let A be a generating system of $\mathcal{PT}_n$. Then this system is irreducible if and only if $A = A_1 \cup \{\alpha, \beta\}$, where A_1 is an irreducible generating system of $\mathcal{S}_n$, α is a total transformation of rank $(n-1)$ and β is a partial permutation of rank $(n-1)$.*

Proof. A composition of total transformations is a total transformation. On the other hand, a composition which consists of nontotal maps only, results in an element which is not total either. Hence from the proof of Lemma 3.1.2, it follows that each generating system of the semigroup $\mathcal{PT}_n$ must contain at least one total transformation, say α, of rank $(n-1)$ and at least one partial transformation, say β, of rank $(n-1)$, which is not total. We have $|\mathrm{dom}(\beta)| \leq$

$(n-1)$ and $|\mathrm{im}(\beta)| = (n-1)$. Moreover, the mapping $\beta : \mathrm{dom}(\beta) \to \mathrm{im}(\beta)$ is surjective. This implies $|\mathrm{dom}(\beta)| = n-1$ and $\beta : \mathrm{dom}(\beta) \to \mathrm{im}(\beta)$ is bijective.

Taking Lemma 3.1.1 and the previous paragraph into account, we have to only show that any $A = A_1 \cup \{\alpha, \beta\}$, where A_1 is an irreducible generating system of $\mathcal{S}_n$, α is a total transformation of rank $(n-1)$, and β is a partial permutation of rank $(n-1)$, generates $\mathcal{PT}_n$.

By Theorems 3.1.3 and 3.1.4 the sets $A_1 \cup \{\alpha\}$ and $A_1 \cup \{\beta\}$ generate $\mathcal{T}_n$ and $\mathcal{IS}_n$, respectively. Each $\gamma \in \mathcal{PT}_n$ can be extended to a total transformation $\gamma' \in \mathcal{T}_n$. Then we have $\gamma = \gamma' \varepsilon_{\mathrm{dom}(\gamma)}$ and $\varepsilon_{\mathrm{dom}(\gamma)} \in \mathcal{IS}_n$. Hence $\gamma \in \langle A \rangle$. This completes the proof. $\qquad\square$

3.2   Addenda and Comments

3.2.1 Theorem 3.1.3 was proved in [Vo2].

3.2.2 Although all groups are semigroups, group theory and semigroup theory are two completely different directions of modern algebra. They differ both in formulations of the principal problems and in methods of their solutions. In particular, a problem in semigroup theory is usually considered as solved if it is reduced to a problem in group theory. Theorems 3.1.3–3.1.5 are good examples of such reduction. They reduce the problem of classification of all irreducible generating systems for the semigroups $\mathcal{PT}_n$, $\mathcal{T}_n$, and $\mathcal{IS}_n$ to the analogous problem for the group $\mathcal{S}_n$. Irreducible generating systems in $\mathcal{S}_n$ have very complicated structure. Two classical irreducible generating systems are: the *Coxeter system* $\{(1,2),(2,3),\ldots,(n-1,n)\}$ and the system $\{(1,2),(1,2,\ldots,n)\}$ consisting of two generators. However, the problem to classify all irreducible generating systems in $\mathcal{S}_n$ is still open. At the same time Theorems 3.1.3–3.1.5 give a satisfactory classification of irreducible generating systems for $\mathcal{PT}_n$, $\mathcal{T}_n$, and $\mathcal{IS}_n$.

3.2.3 In each of the semigroups $\mathcal{PT}_n$, $\mathcal{T}_n$, and $\mathcal{IS}_n$, the set of all noninvertible elements forms a subsemigroup (called the *singular part* of the original semigroup). Hence we have a new natural problem: to study generating systems for the singular parts of our semigroups $\mathcal{PT}_n$, $\mathcal{T}_n$, and $\mathcal{IS}_n$. Here, apart from irreducibility, one could also impose other conditions, for example to consist of idempotents, to consist of elements of defect 1, to consists of elements with prescribed structure of connected components, and so on. One of the first works in this direction was [Ho1] in which it was shown that the singular part of $\mathcal{T}_n$ is generated by idempotents of defect 1. One of the most recent works is [AAH], where the authors reprove some known results (in particular, the above-mentioned result from [Ho1]) and give several new examples of generating systems for the singular part of $\mathcal{T}_n$.

3.3 Additional Exercises

3.3.1 Prove that each element of defect k from $\mathcal{T}_n$ can be written as a product of a permutation and k idempotents of defect 1.

3.3.2 Let $A \subset \mathcal{S}_n$ be a subset consisting of transpositions. Define the (un-oriented) graph Γ_A in the following way: the vertices are $\{1, 2, \ldots, n\}$, and for $i, j \in \{1, 2, \ldots, n\}$ the graph Γ_A contains the edge (i, j) if and only if $(i, j) \in A$. Prove the following:

(a) A is a generating system of $\mathcal{S}_n$ if and only if Γ_A is connected.

(b) A is an irreducible generating system of $\mathcal{S}_n$ if and only if Γ_A is a tree.

(c) $\mathcal{S}_n$ has exactly n^{n-2} irreducible generating systems, consisting of transpositions.

3.3.3 (a) Show that for $n > 2$ the group $\mathcal{S}_n$ does not contain any one-element generating system.

(b) Show that $(1, 2)$, $(2, 3), \ldots, (n-1, n)$ generate $\mathcal{S}_n$.

(c) Show that $(1, 2)$ and $(1, 2, \ldots, n)$ generate $\mathcal{S}_n$.

3.3.4 ([Ho1]) Prove that the semigroup $\mathcal{T}_n \backslash \mathcal{S}_n$ is generated by the set of all idempotents of defect 1.

Chapter 4

Ideals and Green's Relations

4.1 Ideals of Semigroups

Let S be a semigroup. If $A, B \subset S$, we set $AB = \{ab : a \in A; b \in B\}$.

A subset $I \subset S$ is called a *left ideal* provided that for all $a \in S$ and $b \in I$ we have $ab \in I$ (in other words, $SI \subset I$). Analogously, I is called a *right ideal* provided that $IS \subset I$. Left and right ideals are also called *one-sided* ideals. A subset $I \subset S$ is called a *two-sided ideal* or simply an *ideal* provided that it is both a left and a right ideal.

Let I be a (one-sided) ideal of S. Then we have $I \cdot I \subset I$ by definition and hence I is a subsemigroup of S. The converse is not true in the general case. For example, the group $\mathcal{S}_n$ is a subsemigroup of each of the semigroups $\mathcal{T}_n$, $\mathcal{PT}_n$, and $\mathcal{IS}_n$. However, for each (partial) transformation $\alpha \notin \mathcal{S}_n$ both sets $\alpha \mathcal{S}_n$ and $\mathcal{S}_n \alpha$ do not have any common elements with $\mathcal{S}_n$.

It is easy to see that both the intersection and the union of an arbitrary family of left (right, two-sided) ideals of S are in turn a left (resp. right, two-sided) ideal of S.

Exercise 4.1.1 Show that for any $A \subset S$ the set AS is a right ideal of S; the set SA is a left ideal of S; and the set SAS is a two-sided ideal of S.

For each semigroup S we denote the semigroup S by S^1 provided that S contains an identity element, and the semigroup constructed in Proposition 2.2.1 provided that S does not contain any identity element. Analogously we define S^0. We also note that for $k > 1$ the notation S^k means something completely different, namely, $S^k = \{a_1 a_2 \cdots a_k : a_i \in S.\ i = 1, \ldots, k\}$.

A left (resp. right or two-sided) ideal I of S is called *principal* provided that there exists $a \in S$ such that $I = S^1 a$ (resp. $I = aS^1$, $I = S^1 aS^1$). The element a is called the *generator* of the ideal I. Note that $a \in S^1 a$, $a \in aS^1$, and $a \in S^1 aS^1$ by definition.

O. Ganyushkin, V. Mazorchuk, *Classical Finite Transformation Semigroups*, Algebra and Applications 9, DOI: 10.1007/978-1-84800-281-4_4,
© Springer-Verlag London Limited 2009

Proposition 4.1.2 *Each left (right or two-sided) ideal is a union of principal left (resp. right or two-sided) ideals.*

Proof. Since $a \in S^1 a$, if $I \subset S$ is a left ideal, we have $I = \cup_{a \in I} S^1 a$. For other ideals the argument is similar. $\qquad\square$

4.2 Principal Ideals in $\mathcal{T}_n$, $\mathcal{PT}_n$, and $\mathcal{IS}_n$

Since the semigroups $\mathcal{T}_n$, $\mathcal{PT}_n$, and $\mathcal{IS}_n$ are in fact monoids, we have $S = S^1$ for these semigroups.

Theorem 4.2.1 *Let S denote one of the semigroups $\mathcal{T}_n$, $\mathcal{PT}_n$, or $\mathcal{IS}_n$. Then for each $\alpha \in S$ the right principal ideal generated by α has the following form*

$$\alpha S = \{\beta \in S : \operatorname{im}(\beta) \subset \operatorname{im}(\alpha)\}.$$

Proof. Denote $X = \{\beta \in S : \operatorname{im}(\beta) \subset \operatorname{im}(\alpha)\}$. If $\gamma \in S$, we have $\alpha\gamma(x) = \alpha(\gamma(x))$ by definition. Hence $\operatorname{im}(\alpha\gamma) \subset \operatorname{im}(\alpha)$ and thus $\alpha S \subset X$.

To prove the inverse inclusion consider an arbitrary $\beta \in X$. We have $\operatorname{im}(\beta) \subset \operatorname{im}(\alpha)$. For each $b \in \operatorname{im}(\beta)$ choose some a_b such that $\alpha(a_b) = b$. Consider the transformation γ for which $\operatorname{dom}(\gamma) = \operatorname{dom}(\beta)$ and such that for $x \in \operatorname{dom}(\beta)$ we have $\gamma(x) = a_b$ if and only if $\beta(x) = b$. A direct calculation then shows that $\alpha\gamma = \beta$. Note that in the case $S = \mathcal{IS}_n$ the injectivity of α and β gives the following implications for $x_1, x_2 \in \operatorname{dom}(\gamma)$:

$$x_1 \neq x_2 \Rightarrow b_1 = \beta(x_1) \neq \beta(x_2) = b_2 \Rightarrow a_{b_1} \neq a_{b_2} \Rightarrow \gamma(x_1) \neq \gamma(x_2).$$

Hence $\gamma \in \mathcal{IS}_n$. Thus $\beta \in \alpha S$. This implies that $X \subset \alpha S$ and completes the proof. $\qquad\square$

Corollary 4.2.2 (i) *Each of the semigroups $\mathcal{PT}_n$ and $\mathcal{IS}_n$ has exactly 2^n different principal right ideals.*

(ii) *The semigroup $\mathcal{T}_n$ has exactly $2^n - 1$ different principal right ideals.*

Proof. From Theorem 4.2.1 we have that a principal right ideal is uniquely determined by $\operatorname{im}(\alpha)$, that is, by a subset of $\mathbf{N}$, which must be nonempty in the case of $\mathcal{T}_n$. The claim follows. $\qquad\square$

Corollary 4.2.3 *Let S denote one of the semigroups $\mathcal{T}_n$, $\mathcal{PT}_n$, or $\mathcal{IS}_n$ and $\alpha \in S$ be such that $\operatorname{rank}(\alpha) = k$. Then*

$$|\alpha S| = \begin{cases} k^n, & S = \mathcal{T}_n; \\ (k+1)^n, & S = \mathcal{PT}_n; \\ \sum\limits_{i=0}^{k} \binom{n}{i}\binom{k}{i} i!, & S = \mathcal{IS}_n. \end{cases}$$

Proof. Theorem 4.2.1 says that in the case of $\mathcal{T}_n$ the ideal αS is the set of all mappings from **N** to $\text{im}(\alpha)$, where $|\mathbf{N}| = n$ and $|\text{im}(\alpha)| = k$. Hence $|\alpha S| = k^n$.

In the case of $\mathcal{PT}_n$, Theorem 4.2.1 says that αS is the set of all partial mappings from **N** to $\text{im}(\alpha)$, that is, the set of all mappings from **N** to $\text{im}(\alpha) \cup \{\varnothing\}$. Hence $|\alpha S| = (k+1)^n$.

Finally, for $\mathcal{IS}_n$, Theorem 4.2.1 says that αS is the set of all partial injections from **N** to $\text{im}(\alpha)$. Such a partial injection can have rank $i = 0, 1, \ldots, k$. If i is fixed, to define such a partial injection, say β, we have to choose its domain, that is, an i-element subset of **N**, which can be done in $\binom{n}{i}$ different ways; then we have to choose $\text{im}(\beta)$, that is, an i-element subset of $\text{im}(\alpha)$, which can be done in $\binom{k}{i}$ different ways; finally we have to define a bijection from $\text{dom}(\beta)$ to $\text{im}(\beta)$, which can be done in $i!$ different ways. The statement now follows by applying the product rule and the summing up over all i. $\square$

With each $\alpha \in \mathcal{PT}_n$, we associate the binary relation π_α on **N** in the following way

$$x \, \pi_\alpha \, y \Leftrightarrow \big((x, y \in \text{dom}(\alpha) \text{ and } \alpha(x) = \alpha(y)) \text{ or } (x, y \notin \text{dom}(\alpha)) \big) . \qquad (4.1)$$

In other words, $x \, \pi_\alpha \, y$ if either α is not defined on both x and y, or α is defined on both x and y and $\alpha(x) = \alpha(y)$. It is clear that π_α is an equivalence relation and the equivalence classes of π_α are $\overline{\text{dom}(\alpha)}$ and all full preimages of elements from $\text{im}(\alpha)$.

Theorem 4.2.4 *Let S denote one of the semigroups $\mathcal{T}_n$, $\mathcal{PT}_n$, or $\mathcal{IS}_n$. Then for each $\alpha \in S$ the left principal ideal generated by α has the following form*

$$S\alpha = \{\beta \in S \, : \, \text{dom}(\beta) \subset \text{dom}(\alpha) \text{ and } \pi_\alpha \subset \pi_\beta\}. \qquad (4.2)$$

Proof. Denote the right-hand side of (4.2) by X. Let $\beta \in S\alpha$. Then $\beta = \gamma\alpha$ for some $\gamma \in S$. For arbitrary $x \in \text{dom}(\beta)$ we have $\beta(x) = \gamma(\alpha(x))$. Hence $x \in \text{dom}(\alpha)$ and thus $\text{dom}(\beta) \subset \text{dom}(\alpha)$. Furthermore, if $x, y \in \text{dom}(\beta)$ and $\alpha(x) = \alpha(y)$, we have $\beta(x) = \gamma(\alpha(x)) = \gamma(\alpha(y)) = \beta(y)$. Thus $\pi_\alpha \subset \pi_\beta$ and we have $S\alpha \subset X$.

Assume now that $\beta \in X$. Set $M = \alpha(\text{dom}(\beta))$ and define the mapping $\gamma : M \to \mathbf{N}$ as follows:

$$\gamma(\alpha(y)) = \beta(y). \qquad (4.3)$$

The mapping γ is well-defined since if $x = \alpha(y_1) = \alpha(y_2)$, we have $y_1 \, \pi_\alpha \, y_2$, which implies $y_1 \, \pi_\beta \, y_2$ and thus $\beta(y_1) = \beta(y_2)$. Hence the value $\gamma(x)$ is uniquely defined.

If $S = \mathcal{T}_n$, we have $\text{dom}(\beta) = \mathbf{N}$ and $M = \alpha(\mathbf{N}) = \text{im}(\alpha)$. In this case, we can extend γ to a total transformation on **N** in an arbitrary way. Abusing notation we denote the resulting total transformation also by γ. If $S \neq \mathcal{T}_n$, we just consider γ as a partial transformation on **N** with domain M. In particular, if $S = \mathcal{IS}_n$ and $x_1 = \alpha(y_1)$, $x_2 = \alpha(y_2)$ are two different elements in M, we have $y_1 \neq y_2$ and from the injectivity of β it follows that $\gamma(x_1) = \beta(y_1)$ and $\gamma(x_2) = \beta(y_2)$ are also different. This means that in this case $\gamma \in \mathcal{IS}_n$ as well.

From the definition of γ we have that $y \notin \mathrm{dom}(\beta)$ implies either $y \notin \mathrm{dom}(\alpha)$, or $\alpha(y) \notin M$. Indeed, assume that $y \in \mathrm{dom}(\alpha)$ and $\alpha(y) \in M$. Then there exists $y_1 \in \mathrm{dom}(\beta)$ such that $\alpha(y) = \alpha(y_1)$. Hence $y \pi_\alpha y_1$ and thus $y \pi_\beta y_1$, implying $y \in \mathrm{dom}(\beta)$, a contradiction. This means that for $y \notin \mathrm{dom}(\beta)$ the equality $(\gamma\alpha)(y) = \gamma(\alpha(y))$ implies that $y \notin \mathrm{dom}(\gamma\alpha)$.

If $y \in \mathrm{dom}(\beta)$, we have $\alpha(y) \in M$ and the equality (4.3) implies that $(\gamma\alpha)(y) = \gamma(\alpha(y)) = \beta(y)$. This means that $\beta = \gamma\alpha$ and thus $\beta \in S\alpha$. Thus $X \subset S\alpha$ completing the proof.   □

If $S = \mathcal{T}_n$, then $\mathrm{dom}(\alpha) = \mathrm{dom}(\beta) = \mathbf{N}$ and hence the first condition in (4.2) can be omitted. If $S = \mathcal{IS}_n$, then the restriction of π_α to $\mathrm{dom}(\alpha)$ becomes the equality relation. Hence $\mathrm{dom}(\beta) \subset \mathrm{dom}(\alpha)$ automatically implies $\pi_\alpha \subset \pi_\beta$. This means that for the semigroup $S = \mathcal{IS}_n$ the second condition in (4.2) can be omitted.

The number of all unordered partitions $\mathbf{N} = N_1 \cup N_2 \cup \cdots \cup N_k$ of the set $\mathbf{N}$ into disjoint unions of nonempty blocks is called the *nth Bell number* and is denoted by B_n. From the definition of the Stirling numbers $\mathsf{S}(n, k)$ of the second kind we immediately have the equality $\mathsf{B}_n = \sum_{k=1}^{n} \mathsf{S}(n, k)$.

Corollary 4.2.5 *(i) The number of different principal left ideals in the semigroup $\mathcal{T}_n$ equals B_n.*

   (ii) The number of different principal left ideals in the semigroup $\mathcal{PT}_n$ equals B_{n+1}.

   (iii) The number of different principal left ideals in the semigroup $\mathcal{IS}_n$ equals 2^n.

Proof. From Theorem 4.2.4 we have that a principal left ideal of $\mathcal{T}_n$ is uniquely determined by the partition of $\mathbf{N}$ into the equivalence classes of the relation π_α. Since all partitions obviously give rise to some ideals, we obtain that the number of ideals equals B_n. This proves (i).

To prove (ii) we take a new symbol, say $n + 1$. Let $\alpha \in \mathcal{PT}_n$. With π_α we associate the following equivalence relation π'_α on the set $\mathbf{N} \cup \{n + 1\}$: If $\mathrm{dom}(\alpha) \neq \mathbf{N}$, we adjoin $n + 1$ to the block $\overline{\mathrm{dom}}(\alpha)$; if $\mathrm{dom}(\alpha) = \mathbf{N}$, then we say that $\{n + 1\}$ is a new equivalence class. Obviously, π'_α is uniquely determined by π_α. Conversely, any π'_α uniquely determines π_α and $\overline{\mathrm{dom}}(\alpha)$.

By Theorem 4.2.4 each principal left ideal of $\mathcal{PT}_n$ is uniquely determined by the pair $(\pi_\alpha, \overline{\mathrm{dom}}(\alpha))$. Hence the previous paragraph says that the mapping $(\pi_\alpha, \overline{\mathrm{dom}}(\alpha)) \mapsto \pi'_\alpha$ is a bijection from the set of all principal left ideals of $\mathcal{PT}_n$ to the set of all unordered partitions of $\mathbf{N} \cup \{n + 1\}$, the latter containing $n + 1$ element. Thus the number of different principal left ideals of $\mathcal{PT}_n$ equals B_{n+1}. This proves (ii).

Finally, to prove (iii) we note that by Theorem 4.2.4 each principal left ideal of $\mathcal{IS}_n$ is uniquely determined by the set $\mathrm{dom}(\alpha)$, which is just a subset of $\mathbf{N}$. Hence the number of such ideals is 2^n. $\qquad\square$

Remark 4.2.6 Corollary 4.2.5(iii) also follows from Corollary 4.2.2(ii) and 2.9.9.

Corollary 4.2.7 *Let S denote one of the semigroups $\mathcal{T}_n$, $\mathcal{PT}_n$, or $\mathcal{IS}_n$, and $\alpha \in S$ be an element of rank k. Then we have the following:*

$$|S\alpha| = \begin{cases} n^k, & S = \mathcal{T}_n; \\ (n+1)^k, & S = \mathcal{PT}_n; \\ \displaystyle\sum_{i=0}^{k} \binom{n}{i}\binom{k}{i} i!, & S = \mathcal{IS}_n. \end{cases}$$

Proof. We start with the cases $S = \mathcal{T}_n$ and $S = \mathcal{PT}_n$. As $\pi_\alpha \subset \pi_\beta$, each transformation $\beta \in S\alpha$ is uniquely determined by its values on the set of those equivalence classes of π_α which are contained in $\mathrm{dom}(\alpha)$. We have k such classes and for each of them we have to choose the value of β on this class, which is an element from $\mathbf{N}$ (or $\mathbf{N} \cup \{\varnothing\}$ in the case of $\mathcal{PT}_n$). Now for $\mathcal{T}_n$ and $\mathcal{PT}_n$ the statements are obtained by applying the product rule.

In the case $S = \mathcal{IS}_n$ the proof is similar to that of the corresponding part of Corollary 4.2.3. The only difference is that now one has to consider partial injections from $\mathrm{dom}(\alpha)$ to $\mathbf{N}$. $\qquad\square$

Theorem 4.2.8 *Let S denote one of the semigroups $\mathcal{T}_n$, $\mathcal{PT}_n$, or $\mathcal{IS}_n$ and $\alpha \in S$. Then the principal ideal generated by α has the following form:*

$$S\alpha S = \{\beta \in S \ : \ \mathrm{rank}(\beta) \leq \mathrm{rank}(\alpha)\}. \tag{4.4}$$

Proof. Let X denote the right-hand side of (4.4). The inclusion $S\alpha S \subset X$ follows from Exercise 2.1.4(c), so we have only to prove the inclusion $X \subset S\alpha S$.

Let $\mathrm{im}(\alpha) = \{a_1, a_2, \dots, a_k\}$, and $\beta \in X$ be such that $\mathrm{rank}(\beta) = m$, and $\mathrm{im}(\beta) = \{b_1, b_2, \dots, b_m\}$. Then $m \leq k$ and for each $i = 1, \dots, k$ we choose some element c_i in the set $A_i = \{x \in \mathbf{N} \ : \ \alpha(x) = a_i\}$. Define $\gamma \in \mathcal{PT}_n$ in the following way: $\mathrm{dom}(\gamma) = \mathrm{dom}(\beta)$, and for all $y \in B_j = \{z \in \mathbf{N} \ : \ \beta(z) = b_j\}$, $j = 1, \dots, m$, we set $\gamma(y) = c_j$. Note that $\beta \in \mathcal{T}_n$ implies $\gamma \in \mathcal{T}_n$ and $\beta \in \mathcal{IS}_n$ implies $\gamma \in \mathcal{IS}_n$ by construction. Let δ be any permutation satisfying $\delta(a_i) = b_i$, $i = 1, \dots, m$. Then a direct calculation gives $\delta\alpha\gamma = \beta$, which implies $\beta \in S\alpha S$. Thus $X \subset S\alpha S$ and the proof is complete. $\qquad\square$

Corollary 4.2.9 *Each of the semigroups $\mathcal{PT}_n$ and $\mathcal{IS}_n$ contains $(n+1)$ different principal ideals. The semigroup $\mathcal{T}_n$ contains n different principal ideals.*

Proof. By Theorem 4.2.8 a principal ideal is uniquely determined by the rank of the generator. For $\mathcal{PT}_n$ and $\mathcal{IS}_n$ the rank varies between 0 and n, for $\mathcal{T}_n$ it varies between 1 and n. □

Corollary 4.2.10 *Let S denote one of the semigroups $\mathcal{T}_n$, $\mathcal{PT}_n$, or $\mathcal{IS}_n$, and $\alpha \in S$ be an element of rank k. Then we have the following:*

$$|S\alpha S| = \begin{cases} \displaystyle\sum_{i=1}^{k} \mathsf{S}(n,i)\frac{n!}{(n-i)!}, & S = \mathcal{T}_n; \\[2ex] \displaystyle\sum_{i=0}^{k} \mathsf{S}(n+1,i+1)\frac{n!}{(n-i)!}, & S = \mathcal{PT}_n; \\[2ex] \displaystyle\sum_{i=0}^{k} \binom{n}{i}^2 i!, & S = \mathcal{IS}_n. \end{cases}$$

Proof. According to Theorem 4.2.8, to determine $|S\alpha S|$ we have to find the number $\mathbf{t}_i$ of elements of rank $i \in \mathbf{N}$ and then form the sum $\sum_{i \leq k} \mathbf{t}_i$.

If $S = \mathcal{T}_n$, then an element of rank i is uniquely determined by an unordered partition of $\mathbf{N}$ into i nonempty blocks (which can be done in $\mathsf{S}(n,i)$ different ways), and then an injective mapping from the blocks of this partition into $\mathbf{N}$ (this can be done in $n(n-1)\cdots(n-i+1) = \frac{n!}{(n-i)!}$ different ways). Hence in this case $\mathbf{t}_i = \mathsf{S}(n,i)\frac{n!}{(n-i)!}$.

For $S = \mathcal{PT}_n$ the argument is almost the same with the difference that we should consider partitions of $\mathbf{N} \cup \{n+1\}$ instead of those of $\mathbf{N}$. The block containing $n+1$ corresponds to $\overline{\mathrm{dom}}(\beta)$. Hence $\mathbf{t}_i = \mathsf{S}(n+1,i+1)\frac{n!}{(n-i)!}$ in this case.

For $S = \mathcal{IS}_n$ the number of partial permutations of rank k was computed in Theorem 2.5.1. □

4.3   Arbitrary Ideals in $\mathcal{T}_n$, $\mathcal{PT}_n$, and $\mathcal{IS}_n$

Theorem 4.3.1 *Let S denote one of the semigroups $\mathcal{T}_n$, $\mathcal{PT}_n$, or $\mathcal{IS}_n$. Then all two-sided ideals in S are principal and are generated by any element of the ideal, which has the maximal possible rank.*

Proof. Let $I \subset S$ be a two-sided ideal. Choose $\alpha \in I$ of maximal possible rank. As $SIS \subset I$, we have $S\alpha S \subset I$. On the other hand, by Theorem 4.2.8 the set $S\alpha S$ contains all elements whose rank does not exceed that of α. Hence $I \subset S\alpha S$ and thus $I = S\alpha S$. □

Let S denote one of the semigroups $\mathcal{T}_n$, $\mathcal{PT}_n$, or $\mathcal{IS}_n$. For $k \leq n$ set $\mathcal{I}_k = \{\alpha \in S : \mathrm{rank}(\alpha) \leq k\}$. By Theorem 4.3.1, each $\mathcal{I}_k$ is an ideal of S, and each ideal in S is of the form $\mathcal{I}_k$ for some k. In particular, the set of all ideals of S forms a chain with respect to inclusions. For $\mathcal{T}_n$ this chain has the form

$$\mathcal{I}_1 \subset \mathcal{I}_2 \subset \cdots \subset \mathcal{I}_n = \mathcal{T}_n.$$

For $\mathcal{PT}_n$ and $\mathcal{IS}_n$ this chain has the form

$$\{0\} = \mathcal{I}_0 \subset \mathcal{I}_1 \subset \mathcal{I}_2 \subset \cdots \subset \mathcal{I}_n = \mathcal{PT}_n \ (\text{or} \mathcal{IS}_n).$$

Note that $\mathcal{I}_n \backslash \mathcal{I}_{n-1} = \mathcal{S}_n$ in all cases.

The Boolean $\mathcal{B}(X)$ of a set X is partially ordered with respect to inclusions. Recall that an *antichain* of a partially ordered set is a subset such that each two elements in this subset are not comparable. In particular, if $A_1, A_2, \ldots, A_m \subset \mathbf{N}$, the set $\{A_1, \ldots, A_m\}$ is an antichain of $\mathcal{B}(\mathbf{N})$ if and only if for all $i \neq j$ we have both $A_i \not\subset A_j$ and $A_j \not\subset A_i$.

Let $L \subset \mathcal{B}(\mathbf{N})$ be an antichain. Set

$$\mathcal{I}_L = \{\alpha \in S : \text{there exists} A \in L \text{such that} \operatorname{im}(\alpha) \subset A\}.$$

Theorem 4.3.2 *Let S denote one of the semigroups $\mathcal{T}_n$, $\mathcal{PT}_n$, or $\mathcal{IS}_n$.*

(i) For each antichain $L \subset \mathcal{B}(\mathbf{N})$ the set $\mathcal{I}_L$ is a right ideal of S.

(ii) Let L_1 and L_2 be two antichains in $\mathcal{B}(\mathbf{N})$. Then $L_1 \neq L_2$ implies $\mathcal{I}_{L_1} \neq \mathcal{I}_{L_2}$.

(iii) For each right ideal I of S there exists an antichain $L \subset \mathcal{B}(\mathbf{N})$ such that $I = \mathcal{I}_L$.

Proof. For each $\alpha \in \mathcal{I}_L$ and $\mu \in S$ from $\operatorname{im}(\alpha\mu) \subset \operatorname{im}(\alpha)$ (see Exercise 2.1.4(b)) it follows that $\operatorname{im}(\alpha\mu) \in \mathcal{I}_L$. This proves (i).

Let $L_1 \neq L_2$. Without loss of generality we may assume $L_1 \backslash L_2 \neq \varnothing$. Let $A \in L_1 \backslash L_2$. We have to consider two possible cases.

Case 1: A is not comparable with any element of L_2. In this case each $\alpha \in S$ for which $\operatorname{im}(\alpha) = A$ is contained in $\mathcal{I}_{L_1}$ but not in $\mathcal{I}_{L_2}$. Hence $\mathcal{I}_{L_1} \neq \mathcal{I}_{L_2}$.

Case 2: There exists $B \in L_2$ such that A and B are comparable. Then $B \subset L_2 \backslash L_1$. Without loss of generality we may assume $B \subset A$. As $B \neq A$, each element $\alpha \in S$ for which $\operatorname{im}(\alpha) = A$ is contained in $\mathcal{I}_{L_1} \backslash \mathcal{I}_{L_2}$. Hence $\mathcal{I}_{L_1} \neq \mathcal{I}_{L_2}$ in this case as well.

The above proves (ii).

To prove (iii) we first observe that the maximal (with respect to inclusions) elements of the set $\{\operatorname{im}(\alpha) : \alpha \in I\}$ obviously form an antichain, say L, of $\mathcal{B}(\mathbf{N})$. By the definition of L, for each $\alpha \in I$ there exists $A \in L$ such that $\operatorname{im}(\alpha) \subset A$. Hence $\alpha \in \mathcal{I}_L$ and $I \subset \mathcal{I}_L$.

On the other hand, for each $\beta \in \mathcal{I}_L$ there exists $A \in L$ and $\alpha \in I$ such that $\operatorname{im}(\beta) \subset A$ and $\operatorname{im}(\alpha) = A$. By Theorem 4.2.1 we have $\beta \in \alpha S$. But $\alpha S \subset I$ since $\alpha \in I$ and I is a right ideal. Hence $\beta \in I$. This means that $\mathcal{I}_L \subset I$ and thus $\mathcal{I}_L = I$. The proof is complete. $\qquad \square$

The identity mapping from S to the dual semigroup $\overleftarrow{S}$ maps left ideals to right ideals and vice versa. For the inverse semigroup $\mathcal{IS}_n$ we have $\mathcal{IS}_n \cong \overleftarrow{\mathcal{IS}}_n$ via the mapping $a \mapsto a^{-1}$. Hence Theorem 4.3.2 can be used to describe all left ideals in $\mathcal{IS}_n$. The only change one has to make is to substitute im by dom. Indeed, the equalities $\text{im}(\alpha^{-1}) = \text{dom}(\alpha)$ and $\text{im}(\alpha) = \text{dom}(\alpha^{-1})$ follow from the proof of Theorem 2.6.7. Hence we get the following statement:

Theorem 4.3.3 *For each antichain L of $\mathcal{B}(X)$ the set*

$$_L\mathcal{I} = \{\alpha \in S : \text{ there exists} A \in L \text{such that} \text{dom}(\alpha) \subset A\}$$

is a left ideal of $\mathcal{IS}_n$ and the mapping $L \mapsto {}_L\mathcal{I}$ is a bijection from the set of all antichains of $\mathcal{B}(X)$ to the set of all left ideals of $\mathcal{IS}_n$.

As $\mathcal{T}_n$ and $\mathcal{PT}_n$ are not self-dual (see 2.9.9 and Exercise 2.10.23), the description of all left ideals in these semigroups requires more efforts.

Let Part_n denote the set of all (unordered) partitions of $\mathbf{N}$. Define a partial order on Part_n in the following way: Let $\rho = A_1 \cup \cdots \cup A_k$ and $\tau = B_1 \cup \cdots \cup B_m$ be two partitions. We will write $\rho \preceq \tau$ provided that each block A_i is a union of some blocks of the partition τ (i.e., of some B_js). In particular, the maximum partition is the one into one element blocks and the minimum partition is the one with one block.

Let ρ be a partition of $\mathbf{N}$. Denote by π_ρ the corresponding equivalence relation ($x \pi_\rho y$ if and only if x and y belong to the same block of ρ). It is easy to see that $\rho \preceq \tau$ if and only if $\pi_\rho \supset \pi_\tau$.

Example 4.3.4 If we have $A_1 = \{a_1, \ldots, a_p\}$, $A_2 = \{b_1, \ldots, b_q\}, \ldots, A_k = \{c_1, \ldots, c_r\}$, the partition $A_1 \cup A_2 \cup \cdots \cup A_k$ can be written for example as follows:

$$a_1, \ldots, a_p | b_1, \ldots, b_q | \cdots | c_1, \ldots, c_r.$$

Using this notation the Hasse diagram of the partially ordered set of all partitions of $\{1, 2, 3, 4\}$ looks as follows:

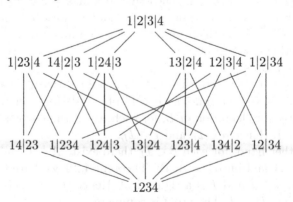

To each element $\alpha \in \mathcal{T}_n$ we associate the partition ρ_α of $\mathbf{N}$ into equivalence classes with respect to the relation π_α. Then $\rho_\alpha \preceq \rho_\beta$ if and only if $\pi_\alpha \supset \pi_\beta$.

Theorem 4.3.5 *For each antichain* L *in* Part_n *the set*

$$_L\mathcal{I} = \{\alpha \in \mathcal{T}_n : \text{there exists } \rho \in L \text{ such that } \rho_\alpha \preceq \rho\}$$

is a left ideal of $\mathcal{T}_n$ *and the mapping* $L \mapsto {_L\mathcal{I}}$ *is a bijection from the set of all antichains of* Part_n *to the set of all left ideals of* $\mathcal{T}_n$.

Proof. For any $\mu, \nu \in \mathcal{T}_n$ the equality $\mu(x) = \mu(y)$ implies the equality $(\nu\mu)(x) = (\nu\mu)(y)$. Hence $\pi_\mu \subset \pi_{\nu\mu}$ and $\rho_{\nu\mu} \preceq \rho_\mu$. Thus for arbitrary $\alpha \in {_L\mathcal{I}}$ and $\beta \in \mathcal{T}_n$, from $\rho_\alpha \preceq \rho$ it follows that $\rho_{\beta\alpha} \preceq \rho_\alpha \preceq \rho$. Hence $\beta\alpha \in {_L\mathcal{I}}$ and $_L\mathcal{I}$ is a left ideal.

It is clear that for each partition $\rho \in \mathrm{Part}_n$ there exists $\alpha \in \mathcal{T}_n$ such that $\rho_\alpha = \rho$. Hence using the same arguments as in the proof of Theorem 4.3.2(ii) (with the obvious substitution of $\mathrm{im}(\alpha)$ by ρ_α) we obtain that for all antichains L_1 and L_2 of Part_n such that $L_1 \neq L_2$ the ideals $_{L_1}\mathcal{I}$ and $_{L_2}\mathcal{I}$ are different.

Finally, to prove that each left ideal of $\mathcal{T}_n$ has the form $_L\mathcal{I}$ for some antichain L in Part_n one should just follow the proof of Theorem 4.3.2(iii): If I is a left ideal of $\mathcal{T}_n$, the antichain L is defined as the set of maximal (with respect to $\preceq$) elements of the set $\{\rho_\alpha : \alpha \in I\}$. One should also use Theorem 4.2.4 instead of Theorem 4.2.1. $\square$

With each element $\alpha \in \mathcal{PT}_n$ we associate the partition ρ_α of $\mathbf{N} \cup \{n+1\}$ as it was done in the proof of Corollary 4.2.5(ii).

Theorem 4.3.6 *For each antichain* L *in* Part_{n+1} *the set*

$$_L\mathcal{I} = \{\alpha \in \mathcal{PT}_n : \text{there exists } \rho \in L \text{ such that } \rho_\alpha \preceq \rho\}$$

is a left ideal of $\mathcal{PT}_n$ *and the mapping* $L \mapsto {_L\mathcal{I}}$ *is a bijection from the set of all antichains of* Part_{n+1} *to the set of all left ideals of* $\mathcal{PT}_n$.

Proof. The proof repeats that of Theorem 4.3.5. $\square$

4.4 Green's Relations

In this section, we introduce several equivalence relations on semigroups, which play a central role in the structure theory. Let S be a semigroup. Elements $a, b \in S$ are called $\mathcal{L}$-*equivalent* provided that they generate the same principal left ideal. In other words, $a\mathcal{L}b$ if and only if $S^1a = S^1b$. Equivalence classes of the relation $\mathcal{L}$ are called $\mathcal{L}$-*classes*. For $a \in S$, the $\mathcal{L}$-class containing a will be denoted by $\mathcal{L}(a)$. In other words, $a\mathcal{L}b$ if and only if $a \in \mathcal{L}(b)$. The following easy but very useful fact follows immediately from the definition.

Proposition 4.4.1 $a\mathcal{L}b$ *if and only if there exist* $x, y \in S^1$ *such that* $a = xb$ *and* $b = ya$.

In the dual way we define the relation $\mathcal{R}$: Elements $a, b \in S$ are called $\mathcal{R}$-*equivalent* provided that they generate the same principal right ideal. In other words, $a\mathcal{R}b$ if and only if $aS^1 = bS^1$. Equivalence classes of the relation $\mathcal{R}$ are called $\mathcal{R}$-*classes*. For $a \in S$ the $\mathcal{R}$-class containing a will be denoted by $\mathcal{R}(a)$.

Proposition 4.4.2 $a\mathcal{R}b$ *if and only if there exist* $x, y \in S^1$ *such that* $a = bx$ *and* $b = ay$.

Note that for every $x \in S^1$ the equality $S^1a = S^1b$ implies the equality $S^1ax = S^1bx$; and the equality $aS^1 = bS^1$ implies the equality $xaS^1 = xbS^1$. In other words, the relation $\mathcal{L}$ is *compatible* with the right multiplication by elements of S^1 (this is also called *right compatible*), that is, $a\mathcal{L}b$ implies $ax\mathcal{L}bx$. Dually, the relation $\mathcal{R}$ is *left compatible*, that is, $a\mathcal{R}b$ implies $xa\mathcal{R}xb$. A relation which is both left and right compatible is simply called *compatible*.

If X is a set and ξ, η are two binary relations on X, then the *product* $\xi \circ \eta$ is defined in the following way:

$$\xi \circ \eta = \{(a,b) : \text{there exists } c \in X \text{ such that } (a,c) \in \xi \text{ and } (c,b) \in \eta\}.$$

Exercise 4.4.3 Show that the set of all binary relations on X with respect to the product defined above is a monoid.

Lemma 4.4.4 *The relations* $\mathcal{L}$ *and* $\mathcal{R}$ *commute, that is,* $\mathcal{L} \circ \mathcal{R} = \mathcal{R} \circ \mathcal{L}$.

Proof. Let $(a,b) \in \mathcal{L} \circ \mathcal{R}$. Then there exists $c \in S$ such that $a\mathcal{L}c$ and $c\mathcal{R}b$. By Propositions 4.4.1 and 4.4.2, there exist $x, y \in S^1$ such that $a = xc$, $b = cy$. Moreover, $c\mathcal{R}b$ implies $xc\mathcal{R}xb$, and $a\mathcal{L}c$ implies $ay\mathcal{L}cy$. But $xc = a$, $xb = xcy = ay$, and $cy = b$. Hence $a\mathcal{R}xcy$ and $xcy\mathcal{L}b$, which implies $(a,b) \in \mathcal{R} \circ \mathcal{L}$, and thus $\mathcal{L} \circ \mathcal{R} \subset \mathcal{R} \circ \mathcal{L}$.

Analogously one shows that $\mathcal{R} \circ \mathcal{L} \subset \mathcal{L} \circ \mathcal{R}$ and hence $\mathcal{L} \circ \mathcal{R} = \mathcal{R} \circ \mathcal{L}$. $\square$

Exercise 4.4.5 Let ξ and η be two equivalence relations on X. Show that $\xi \circ \eta = \eta \circ \xi$ implies that $\xi \circ \eta$ is again an equivalence relation. Moreover, show that this product is the minimum equivalence relation which contains both ξ and η.

The minimum equivalence relation on S which contains both $\mathcal{R}$ and $\mathcal{L}$ is denoted by $\mathcal{D}$ and is called the $\mathcal{D}$-*relation*. From Lemma 4.4.4 it follows that $\mathcal{D} = \mathcal{L} \circ \mathcal{R} = \mathcal{R} \circ \mathcal{L}$. All other notions and notation for $\mathcal{D}$ are similar to the ones used for $\mathcal{L}$. In particular, $\mathcal{D}(a)$ denotes the $\mathcal{D}$-class of an element a.

The intersection of two equivalence relations is always an equivalence relation. We define the $\mathcal{H}$-*relation* as the intersection of $\mathcal{R}$ and $\mathcal{L}$. All other notions and notation for $\mathcal{H}$ are similar to the ones used for $\mathcal{L}$. In particular, $\mathcal{H}(a)$ denotes the $\mathcal{H}$-class of an element a.

Finally, we will say that the elements a and b are $\mathcal{J}$-*equivalent* provided that they generate the same principal two-sided ideal, that is, $S^1 a S^1 = S^1 b S^1$. All other notions and notation for $\mathcal{J}$ are similar to the ones used for $\mathcal{L}$. In particular, $\mathcal{J}(a)$ denotes the $\mathcal{J}$-class of an element a.

It is obvious that $\mathcal{R} \subset \mathcal{J}$ and $\mathcal{L} \subset \mathcal{J}$. In particular, it follows that $\mathcal{D} \subset \mathcal{J}$. Hence we have the following diagram depicting the introduced relations on S:

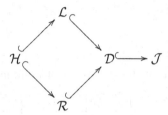

The relations $\mathcal{L}$, $\mathcal{R}$, $\mathcal{H}$, $\mathcal{D}$, and $\mathcal{J}$ on the semigroup S are called *Green's relations* after J. A. Green, who introduced them in 1951 in [Gr].

Lemma 4.4.6 *The following conditions are equivalent:*

(a) $a\mathcal{D}b$.

(b) $\mathcal{R}(a) \cap \mathcal{L}(b) \neq \varnothing$.

(c) $\mathcal{L}(a) \cap \mathcal{R}(b) \neq \varnothing$.

Proof. Since the relation $\mathcal{D}$ is symmetric, it is enough to show that (a)⇔(b).

Let $a\mathcal{D}b$. As $\mathcal{D} = \mathcal{R} \circ \mathcal{L}$, there exists $c \in S^1$ such that $a\mathcal{R}c$ and $c\mathcal{L}b$. But this means that $c \in \mathcal{R}(a)$ and $c \in \mathcal{L}(b)$. Thus $\mathcal{R}(a) \cap \mathcal{L}(b) \neq \varnothing$, proving the implication (a)⇒(b).

Assume that $c \in \mathcal{R}(a) \cap \mathcal{L}(b)$. Then $c \in \mathcal{R}(a)$ and $c \in \mathcal{L}(b)$, which means that $a\mathcal{R}c$ and $c\mathcal{L}b$. By definition, this implies that $(a, b) \in \mathcal{R} \circ \mathcal{L} = \mathcal{D}$. □

Lemma 4.4.6 says that every $\mathcal{L}$-class and every $\mathcal{R}$-class which belong to the same $\mathcal{D}$-class have a nonempty intersection. Hence it is convenient to think of a $\mathcal{D}$-class as a rectangular table in which the rows correspond to, say, $\mathcal{R}$-classes and the columns correspond to $\mathcal{L}$-classes:

$\mathcal{R}(a_1)$			$\dots$		
$\mathcal{R}(a_2)$			$\dots$		
$\vdots$	$\vdots$	$\vdots$		$\vdots$	
$\mathcal{R}(a_k)$			$\dots$		
	$\mathcal{L}(b_1)$	$\mathcal{L}(b_2)$	$\dots$	$\mathcal{L}(b_m)$	

The cells in this table give all $\mathcal{H}$-classes, contained in our $\mathcal{D}$-class. From Lemma 4.4.6 it follows that all cells are nonempty. The rectangular table above is called the *egg-box diagram* of the $\mathcal{D}$-class.

Lemma 4.4.7 (Green) *Let S be a semigroup and $a, b \in S$ be such that $a\mathcal{R}b$. Let further $u, v \in S^1$ be such that $au = b$ and $bv = a$.*

(i) *The mapping $\lambda_u : x \mapsto xu$ maps $\mathcal{L}(a)$ to $\mathcal{L}(b)$, and the mapping $\lambda_v : y \mapsto yv$ maps $\mathcal{L}(b)$ to $\mathcal{L}(a)$.*

(ii) *$\lambda_u : \mathcal{L}(a) \to \mathcal{L}(b)$ and $\lambda_v : \mathcal{L}(b) \to \mathcal{L}(a)$ are mutually inverse bijections.*

(iii) *Both λ_u and λ_v preserve $\mathcal{R}$-classes, that is, $x\mathcal{R}\lambda_u(x)$ and $y\mathcal{R}\lambda_v(y)$ for all $x \in \mathcal{L}(a)$ and $y \in \mathcal{L}(b)$.*

Here is an illustration of the statement of Lemma 4.4.7:

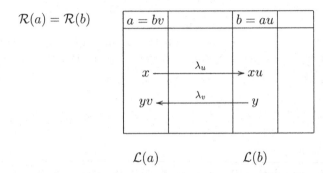

Proof. The statement (i) follows from the fact that $\mathcal{L}$ is right compatible. Let $x \in \mathcal{L}(a)$. Then there exists $p \in S^1$ such that $x = pa$. Hence

$$(\lambda_v\lambda_u)(x) = \lambda_v(\lambda_u(x)) = xuv = pauv = pa = x.$$

This means that $\lambda_v\lambda_u$ is the identity transformation of $\mathcal{L}(a)$. Analogously one shows that $\lambda_u\lambda_v$ is the identity transformation of $\mathcal{L}(b)$. This proves the statement (ii).

Let $x \in \mathcal{L}(a)$. Using (ii) we have

$$\lambda_u(x) = xu \quad \text{and} \quad x = \lambda_v(xu) = xu \cdot v = \lambda_u(x) \cdot v.$$

Using Proposition 4.4.2 we have $x\mathcal{R}\lambda_u(x)$. Hence λ_u preserves $\mathcal{R}$. Since λ_v is inverse to λ_u, it must preserve $\mathcal{R}$ as well. This completes the proof. □

From Green's Lemma it follows that the mappings λ_u and λ_v induce mutually inverse bijections between the sets of $\mathcal{H}$-classes in $\mathcal{L}(a)$ and $\mathcal{L}(b)$. Moreover, λ_u and λ_v also induce mutually inverse bijections between the corresponding $\mathcal{H}$-classes.

Applying Green's Lemma to the opposite semigroup $\overleftarrow{S}$ we obtain the following dual version of this lemma.

Lemma 4.4.8 *Let S be a semigroup and $a, b \in S$ be such that $a\mathcal{L}b$. Let $u, v \in S^1$ be such that $ua = b$ and $vb = a$.*

(i) *The mapping $\mu_u : x \mapsto ux$ maps $\mathcal{R}(a)$ to $\mathcal{R}(b)$, and the mapping $\mu_v : y \mapsto vy$ maps $\mathcal{R}(b)$ to $\mathcal{R}(a)$.*

(ii) *$\mu_u : \mathcal{R}(a) \to \mathcal{R}(b)$ and $\mu_v : \mathcal{R}(b) \to \mathcal{R}(a)$ are mutually inverse bijections.*

(iii) *Both μ_u and μ_v preserve $\mathcal{L}$-classes, that is, $x\mathcal{L}\mu_u(x)$ and $y\mathcal{L}\mu_v(y)$ for all $x \in \mathcal{R}(a)$ and $y \in \mathcal{R}(b)$.*

We note that the existence of the elements u and v, which are essentially used in Lemmas 4.4.7 and 4.4.8, follows from Propositions 4.4.1 and 4.4.2.

Corollary 4.4.9 *All $\mathcal{H}$-classes inside the same $\mathcal{D}$-class are of the same cardinality.*

Proof. Let $a\mathcal{D}b$. As $\mathcal{D} = \mathcal{L} \circ \mathcal{R}$, there exists $c \in S^1$ such that $a\mathcal{R}c$ and $c\mathcal{L}b$. By Propositions 4.4.1 and 4.4.2, there exist $u, v \in S$ such that $c = au$, $b = vc$. By Green's Lemma, the mapping λ_u is a bijection from $\mathcal{H}(a)$ to $\mathcal{H}(c)$. By the dual of Green's Lemma the mapping μ_v is a bijection from $\mathcal{H}(c)$ to $\mathcal{H}(b)$.

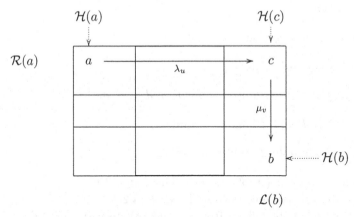

Hence the composition $\mu_v\lambda_u$ is a bijection from $\mathcal{H}(a)$ to $\mathcal{H}(b)$. $\square$

Exercise 4.4.10 Let S be a semigroup such that for each $a \in S$ we have $aS = Sa = S$. Show that S is a group.

Theorem 4.4.11 *Let H be an $\mathcal{H}$-class of S. Then the following conditions are equivalent:*

(a) *H is a group.*

(b) *H contains an idempotent.*

(c) *There exist $a, b \in H$ such that $ab \in H$.*

Proof. Assume that H is a group. Let $e \in H$ be the identity element. Then $e \cdot e = e$ and hence (a)$\Rightarrow$(b). The implication (b)$\Rightarrow$(c) is obvious.

To complete the proof we have to prove the implication (c)$\Rightarrow$(a). Let $a, b, ab \in H$. As $ab\mathcal{L}b$ and $a\mathcal{R}ab$, by Green's Lemma we have

$$aH = H = Hb. \tag{4.5}$$

Let now $x \in H$ be arbitrary. From (4.5) we have $ax \in H$ and $xb \in H$. Analogously to the proof of (4.5) we have $Hx = xH = H$. As x is arbitrary, $H \cdot H = H$ and thus H is a semigroup. The statement now follows from Exercise 4.4.10. □

4.5   Green's Relations on $\mathcal{T}_n$, $\mathcal{PT}_n$, and $\mathcal{IS}_n$

The description of principal ideals in $\mathcal{T}_n$, $\mathcal{PT}_n$, and $\mathcal{IS}_n$ obtained in Sect. 4.2 allows us to describe Green's relations in these semigroups. As before for a (partial) transformation α we denote by ρ_α the partition of $\mathbf{N}$ (in the case of $\mathcal{T}_n$) or $\mathbf{N} \cup \{n+1\}$ (in the case of $\mathcal{PT}_n$, or $\mathcal{IS}_n$) generated by the relation π_α.

Theorem 4.5.1 *Let S be one of the semigroups $\mathcal{T}_n$, $\mathcal{PT}_n$, or $\mathcal{IS}_n$, and $\alpha, \beta \in S$. Then*

(i) $\alpha\mathcal{R}\beta$ if and only if $\mathrm{im}(\alpha) = \mathrm{im}(\beta)$

(ii) $\alpha\mathcal{L}\beta$ if and only if $\rho_\alpha = \rho_\beta$

(iii) $\alpha\mathcal{H}\beta$ if and only if $\mathrm{im}(\alpha) = \mathrm{im}(\beta)$ and $\rho_\alpha = \rho_\beta$

(iv) $\alpha\mathcal{D}\beta$ if and only if $\mathrm{rank}(\alpha) = \mathrm{rank}(\beta)$

(v) $\alpha\mathcal{J}\beta$ if and only if $\mathrm{rank}(\alpha) = \mathrm{rank}(\beta)$

Proof. The statement (i) follows from Theorem 4.2.1.

The statement (ii) follows from Theorem 4.2.4.

The statement (iii) follows from (i) and (ii) since $\mathcal{H} = \mathcal{R} \cap \mathcal{L}$.

The statement (v) follows from Theorem 4.2.8.

As $\mathcal{D} \subset \mathcal{J}$, from (v) it follows that $\alpha\mathcal{D}\beta$ implies $\mathrm{rank}(\alpha) = \mathrm{rank}(\beta)$. Conversely, let $\mathrm{rank}(\alpha) = \mathrm{rank}(\beta)$. Set $\mathrm{im}(\alpha) = \{a_1, a_2, \ldots, a_k\}$ and $\mathrm{im}(\beta) = \{b_1, b_2, \ldots, b_k\}$ and consider the element γ, defined in the following way:

$$\gamma(x) = b_j \text{ if and only if } \alpha(x) = a_j, j = 1, \ldots, k.$$

Obviously $\rho_\gamma = \rho_\alpha$ and hence $\alpha\mathcal{L}\gamma$ by (ii). On the other hand, $\mathrm{im}(\gamma) = \mathrm{im}(\beta)$ and hence $\gamma\mathcal{L}\beta$ by (i). This means that $(\alpha, \beta) \in \mathcal{L} \circ \mathcal{R} = \mathcal{D}$ and proves (iv). □

Corollary 4.5.2 *In the semigroups $\mathcal{T}_n$, $\mathcal{PT}_n$, and $\mathcal{IS}_n$ we have $\mathcal{D} = \mathcal{J}$.*

As we have already mentioned after Theorem 4.2.4, for the semigroup $\mathcal{IS}_n$ the condition $\rho_\alpha = \rho_\beta$ is equivalent to the condition $\mathrm{dom}(\alpha) = \mathrm{dom}(\beta)$.

Proposition 4.5.3 *Let S be one of the semigroups $\mathcal{T}_n$, $\mathcal{PT}_n$, or $\mathcal{IS}_n$, and $\alpha, \beta \in S$. Then $\alpha \mathcal{L} \beta$ if and only if there exists $\mu \in \mathcal{S}_n$ such that $\alpha = \mu\beta$.*

Proof. Let $\alpha = \mu\beta$, where $\mu \in \mathcal{S}_n$. Since μ is invertible, we have $\beta = \mu^{-1}\alpha$. Hence $\alpha \mathcal{L} \beta$ by Proposition 4.4.1.

Conversely, let $\alpha \mathcal{L} \beta$. Then $\rho_\alpha = \rho_\beta$. Let $A_1, \ldots, A_k$ be those blocks of ρ_α which are contained in $\mathrm{dom}(\alpha)$ and let $\alpha(x) = a_i$ and $\beta(x) = b_i$ for all $x \in A_i$, $i = 1, \ldots, k$. Take any permutation $\mu \in \mathcal{S}_n$ such that $\mu(b_i) = a_i$, $i = 1, \ldots, k$. Then a direct calculation shows that $\alpha = \mu\beta$. $\square$

Proposition 4.5.4 *Let $\alpha, \beta \in \mathcal{IS}_n$. Then*

(i) $\alpha \mathcal{R} \beta$ if and only if there exists $\mu \in \mathcal{S}_n$ such that $\alpha = \beta\mu$.

(ii) $\alpha \mathcal{D} \beta$ if and only if there exists $\mu, \nu \in \mathcal{S}_n$ such that $\alpha = \nu\beta\mu$.

Proof. The statement (i) follows from Proposition 4.5.3 and the fact that $\mathcal{IS}_n \cong \overleftarrow{\mathcal{IS}_n}$.

If $\alpha = \nu\beta\mu$ for some $\mu, \nu \in \mathcal{S}_n$, then $\beta = \nu^{-1}\alpha\mu^{-1}$ and hence the principal ideals generated by α and β coincide. This means $\alpha \mathcal{J} \beta$. The latter implies $\alpha \mathcal{D} \beta$ using Theorem 4.5.1(iv) and (v).

Conversely, if $\alpha \mathcal{D} \beta$, then there exists γ such that $\alpha \mathcal{L} \gamma$ and $\gamma \mathcal{L} \beta$. By Proposition 4.5.3 we have $\alpha = \nu\gamma$ for some $\nu \in \mathcal{S}_n$, and by (i) we have $\gamma = \beta\mu$ for some $\mu \in \mathcal{S}_n$. Thus $\alpha = \nu\beta\mu$. $\square$

Example 4.5.5 Here we present the egg-box diagrams for all $\mathcal{D}$-classes of the semigroup $\mathcal{PT}_3$. This semigroup has four $\mathcal{D}$-classes $\mathcal{D}_i$, $i = 0, 1, 2, 3$, indexed by the rank of elements inside the $\mathcal{D}$-class. The transformation $\begin{pmatrix} 1 & 2 & 3 \\ a_1 & a_2 & a_3 \end{pmatrix}$ from $\mathcal{PT}_3$ is given by the second row $(a_1 a_2 a_3)$ (do not mix this with our notation for cycles). To accommodate the picture to the page size we are forced to transpose it. So, in our picture rows are $\mathcal{L}$-classes and columns are $\mathcal{R}$-classes. For each $\mathcal{L}$-class L we show on the left of the diagram the partition ρ_α of the domain of $\alpha \in L$, and for each $\mathcal{R}$-class R we show above the diagram the set $\mathrm{im}(\alpha)$, where $\alpha \in R$. The elements marked with $*$ are idempotents.

$\{1,2,3\}$

$\mathcal{D}_3:$ 1|2|3 | $(123)^*, (132), (213), (231), (312), (321)$ |

$\mathcal{D}_2:$

	$\{1,2\}$	$\{1,3\}$	$\{2,3\}$
12|3	$(112), (221)$	$(113)^*, (331)$	$(223)^*, (332)$
13|2	$(121)^*, (212)$	$(131), (313)$	$(232), (323)^*$
1|23	$(122)^*, (211)$	$(133)^*, (311)$	$(233), (322)$
1|2	$(12\varnothing)^*, (21\varnothing)$	$(13\varnothing), (31\varnothing)$	$(23\varnothing), (32\varnothing)$
1|3	$(1\varnothing 2), (2\varnothing 1)$	$(1\varnothing 3)^*, (3\varnothing 1)$	$(2\varnothing 3), (3\varnothing 2)$
2|3	$(\varnothing 12), (\varnothing 21)$	$(\varnothing 13), (\varnothing 31)$	$(\varnothing 23)^*, (\varnothing 32)$

$\mathcal{D}_1:$

	$\{1\}$	$\{2\}$	$\{3\}$
123	$(111)^*$	$(222)^*$	$(333)^*$
12	$(11\varnothing)^*$	$(22\varnothing)^*$	$(33\varnothing)$
13	$(1\varnothing 1)^*$	$(2\varnothing 2)$	$(3\varnothing 3)^*$
23	$(\varnothing 11)$	$(\varnothing 22)^*$	$(\varnothing 33)^*$
1	$(1\varnothing \varnothing)^*$	$(2\varnothing \varnothing)$	$(3\varnothing \varnothing)$
2	$(\varnothing 1\varnothing)$	$(\varnothing 2\varnothing)^*$	$(\varnothing 3\varnothing)$
3	$(\varnothing \varnothing 1)$	$(\varnothing \varnothing 2)$	$(\varnothing \varnothing 3)^*$

$\varnothing$

$\mathcal{D}_0:$ | $(\varnothing \varnothing \varnothing)^*$ |

If from the above tables we delete all elements containing the symbol $\varnothing$, we obtain the egg-box diagram for the semigroup $\mathcal{T}_3$. If from the above tables we delete all elements containing some repetitions of 1, 2, or 3, we obtain the egg-box diagram for the semigroup $\mathcal{IS}_3$.

4.6   Combinatorics of Green's Relations in the Semigroups $\mathcal{T}_n$, $\mathcal{PT}_n$, and $\mathcal{IS}_n$

From the definition of Green's relations we have that the number of $\mathcal{L}$-, $\mathcal{R}$- and $\mathcal{J}$-classes in a semigroup coincides with the number of principal left, right and two-sided ideals, respectively. Hence from Corollaries 4.2.2, 4.2.5, and 4.2.9 we get:

Proposition 4.6.1   (i) *The semigroup $\mathcal{T}_n$ contains B_n different $\mathcal{L}$-classes, $(2^n - 1)$ different $\mathcal{R}$-classes and n different $\mathcal{J}$-classes.*

(ii) *The semigroup $\mathcal{PT}_n$ contains B_{n+1} different $\mathcal{L}$-classes, 2^n different $\mathcal{R}$- classes and $n + 1$ different $\mathcal{J}$-classes.*

(iii) *The semigroup $\mathcal{IS}_n$ contains 2^n different $\mathcal{L}$-classes, 2^n different $\mathcal{R}$- classes and $n + 1$ different $\mathcal{J}$-classes.*

Since in $\mathcal{T}_n$, $\mathcal{PT}_n$, and $\mathcal{IS}_n$ we have $\mathcal{J} = \mathcal{D}$ by Theorem 4.5.1, in what follows we will not consider $\mathcal{J}$-classes.

For each semigroup $\mathcal{T}_n$, $\mathcal{PT}_n$, and $\mathcal{IS}_n$ and each $k = 0, \ldots, n$ we denote by $\mathcal{D}_k$ the $\mathcal{D}$-class which consists of elements of rank k. Then $\mathcal{T}_n$ has $\mathcal{D}$-classes $\mathcal{D}_k$, $k = 1, \ldots, n$, and $\mathcal{PT}_n$ and $\mathcal{IS}_n$ have an additional $\mathcal{D}$-class $\mathcal{D}_0$. Note that in each of these semigroups the ideal $\mathcal{I}_k$ has the form $\mathcal{I}_k = \cup_{i \leq k} \mathcal{D}_k$ and hence $\mathcal{D}_k = \mathcal{I}_k \backslash \mathcal{I}_{k-1}$.

The cardinality of $\mathcal{D}_k$ was already determined during the proof of Corollary 4.2.10:

Proposition 4.6.2 *Let S be one of the semigroups $\mathcal{T}_n$, $\mathcal{PT}_n$, or $\mathcal{IS}_n$. We have*

$$
|\mathcal{D}_k| = \begin{cases} \mathsf{S}(n,k)\frac{n!}{(n-k)!}, & S = \mathcal{T}_n; \\ \mathsf{S}(n+1,k+1)\frac{n!}{(n-k)!}, & S = \mathcal{PT}_n; \\ \binom{n}{k}^2 k!, & S = \mathcal{IS}_n. \end{cases}
$$

Proposition 4.6.3 *For each of the semigroups $\mathcal{T}_n$, $\mathcal{PT}_n$, and $\mathcal{IS}_n$ there is a natural bijection between the k-element subsets of $\mathbf{N}$ and the $\mathcal{R}$-classes inside the $\mathcal{D}$-class $\mathcal{D}_k$. In particular, $\mathcal{D}_k$ contains exactly $\binom{n}{k}$ different $\mathcal{R}$-classes.*

Proof. By Theorem 4.5.1 each $\mathcal{R}$-class is uniquely defined by the image of elements from this class. For $\mathcal{R}$-classes inside $\mathcal{D}_k$ the image can be an arbitrary k-element subset of $\mathbf{N}$. The statement follows. □

Proposition 4.6.4 *(i) For the semigroup $\mathcal{T}_n$ there is a natural bijection between partitions of $\mathbf{N}$ into k blocks and the $\mathcal{L}$-classes inside the $\mathcal{D}$-class $\mathcal{D}_k$. In particular, $\mathcal{D}_k$ contains $\mathsf{S}(n,k)$ different $\mathcal{L}$-classes.*

(ii) For the semigroup $\mathcal{PT}_n$ there is a natural bijection between partitions of $\mathbf{N} \cup \{n+1\}$ into $k+1$ blocks and the $\mathcal{L}$-classes inside the $\mathcal{D}$-class $\mathcal{D}_k$. In particular, $\mathcal{D}_k$ contains $\mathsf{S}(n+1, k+1)$ different $\mathcal{L}$-classes.

(iii) For the semigroup $\mathcal{IS}_n$ there is a natural bijection between k-element subsets of $\mathbf{N}$ and the $\mathcal{L}$-classes inside the $\mathcal{D}$-class $\mathcal{D}_k$. In particular, $\mathcal{D}_k$ contains exactly $\binom{n}{k}$ different $\mathcal{L}$-classes.

Proof. By Theorem 4.5.1 each $\mathcal{L}$-class L of $\mathcal{T}_n$ is uniquely determined by some partition ρ_α of the set $\mathbf{N}$. This partition is the same for all elements of L. For $\mathcal{L}$-classes inside the $\mathcal{D}_k$ such partition can be an arbitrary partition of $\mathbf{N}$ into k blocks. The statement (i) follows.

The proof of (ii) is analogous to that of (i) with substitution of $\mathbf{N}$ by $\mathbf{N} \cup \{n+1\}$.

The proof of (iii) repeats that of Proposition 4.6.3 considering domains instead of images. □

Proposition 4.6.5   *(i) In each of the semigroups $\mathcal{T}_n$, $\mathcal{PT}_n$ and $\mathcal{IS}_n$ every $\mathcal{L}$-class inside the $\mathcal{D}_k$ contains exactly $\binom{n}{k}$ different $\mathcal{H}$-classes.*

(ii) In semigroups $\mathcal{T}_n$, $\mathcal{PT}_n$ and $\mathcal{IS}_n$ every $\mathcal{R}$-class inside the $\mathcal{D}_k$ contains $\mathsf{S}(n,k)$, $\mathsf{S}(n+1,k+1)$ and $\binom{n}{k}$ different $\mathcal{H}$-classes, respectively.

(iii) In semigroups $\mathcal{T}_n$, $\mathcal{PT}_n$ and $\mathcal{IS}_n$ every $\mathcal{D}$-class $\mathcal{D}_k$ contains $\mathsf{S}(n,k)\binom{n}{k}$, $\mathsf{S}(n+1,k+1)\binom{n}{k}$ and $\binom{n}{k}^2$ different $\mathcal{H}$-classes, respectively.

Proof. Let D be a $\mathcal{D}$-class. By Lemma 4.4.6, there is a bijection between the $\mathcal{H}$-classes inside D and pairs (R, L), where L and R are an $\mathcal{L}$-class and an $\mathcal{R}$-class inside D, respectively. The bijection is given by taking $R \cap L$. Hence the number of $\mathcal{H}$-classes inside R equals the number of $\mathcal{L}$-classes inside D, and vice versa. The statement now follows from Propositions 4.6.3 and 4.6.4.   □

Theorem 4.6.6 *For each of the semigroups $\mathcal{T}_n$, $\mathcal{PT}_n$, and $\mathcal{IS}_n$ the cardinality of any $\mathcal{H}$-class inside the $\mathcal{D}$-class $\mathcal{D}_k$ equals $k!$.*

Proof. Let H be an $\mathcal{H}$-class inside the $\mathcal{D}$-class $\mathcal{D}_k$. By Theorem 4.5.1(iii) it consists of all elements α for which the image $\mathrm{im}(\alpha)$ and the partition ρ_α are fixed. Let $\mathrm{im}(\alpha) = \{a_1, \ldots, a_k\}$ and $B_1, \ldots, B_k$ be those blocks of ρ_α which are contained in $\mathrm{dom}(\alpha)$. Then each element from H is completely determined by a surjective function from the set $\{B_1, \ldots, B_k\}$ to the set $\{a_1, \ldots, a_k\}$. The number of such functions is obviously $k!$.   □

From Theorem 4.6.6 and Proposition 4.6.5 we obtain:

Corollary 4.6.7   *(i) In each of the semigroups $\mathcal{T}_n$, $\mathcal{PT}_n$, and $\mathcal{IS}_n$ each $\mathcal{L}$-class inside the $\mathcal{D}$-class $\mathcal{D}_k$ contains exactly $\binom{n}{k}k!$ elements.*

(ii) In the semigroups $\mathcal{T}_n$, $\mathcal{PT}_n$, and $\mathcal{IS}_n$ each $\mathcal{R}$-class inside the $\mathcal{D}$-class $\mathcal{D}_k$ contains $\mathsf{S}(n,k)k!$, $\mathsf{S}(n+1,k+1)k!$ and $\binom{n}{k}k!$ elements, respectively.

4.7   Addenda and Comments

4.7.1 Two-sided ideals of $\mathcal{T}_n$ were independently described by Mal'cev in [Ma1] and Vorob'ev in [Vo1]. However, principal one-sided ideals of $\mathcal{T}_n$ had been described already by A. Suschkewitsch in [Su2]. Two-sided ideals of $\mathcal{IS}_n$ and $\mathcal{PT}_n$ were described by Liber in [Lib] and Sutov in [Sut2], respectively. The cardinalities of all principal one-sided ideals of $\mathcal{T}_n$ were determined in [HM]. We did not manage to find any description of all one-sided ideals for $\mathcal{T}_n$, $\mathcal{PT}_n$, or $\mathcal{IS}_n$ in the literature.

4.7.2 Green's relations were introduced in [Gr]. They play a very important role in the study of the structure of semigroups. One of the main reasons is that Green's Lemma 4.4.7 in some sense allows us to define a kind of Cartesian coordinate system on each $\mathcal{D}$-class.

4.7.3 Green's relations for $\mathcal{T}_n$ were described by Miller and Doss, see [Do]. A nice presentation of this description can be found in [CP1, Sect. 2.2]. Green's relations for $\mathcal{IS}_n$ and $\mathcal{PT}_n$ were described by Reilly in [Re1] and FitzGerald and Preston in [FP], respectively.

4.7.4 In Sect. 2.7 we already mentioned a very close connection between regular elements and idempotents. Another aspect of this connection is the following:

Proposition 4.7.1 *Let S be a semigroup and $a \in S$. Then the following conditions are equivalent:*

(a) a is regular.

(b) There exists an idempotent $e \in S$ such that $a\mathcal{R}e$.

(c) There exists an idempotent $e \in S$ such that $a\mathcal{L}e$.

Proof. Assume that a is regular and b is an inverse of a. Then $ab = a \cdot b$ and $a = ab \cdot a$. Hence $a\mathcal{R}ab$ by Proposition 4.4.2. At the same time $e = ab = (aba)b = (ab)^2$ is an idempotent. This proves the implication (a)$\Rightarrow$(b).

Let $a\mathcal{R}e$ for some idempotent e. Then, by Proposition 4.4.2, there exist $x, y \in S^1$ such that $ax = e$ and $ey = a$. If $x = 1$, then $a = e$ is an idempotent, hence regular. If $x \neq 1$, we have $x \in S$ and $axa = ea = e \cdot ey = ey = a$. Hence a is regular, which proves the implication (b)$\Rightarrow$(a).

The equivalence (a)$\Leftrightarrow$(c) is proved similarly. $\square$

Corollary 4.7.2 *Let S be a semigroup and $s \in S$. If the element a is regular, then every element from $\mathcal{D}(a)$ is regular as well.*

A $\mathcal{D}$-class containing a regular element is called a *regular $\mathcal{D}$-class*. As the semigroups $\mathcal{T}_n$, $\mathcal{PT}_n$, and $\mathcal{IS}_n$ are regular, we have the following:

Corollary 4.7.3   *(i) In the semigroups $\mathcal{T}_n$, $\mathcal{PT}_n$, and $\mathcal{IS}_n$ all principal (right, left, two-sided) ideals are generated by idempotents.*

(ii) In the semigroups $\mathcal{T}_n$, $\mathcal{PT}_n$, and $\mathcal{IS}_n$ each $\mathcal{L}$-class and each $\mathcal{R}$-class contain an idempotent.

4.7.5 The $\mathcal{D}$-class $\mathcal{D}_k$ of the semigroup $\mathcal{T}_n$ contains $\binom{n}{k}k^{n-k}$ idempotents. This is shown in the proof of Corollary 2.7.4. The $\mathcal{D}$-class $\mathcal{D}_k$ of the semigroup $\mathcal{PT}_n$ contains $\binom{n}{k}(k+1)^{n-k}$ idempotents. This is shown in the proof of Corollary 2.7.5.

4.7.6 Let G be a group and $a, b \in G$. Then each of the equations $ax = b$ and $ya = b$ has a (unique) solution. Hence if G is a subgroup of some semigroup S, then any two elements from G are both $\mathcal{L}$- and $\mathcal{R}$-equivalent in S. Hence they are $\mathcal{H}$-equivalent. This means that any subgroup of S is contained in some $\mathcal{H}$-class, say H. In particular, H must contain an idempotent: the identity element of this group. By Theorem 4.4.11, in this case H itself is a group.

Theorem 4.7.4 *Let H be an $\mathcal{H}$-class of $\mathcal{T}_n$, $\mathcal{PT}_n$, or $\mathcal{IS}_n$, which contains an idempotent of rank k. Then H is isomorphic to the symmetric group $\mathcal{S}_k$.*

Proof. Let $\epsilon \in H$ be an idempotent. Let further $\mathrm{im}(\epsilon) = \{a_1, \ldots, a_k\}$ and for $i = 1, \ldots, k$ let B_i be the full preimage of a_i. From $\epsilon^2 = \epsilon$ we have that $a_i \in B_i$ for all i. Then, by Theorem 4.5.1(iii) the elements from H are given by all possible injections from the set $\{B_1, \ldots, B_k\}$ to the set $\{a_1, \ldots, a_k\}$. In particular, the restriction to $\{a_1, \ldots, a_k\}$ defines a bijective mapping from H to the symmetric group on $\{a_1, \ldots, a_k\}$. It is easy to see that this mapping is compatible with the composition and hence is an isomorphism. The image of this mapping is obviously isomorphic to $\mathcal{S}_k$. $\qquad\square$

4.7.7 Let S be one of the semigroups $\mathcal{T}_n$, $\mathcal{PT}_n$, or $\mathcal{IS}_n$. Let further H_1 and H_2 be two different $\mathcal{H}$-classes of S inside the same $\mathcal{D}$-class. If both H_1 and H_2 are groups, then from Theorem 4.7.4 it follows that $H_1 \cong H_2$. This is a special case of the following more general statement.

Theorem 4.7.5 *Let S be a semigroup and H_1 and H_2 be two different $\mathcal{H}$-classes of S inside the same $\mathcal{D}$-class. If both H_1 and H_2 are groups, then $H_1 \cong H_2$ as groups.*

Proof. Denote the idempotents of H_1 and H_2 by e and f, respectively. Then e and f are the identities in the groups H_1 and H_2, respectively. Let $a \in \mathcal{R}(e) \cap \mathcal{L}(f)$. As $e\mathcal{R}a$, there exists $b \in S^1$ such that $a = eb$. Hence $ea = e \cdot eb = eb = a$. Analogously one shows that $af = a$.

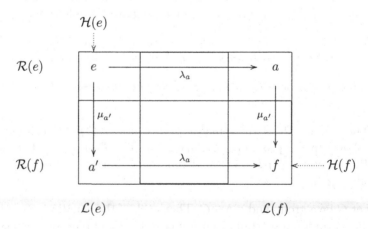

By Green's Lemma the mapping $\lambda_a : x \mapsto xa$ is a bijection of the $\mathcal{L}$-class $\mathcal{L}(e)$ onto the $\mathcal{L}$-class $\mathcal{L}(a) = \mathcal{L}(f)$. Let $a' \in \mathcal{L}(e)$ be such that $\lambda_a(a') = f$.

As $a'e = a'$, the dual Green's lemma says that the mapping $\mu_{a'} : y \to a'y$ is a bijection from $\mathcal{R}(e)$ to $\mathcal{R}(f)$. In particular, $\mu_{a'}(a) = a'a = \lambda_a(a') = f$.

The composition $\varphi = \mu_{a'}\lambda_a : x \mapsto a'xa$ is a bijection from $\mathcal{H}(e)$ to $\mathcal{H}(f)$. Since $f = a'a$ and $a = af$, by Green's Lemma the mapping $\mu_a : \mathcal{R}(f) \to \mathcal{R}(e)$ is inverse to the mapping $\mu_{a'}$. Analogously $\lambda_{a'} : \mathcal{L}(f) \to \mathcal{L}(e)$ is inverse to λ_a. Hence $(\lambda_{a'}\mu_a)(f) = e$. On the other hand,

$$(\lambda_{a'}\mu_a)(f) = afa' = af \cdot a' = aa'.$$

This shows that $aa' = e$, which implies the following:

$$\varphi(xy) = a'xya = a'xeya = a'xa \cdot a'ya = \varphi(x)\varphi(y).$$

Hence φ is an isomorphism from H_1 to H_2. $\qquad\square$

4.7.8 If S is a semigroup with the zero element 0, then for each nonempty subset $A \subset S$ we can define the *left annihilator* of A as follows:

$$\mathrm{Ann}_l(A) = \{x \in S : xa = 0 \text{ for all } a \in A\}.$$

Analogously one defines the *right annihilator* of A as follows:

$$\mathrm{Ann}_r(A) = \{x \in S : ax = 0 \text{ for all } a \in A\}.$$

It is easy to see that $\mathrm{Ann}_l(A)$ is a left ideal of S and $\mathrm{Ann}_r(A)$ is a right ideal of S for each A.

If $A = \{a\}$, one usually uses the notation $\mathrm{Ann}_l(a)$, or $\mathrm{Ann}_r(a)$ instead of $\mathrm{Ann}_l(\{a\})$, or $\mathrm{Ann}_r(\{a\})$, respectively.

4.8 Additional Exercises

4.8.1 Prove the identity

$$B_{n+1} = \sum_{k=1}^{n} (k+1)S(n,k) = \sum_{k=0}^{n} \binom{n}{k} B_k.$$

4.8.2 Let I be an ideal of one of the semigroups $\mathcal{T}_n$, $\mathcal{PT}_n$, or $\mathcal{IS}_n$. Prove that $I^2 = I$.

4.8.3 Prove that in each of the semigroups $\mathcal{T}_n$, $\mathcal{PT}_n$, or $\mathcal{IS}_n$ the number a_R of right ideals satisfies the inequality

$$2^{2^n} > a_R > \sum_{k=1}^{n} 2^{\binom{n}{k}} - n.$$

4.8.4 Find the number of left ideals in $\mathcal{IS}_n$ for

(a) $n = 2$,

(b) $n = 3$,

(c) $n = 4$.

4.8.5 Let ξ be an equivalence relation on some set X. Prove that $\xi \circ \xi = \xi$.

4.8.6 (a) Let $S = \mathcal{T}_n$, or $S = \mathcal{PT}_n$. Prove that $\alpha \mathcal{R} \alpha \eta$ for each $\eta \in \mathcal{S}_n$ and each $\alpha \in S$.

(b) Let $S = \mathcal{T}_n$, $n > 3$, or $S = \mathcal{PT}_n$, $n > 1$. Show that for $\alpha, \beta \in S$ the relation $\alpha \mathcal{R} \beta$ in general does not imply the existence of $\eta \in \mathcal{S}_n$ such that $\alpha = \beta \eta$.

4.8.7 ([HM]) Let $n > 3$, $1 < k < n$, and $\alpha \in \mathcal{T}_n$ be an element of rank k. Prove that $|\alpha \mathcal{T}_n| \geq |\mathcal{T}_n \alpha|$, moreover, the equality is possible only if $n = 4$ and $k = 2$.

4.8.8 Compute the number of elements in each $\mathcal{D}$-class of the semigroups

(a) $\mathcal{IS}_5$,

(b) $\mathcal{T}_5$,

(c) $\mathcal{PT}_5$.

4.8.9 Prove that a regular semigroup is inverse if and only if for each principal one-sided ideal I there exists a unique idempotent which generates I.

4.8.10 An ideal $I \subset S$ is called *prime* provided that $ab \in I$ implies $a \in I$, or $b \in I$ for all $a, b \in S$. Find all prime ideals in $\mathcal{T}_n$, $\mathcal{PT}_n$, and $\mathcal{IS}_n$.

4.8.11 An ideal $I \subset S$ is called *semiprime* provided that $a^2 \in I$ implies $a \in I$ for all $a \in S$. Find all semiprime ideals in $\mathcal{T}_n$, $\mathcal{PT}_n$, and $\mathcal{IS}_n$.

4.8.12 An ideal $I \subset S$ is called *reflexive* provided that $ab \in I$ implies $ba \in I$ for all $a, b \in S$. Find all reflexive ideals in $\mathcal{T}_n$, $\mathcal{PT}_n$, and $\mathcal{IS}_n$.

4.8.13 Find the number of idempotents in each $\mathcal{R}$-class inside $\mathcal{D}_k$ for the semigroup

(a) $\mathcal{T}_n$

(b) $\mathcal{PT}_n$

4.8.14 (a) Let $\alpha \in \mathcal{T}_n$ be a transformation of rank k and $n_1, \ldots, n_k$ be the cardinalities of the full preimages of elements from $\operatorname{im}(\alpha)$. Find the number of idempotents in $\mathcal{R}(\alpha)$.

(b) The same problem for $\mathcal{PT}_n$.

4.8.15 Prove that the class $\mathcal{H}(\alpha)$ of the semigroup $\mathcal{T}_n$ contains an idempotent if and only if for each block B of the partition ρ_α we have $B \cap \mathrm{im}(\alpha) \neq \varnothing$.

4.8.16 (a) Let R be an $\mathcal{R}$-class of some semigroup S and $e \in R$ be an idempotent. Show that e is a left identity on R, that is, $ex = x$ for all $x \in R$.

(b) Let L be an $\mathcal{L}$-class of some semigroup S and $e \in L$ be an idempotent. Show that e is a right identity on L, that is, $xe = x$ for all $x \in L$.

4.8.17 For each $\alpha \in \mathcal{IS}_n$ prove that $|\mathrm{Ann}_l(\alpha)| = |\mathrm{Ann}_r(\alpha)|$.

4.8.18 Let $\alpha \in \mathcal{PT}_n$. Determine:

(a) $|\mathrm{Ann}_l(\alpha)|$

(b) $|\mathrm{Ann}_r(\alpha)|$

4.8.19 Prove that in the semigroups $\mathcal{PT}_n$ and $\mathcal{IS}_n$ none of the ideals $\mathcal{I}_k$, $0 < k < n$, is a (left or right) annihilator of some set.

Chapter 5

Subgroups and Subsemigroups

5.1 Subgroups

Let S be a semigroup and G be a subgroup of S. Then G contains the unique idempotent e (the identity element of G). Conversely, if $e \in S$ is an idempotent, then $\{e\}$ forms a trivial subgroup of S. This shows that there is a close connection between the subgroups of S and idempotents of S. In this section we would like to illustrate this connection. We start from the following two obvious statements:

Lemma 5.1.1 *Let $e \in \mathcal{E}(S)$. Then eSe is a submonoid of S with the identity element e.*

Lemma 5.1.2 *Let G be a subgroup of S with the identity element $e \in \mathcal{E}(S)$. Then G is a subgroup of $(eSe)^*$.*

A subgroup G of S is called *maximal* provided that G is not properly contained in any other subgroup of S.

Theorem 5.1.3   *(i) For each $e \in \mathcal{E}(S)$ there is a unique maximal subgroup G_e of S in which e is the identity element.*

(ii) If $e, f \in \mathcal{E}(S)$ and $e \neq f$, then $G_e \cap G_f = \varnothing$.

Proof. From Lemma 5.1.2 it follows that any subgroup of S in which e is the identity element is contained in $(eSe)^*$. The latter is a group by Proposition 2.2.3. Hence $G_e = (eSe)^*$. This proves (i).

Let $a \in G_e \cap G_f$ and let b and c denote the inverses to a in G_e and G_f, respectively. We have $ab = e$ and $ca = f$. From this and the facts that $fa = a$ (since $a \in G_f$) and $ae = e$ (since $a \in G_e$) we have

$$e = ab = fab = fe = cae = ca = f,$$

a contradiction. $\square$

O. Ganyushkin, V. Mazorchuk, *Classical Finite Transformation Semigroups*, Algebra and Applications 9, DOI: 10.1007/978-1-84800-281-4_5, © Springer-Verlag London Limited 2009

In 4.7.6 it is shown that maximal subgroups of S are exactly the $\mathcal{H}$-classes $\mathcal{H}(e)$, $e \in \mathcal{E}(S)$. As each $\mathcal{H}$-class contains at most one idempotent and different $\mathcal{H}$-classes do not have common elements, we get an alternative proof of Theorem 5.1.3. In what follows for $e \in \mathcal{E}(S)$ we shall denote the corresponding maximal subgroup by G_e. For the semigroups $\mathcal{T}_n$, $\mathcal{PT}_n$, and $\mathcal{IS}_n$ the following statement is a direct consequence of Theorems 4.5.1 and 4.7.4:

Theorem 5.1.4 *Let S denote one of the semigroups $\mathcal{T}_n$, $\mathcal{PT}_n$, or $\mathcal{IS}_n$. For $\epsilon \in \mathcal{E}(S)$ we have:*

(i)
$$G_\epsilon = \{\alpha \in S : \operatorname{im}(\alpha) = \operatorname{im}(\epsilon), \rho_\alpha = \rho_\epsilon\}.$$

(ii) *If* $\operatorname{rank}(\epsilon) = k$, *then* $G_\epsilon \cong S_k$.

An element $g \in S$ is called a *group element* provided that $g \in G_e$ for some $e \in \mathcal{E}(S)$.

Exercise 5.1.5 Let S denote one of the semigroups $\mathcal{T}_n$, $\mathcal{PT}_n$, or $\mathcal{IS}_n$. Show that an element $\alpha \in S$ is a group element if and only if the full subgraph of Γ_α with the vertex set $\mathbf{N}\backslash\operatorname{stim}(\alpha)$ is an *empty* graph, that is, it does not contain any arrow.

Proposition 5.1.6   (i) *The semigroup $\mathcal{T}_n$ contains $\sum_{k=1}^{n} \binom{n}{k} k^{n-k} k!$ group elements.*

(ii) *The semigroup $\mathcal{PT}_n$ contains $\sum_{k=1}^{n} \binom{n}{k}(k+1)^{n-k} k!$ group elements.*

(iii) *The semigroup $\mathcal{IS}_n$ contains $\sum_{k=1}^{n} \binom{n}{k} k!$ group elements.*

Proof. In the proofs of Corollaries 2.7.4, 2.7.5, and 2.7.3, it was shown that the semigroups $\mathcal{T}_n$, $\mathcal{PT}_n$, and $\mathcal{IS}_n$ contain

$$\binom{n}{k}k^{n-k}, \quad \binom{n}{k}(k+1)^{n-k}, \quad \text{and} \quad \binom{n}{k}$$

idempotents of rank k, respectively. After this the claim follows directly from Theorems 5.1.3(ii) and 5.1.4(ii).   □

5.2   Cyclic Subsemigroups

A semigroup S is called *cyclic* provided that it has a generating system consisting of one element. If this element is a, one writes $S = \langle a \rangle$.

Let $S = \langle a \rangle$ be a finite cyclic semigroup. Then the sequence of elements $a^1 = a$, a^2, $a^3, \ldots$ must contain repeating elements. Assume that the elements $a, a^2, \ldots, a^l$ are pairwise different and $a^{l+1} = a^k$, where $k \leq l$. The number l is called the *order* of a and is denoted by $|a|$, the number k is called the *index* of a and the number $m = (l+1) - k$ is called the *period* of a. The pair (k, m) is called the *type* of a.

From $a^{k+m} = a^k$ we get that the different elements in the sequence $a = a$, a^2, a^3, ... are

$$\langle a \rangle = \{a, a^2, \ldots, a^{k+m-1}\}.$$

Lemma 5.2.1 *If a and b are of the same type, the semigroups $\langle a \rangle$ and $\langle b \rangle$ are isomorphic.*

To prove Lemma 5.2.1 we have to recall some notation. Recall that for $x \in \mathbb{Z}$ and $y \in \mathbb{N}$ the expression $x \bmod y$ denotes the unique number $0 \leq z < y$ such that $x - z$ is divisible by y. This number is called the *residue* of x modulo y. The set $\{x + sy : s \in \mathbb{Z}\}$ is called the *residue class* of x modulo y and is denoted by $\bar{x}$. The set $\mathbb{Z}_y$ of all residue classes modulo y forms a group with respect to the addition $\overline{+}$ of residues, defined as follows:

$$\bar{a} \,\overline{+}\, \bar{b} = \overline{a + b}. \tag{5.1}$$

This group is usually referred to as the group of *residue classes* modulo y.

Exercise 5.2.2 Show that the addition of residue classes, defined in (5.1), is well defined, that is, $\bar{a} = \bar{a'}$ and $\bar{b} = \bar{b'}$ imply $\overline{a + b} = \overline{a' + b'}$.

Exercise 5.2.3 Check that $(\mathbb{Z}_y, \overline{+})$ is indeed a group.

Proof of Lemma 5.2.1. If a is an element of type (k, m), then

$$a^s \cdot a^t = \begin{cases} a^{s+t}, & s+t < k+m \\ a^{k+(s+t-k)\bmod m} & s+t \geq k+m. \end{cases} \tag{5.2}$$

Hence in the semigroup $\langle a \rangle$ the multiplication is completely determined by the type of a. Thus if b has the same type as a, the mapping $a^i \mapsto b^i$, $1 \leq i \leq k + m - 1$, is an isomorphism of semigroups. □

Lemma 5.2.4 *For each pair $(k, m) \in \mathbb{N}^2$ there exists a cyclic semigroup, generated by an element of type (k, m).*

Proof. A direct calculation shows that the element

$$\alpha = [1, 2, \ldots, k](k+1, k+2, \ldots, k+m) \in \mathcal{IS}_{k+m}$$

has type (k, m) and hence generates the necessary cyclic semigroup. □

Lemma 5.2.5 *Let $\langle a \rangle$ be a cyclic semigroup, generated by the element a of type (k, m). Then the set $P = \{a^k, \ldots, a^{k+m-1}\}$ is a subgroup of $\langle a \rangle$. Moreover, this subgroup is isomorphic to $\mathbb{Z}_m$.*

Proof. That P is a subsemigroup of $\langle a \rangle$ is obvious. Consider the mapping

$$\varphi: \quad \begin{array}{ccc} P & \to & \mathbb{Z}_m \\ a^{k+i} & \mapsto & \overline{k+i}, \end{array}$$

where $i = 0, 1, \ldots, m-1$. The mapping φ is obviously a bijection. On the other hand, we have

$$\varphi(a^{k+i} \cdot a^{k+j}) = \varphi(a^{2k+i+j}) = \overline{2k+i+j} = \overline{k+i} + \overline{k+j} = \varphi(a^{k+i}) \mp \varphi(a^{k+j}).$$

Hence φ is an isomorphism. $\qquad\qquad\qquad\qquad\qquad\qquad\qquad\qquad$ $\square$

Corollary 5.2.6 *Let S be a semigroup and $a \in S$ be an element of finite order. Then the semigroup $\langle a \rangle$ contains an idempotent. In particular, each finite semigroup contains an idempotent.*

Proof. To prove the first statement we just have to note that the identity element of the group P from Lemma 5.2.5 is an idempotent. The proof is then completed by observing that in a finite semigroup all elements have finite order. $\qquad\qquad\qquad\qquad\qquad\qquad\qquad\qquad\qquad\qquad$ $\square$

The identity element of P is the a^{k+i} for which $\varphi(a^{k+i}) = \overline{0}$, that is, $i \equiv -k \pmod{m}$. It is worth noting that $a^{k+(-k \bmod m)}$ is the unique idempotent of $\langle a \rangle$, as for each $b \in \langle a \rangle \backslash P$ we have $b^k \in P$ and hence b cannot be an idempotent.

The cyclic semigroup $\langle a \rangle$, generated by an element of type (k, m), can be depicted via the following diagram.

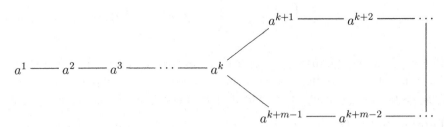

The results obtained above can be summarized in the following statement.

Theorem 5.2.7 (Structure theorem for finite cyclic semigroups)

(i) *Let S be a finite cyclic semigroup, generated by the element a. Then there exist $k, m \in \mathbb{N}$ such that $S = \langle a \rangle = \{a^1, a^2, \ldots, a^{k+m-1}\}$ and $a^{k+m} = a^k$.*

(ii) *For each $k, m \in \mathbb{N}$ there exists a finite cyclic semigroup, generated by an element of type (k, m).*

(iii) *Two finite cyclic semigroups are isomorphic if and only if they are generated by elements of the same type.*

(iv) *If a has type (k, m), then the set $\{a^k, a^{k+1}, \ldots, a^{k+m-1}\}$ is a subgroup of $\langle a \rangle$, which is isomorphic to the group $(\mathbb{Z}_m, +)$.*

(v) *Every finite cyclic semigroup contains a unique idempotent.*

Proof. Almost everything was already proved in Lemmas 5.2.1, 5.2.4, 5.2.5 and Corollary 5.2.6. The only thing which we have to show is that if the semigroups $\langle a \rangle$ and $\langle b \rangle$ are isomorphic, then the types of a and b coincide. Let (k, m) be the type of a and (k', m') be the type of b. The fact that $\langle a \rangle$ and $\langle b \rangle$ are isomorphic implies $k + m = k' + m'$. From the observation after Corollary 5.2.6 we have that an element of $\langle a \rangle$ is a group element if and only if it belongs to P. In particular, $\langle a \rangle$ contains exactly m group elements. As $\langle a \rangle$ and $\langle b \rangle$ are isomorphic, we get that $\langle b \rangle$ contains exactly m group elements as well. Hence $m = m'$ and thus $k = k'$ as well, since we already know that $k + m = k' + m'$. This completes the proof. $\square$

Proposition 5.2.8 *Let $\alpha \in \mathcal{PT}_n$. Then the index of α equals the minimum k for which $\mathrm{im}(\alpha^k) = \mathrm{stim}(\alpha)$, and the period of α equals the order of the permutation $\alpha|_{\mathrm{stim}(\alpha)}$.*

Proof. For every $i > 0$ the set $\mathrm{im}(\alpha^i)$ is invariant with respect to α. Hence we have the following chain of invariant sets:

$$\mathrm{im}(\alpha) \supset \mathrm{im}(\alpha^2) \supset \mathrm{im}(\alpha^3) \supset \cdots . \tag{5.3}$$

As $\mathrm{im}(\alpha^i) = \mathrm{im}(\alpha^{i+1})$ implies $\mathrm{im}(\alpha^i) = \mathrm{im}(\alpha^{i+j}) = \mathrm{stim}(\alpha)$ for all $j > 0$, the chain (5.3) has the form

$$\mathrm{im}(\alpha) \supsetneq \mathrm{im}(\alpha^2) \supsetneq \cdots \supsetneq \mathrm{im}(\alpha^k) = \mathrm{im}(\alpha^{k+1}) = \cdots .$$

By Proposition 2.9.3, the restriction of α to $\mathrm{stim}(\alpha)$ is a permutation. Let m be the order of this permutation. Then the elements $\alpha^1, \alpha^2, \ldots, \alpha^{k+m-1}$ are different. Indeed, the first k elements have different images, the last m elements have the same image, but their restrictions to this image are permutations $\alpha^k|_{\mathrm{stim}(\alpha)}, \alpha^{k+1}|_{\mathrm{stim}(\alpha)}, \ldots, \alpha^{k+m-1}|_{\mathrm{stim}(\alpha)}$, which are all different since the order of $\alpha|_{\mathrm{stim}(\alpha)}$ is m. It is obvious that α^{k+m} cannot be equal to any of α^i, $i < k$, because the image of α^{k+m} is different from that of each such α^i. At the same time, we claim that $\alpha^{k+m} = \alpha^k$. Indeed, for every $x \in \mathrm{dom}(\alpha^k)$ we have

$$\alpha^{k+m}(x) = \alpha^m(\alpha^k(x)) = \alpha^k(x)$$

as $\alpha^k(x) \in \mathrm{stim}(\alpha)$ and α^m acts as the identity transformation on $\mathrm{stim}(\alpha)$. Since $\mathrm{dom}(\alpha^{k+m}) \subset \mathrm{dom}(\alpha^k)$, we get $\alpha^{k+m} = \alpha^k$. This means that α has index k and period m, as claimed. The statement is proved. $\square$

Exercise 5.2.9 For each $\alpha \in \mathcal{T}_n$ the period of α equals the least common multiple of the cardinalities of the kernels of all orbits of α.

Exercise 5.2.10 For each $\alpha \in \mathcal{T}_n$ the index of α is the maximum over all cardinalities of trajectories for the vertices of the full subgraph of Γ_α with the vertex set $\mathbf{N}\backslash\mathrm{stim}(\alpha)$.

5.3   Isolated and Completely Isolated Subsemigroups

A subsemigroup T of a semigroup S is called *completely isolated* provided that $ab \in T$ implies $a \in T$, or $b \in T$ for all $a, b \in S$.

Exercise 5.3.1 Show that a proper subsemigroup $T \subsetneq S$ is completely isolated if and only if its complement $\overline{T} = S\backslash T$ is a subsemigroup. In particular, if T is completely isolated, then $\overline{T}$ is completely isolated as well.

A subsemigroup T of a semigroup S is called *isolated* provided that $a^n \in T$ implies $a \in T$ for all $a \in S$ and $n \in \mathbb{N}$. It is obvious that each completely isolated subsemigroup is isolated. Each semigroup is a completely isolated subsemigroup of itself.

Exercise 5.3.2 Show that a proper subsemigroup $T \subsetneq S$ is isolated if and only if $\overline{T}$ is a union of subsemigroups.

Immediately from the definitions, Exercises 5.3.1 and 5.3.2 we have

Proposition 5.3.3 *Let S be a semigroup.*

(i) *The intersection of any family of isolated subsemigroups of S is an isolated subsemigroup. In particular, for each $a \in S$ there exists a minimum isolated subsemigroup of S, containing a.*

(ii) *If a union of some family of (completely) isolated subsemigroups of S is a semigroup, then it is a (completely) isolated subsemigroup of S.*

For each $e \in \mathcal{E}(S)$ set

$$\sqrt{e} = \{x \in S \ : \ x^m = e \text{ for some } m > 0\}.$$

Lemma 5.3.4 *Let S be a semigroup.*

(i) *If $T \subset S$ is an isolated subsemigroup, then $\sqrt{e} \subset T$ for all $e \in T$.*

(ii) *If $T \subset S$ is isolated and S is finite, then $T = \cup_{e \in \mathcal{E}(T)} \sqrt{e}$.*

Proof. The statement (i) is obvious. From (i) it follows that $\cup_{e \in \mathcal{E}(T)} \sqrt{e} \subset T$. On the other hand, S is finite and hence if $a \in T$, then $\langle a \rangle$ contains a unique idempotent, say e, by Theorem 5.2.7. This means that $a \in \sqrt{e}$ and thus $T \subset \cup_{e \in \mathcal{E}(T)} \sqrt{e}$. This proves (ii) and completes the proof.   □

Remark 5.3.5 If S is finite, then $\sqrt{e}$ always contains G_e, that is, the maximal subgroup, corresponding to e.

Remark 5.3.6 For the semigroups $\mathcal{T}_n$, $\mathcal{PT}_n$, and $\mathcal{IS}_n$ we have $\sqrt{\varepsilon} = \mathcal{S}_n$.

Exercise 5.3.7 Show that $\sqrt{0}$ is not a subsemigroup of $\mathcal{IS}_n$ if $n > 1$.

To describe all (completely) isolated subsemigroups of $\mathcal{T}_n$, $\mathcal{PT}_n$, and $\mathcal{IS}_n$ we will need a number of auxiliary statements. For simplicity we shall call noninvertible elements of semigroups *singular*. Also for simplicity, we shall use the following notation for the elements from $\mathcal{PT}_n$:

$$
\alpha = \left(\begin{array}{cccc} A_1 & A_2 & \cdots & A_m \\ a_1 & a_2 & \cdots & a_m \end{array} \right),
$$

which means that $\alpha(a) = a_i$ for all $a \in A_i$, $i = 1, \ldots, m$, and that on the complement $B = \mathbf{N} \backslash (A_1 \cup \cdots \cup A_m)$ the element α acts as the identity on all elements, on which it is defined. If we consider several elements at the same time, we assume that they act in the same way on the complement B.

Lemma 5.3.8 *Let S be one of the semigroups $\mathcal{T}_n$, $\mathcal{PT}_n$, or $\mathcal{IS}_n$, and T be a proper isolated subsemigroup of S. If T contains all singular idempotents of S, then $T = \mathcal{I}_{n-1}$.*

Proof. The ideal $\mathcal{I}_{n-1}$ coincides with the set of all singular elements in S. The cyclic subsemigroup of S, generated by a singular element a, contains a singular idempotent, say e. But $a \in \sqrt{e}$ by definition, and hence $\sqrt{e} \subset T$ by Lemma 5.3.4(i). Hence $\mathcal{I}_{n-1} \subset T$. On the other hand, if T would contain some invertible element, T would contain the identity transformation as a power of this element, and hence the whole $\mathcal{S}_n$ by Remark 5.3.6 and Lemma 5.3.4(i). This would imply $T = S$, a contradiction. Hence $T = \mathcal{I}_{n-1}$.   □

Lemma 5.3.9 *Let T be an isolated subsemigroup of $\mathcal{T}_n$. If T contains all left zeros of $\mathcal{T}_n$, then T contains all singular idempotents of $\mathcal{T}_n$.*

Proof. Since in $\mathcal{T}_2$ each singular idempotent is a left zero, we may assume $n > 2$. Let $A \subset \mathbf{N}$ and let $a \in A$, $b, c \notin A$, $b \neq c$. We have

$$
\left(\begin{array}{ccc} b & c & A \\ c & a & a \end{array} \right)^2 = \left(\begin{array}{ccc} b & c & A \\ a & a & a \end{array} \right) = \left(\begin{array}{ccc} b & c & A \\ a & b & a \end{array} \right)^2,
$$

$$\begin{pmatrix} b & c & A \\ a & b & a \end{pmatrix} \begin{pmatrix} b & c & A \\ c & a & a \end{pmatrix} = \begin{pmatrix} b & c & A \\ b & a & a \end{pmatrix}.$$

Hence if an isolated subsemigroup T contains the element $\begin{pmatrix} b & c & A \\ a & a & a \end{pmatrix}$, then T contains the element $\begin{pmatrix} b & c & A \\ b & a & a \end{pmatrix}$ as well. Applying the last statement inductively starting from left zeros, we get that T contains all idempotents of the form $\epsilon_A = \begin{pmatrix} A \\ a \end{pmatrix}$, where $A \subset \mathbf{N}$, $|A| > 1$, and $a \in A$. For the element ϵ_A we have that the corresponding partition ρ_{ϵ_A} contains a unique block A of cardinality $|A| \geq 2$, and each of the remaining blocks of this partition consists of one element. Let now

$$\epsilon = \begin{pmatrix} A_1 & A_2 & \cdots & A_k \\ a_1 & a_2 & \cdots & a_k \end{pmatrix}$$

be a singular idempotent, where $A_1, \ldots, A_k$ are all blocks of ρ_ϵ, which consists of more than one element. The proof is now completed by observing that $\epsilon = \epsilon_{A_1} \epsilon_{A_2} \cdots \epsilon_{A_k}$. $\square$

Let $\epsilon \in \mathcal{E}(\mathcal{T}_n)$ be a singular idempotent and $A \subset \mathbf{N}$. We will say that ϵ is *singular on* A provided that A is invariant with respect to ϵ and ϵ acts as the identity transformation on the complement $\mathbf{N}\backslash A$.

If A is the unique block of ρ_ϵ, which contains more than one element, we will say that the idempotent ρ_ϵ is a *constant on* A. Clearly, via restriction, all constants on A can be considered as left zeros of the semigroup $\mathcal{T}(A)$.

Lemma 5.3.10 *Let T be an isolated subsemigroup of $\mathcal{T}_n$. Assume that for some $A \subset \mathbf{N}$, $|A| \geq 3$, the semigroup T contains all idempotents, singular on A. Then for each $k \in \mathbf{N}\backslash A$ the semigroup T contains all idempotents, singular on $A \cup \{k\}$.*

Proof. Taking Lemma 5.3.9 into account, it is enough to show that T contains all constants on $A \cup \{k\}$. Let $a \in A$. Fix a partition $A\backslash\{a\} = A_1 \cup A_2$ and $a_i \in A_i$, $i = 1, 2$. The element

$$\omega_a = \begin{pmatrix} A_1 & a & A_2 & k \\ a & a & a & k \end{pmatrix}$$

is a constant on A and hence belongs to T by our assumptions. For the elements

$$\beta = \begin{pmatrix} A_1 & a & A_2 & k \\ a_1 & a_1 & k & a_2 \end{pmatrix}, \quad \gamma = \begin{pmatrix} A_1 & a & A_2 & k \\ a_1 & k & k & a \end{pmatrix},$$

$$\delta = \begin{pmatrix} A_1 & a & A_2 & k \\ k & k & k & a \end{pmatrix}$$

we have that the elements

$$\beta^2 = \left(\begin{array}{cccc} A_1 & a & A_2 & k \\ a_1 & a_1 & a_2 & k \end{array} \right), \quad \gamma^2 = \left(\begin{array}{cccc} A_1 & a & A_2 & k \\ a_1 & a & a & k \end{array} \right), \quad \delta^2 = \omega_a$$

are singular idempotents on A. Hence $\beta, \gamma, \delta \in T$. The latter implies that the following elements belong to T as well:

$$\mu_a = \omega_a \beta \gamma = \left(\begin{array}{cccc} A_1 & a & A_2 & k \\ a & a & a & a \end{array} \right), \quad \mu_k = \delta \mu_a = \left(\begin{array}{cccc} A_1 & a & A_2 & k \\ k & k & k & k \end{array} \right).$$

The statement follows. $\qquad\qquad\square$

Corollary 5.3.11 *Let T be an isolated subsemigroup of $\mathcal{T}_n$. Assume that for some $A \subset \mathbf{N}$, $|A| \geq 3$, the semigroup T contains all idempotents, singular on A. Then T contains $\mathcal{I}_{n-1}$.*

Proof. Inductively applying Lemma 5.3.10 we obtain that T contains all singular idempotents and the statement follows from Lemma 5.3.8. $\qquad\square$

For $k, m \in \mathbf{N}$, $m \neq k$, let $\varepsilon_{m,k}$ denote the unique idempotent of $\mathcal{T}_n$ of rank $(n-1)$ satisfying $\varepsilon_{m,k}(m) = \varepsilon_{m,k}(k) = m$.

Lemma 5.3.12 *Let T be an isolated subsemigroup of $\mathcal{T}_n$. If for some $k_1 \neq k_2$ and m the semigroup T contains ε_{m,k_1} and ε_{m,k_2}, then T contains $\mathcal{I}_{n-1}$.*

Proof. Without loss of generality we may assume $k_1 = 1$, $m = 2$, and $k_2 = 3$. Let us show that T contains all constants on the set $\{1, 2, 3\}$. For the element $\alpha = \left(\begin{array}{ccc} 1 & 2 & 3 \\ 3 & 3 & 2 \end{array} \right)$ we have $\alpha^2 = \varepsilon_{2,1}$ and hence $\alpha \in T$. Then

$$\alpha \varepsilon_{2,3} = \left(\begin{array}{ccc} 1 & 2 & 3 \\ 3 & 3 & 3 \end{array} \right) \in T, \quad \varepsilon_{2,3}\alpha = \left(\begin{array}{ccc} 1 & 2 & 3 \\ 2 & 2 & 2 \end{array} \right) \in T.$$

Analogously one also shows that $\left(\begin{array}{ccc} 1 & 2 & 3 \\ 1 & 1 & 1 \end{array} \right) \in T$. By Lemma 5.3.9 we thus have that T contains all idempotents, singular on A. The statement now follows from Corollary 5.3.11. $\qquad\square$

Lemma 5.3.13 *Let T be an isolated subsemigroup of $\mathcal{T}_n$. Assume that T contains ε_{m_1,k_1} and ε_{m_2,k_2} for some pairwise different k_1, k_2, m_1, m_2. Then T contains $\mathcal{I}_{n-1}$.*

Proof. Without loss of generality we may assume $m_1 = 1$, $k_1 = 2$, $m_2 = 3$, $k_2 = 4$. By the same arguments as in the proof of Lemma 5.3.12, it is enough to show that T contains all constants on $\{1, 2, 3, 4\}$. For the element $\alpha = \left(\begin{array}{cccc} 1 & 2 & 3 & 4 \\ 2 & 3 & 1 & 1 \end{array} \right)$ we have $\alpha^3 = \varepsilon_{3,4}$ and hence $\alpha \in T$. Thus $\beta = \varepsilon_{1,2}\alpha\varepsilon_{1,2} \in T$

and a direct commutation shows that β is the constant on $\{1,2,3,4\}$ with image $\{1\}$. At the same time $\alpha\beta \in T$ and $\alpha\beta$ is the constant on $\{1,2,3,4\}$ with image $\{2\}$. Analogously one shows that the remaining constants are also in T. $\qquad\square$

Lemma 5.3.14 *Let T be an isolated subsemigroup of $\mathcal{T}_n$. Assume that T contains some idempotent ϵ of rank $r < n - 1$. Then T contains an idempotent of rank $(r + 1)$.*

Proof. If the partition ρ_ϵ has some block B of cardinality $|B| \geq 3$, then an idempotent of rank $(r + 1)$ in T can be easily obtained using the same construction as in the proof of Lemma 5.3.9. If all blocks of ρ_ϵ contain at most two elements, there should be more than one two-element block. Hence without loss of generality we may assume that

$$\epsilon = \begin{pmatrix} 1 & 2 & 3 & 4 & 5 & \cdots & n \\ 1 & 1 & 3 & 3 & a_5 & \cdots & a_n \end{pmatrix}.$$

For the elements

$$\alpha = \begin{pmatrix} 1 & 2 & 3 & 4 & 5 & \cdots & n \\ 3 & 4 & 1 & 1 & a_5 & \cdots & a_n \end{pmatrix}, \quad \beta = \begin{pmatrix} 1 & 2 & 3 & 4 & 5 & \cdots & n \\ 3 & 3 & 1 & 2 & a_5 & \cdots & a_n \end{pmatrix}$$

we have $\alpha^2 = \beta^2 = \epsilon$ and hence $\alpha, \beta \in T$. Thus

$$\beta\alpha = \begin{pmatrix} 1 & 2 & 3 & 4 & 5 & \cdots & n \\ 1 & 2 & 3 & 3 & a_4 & \cdots & a_n \end{pmatrix} \in T,$$

and the element $\beta\alpha$ is an idempotent of rank $(r + 1)$. $\qquad\square$

Lemma 5.3.15 *Let T be an isolated subsemigroup of $\mathcal{T}_n$. Assume that T contains some idempotent ϵ of rank $(n - 2)$. Then T contains $\mathcal{I}_{n-1}$.*

Proof. Without loss of generality we may assume that $\epsilon = \begin{pmatrix} 1 & 2 & 3 \\ 1 & 1 & 1 \end{pmatrix}$, or $\epsilon = \begin{pmatrix} 1 & 2 & 3 & 4 \\ 1 & 1 & 3 & 3 \end{pmatrix}$.

In the case $\epsilon = \begin{pmatrix} 1 & 2 & 3 \\ 1 & 1 & 1 \end{pmatrix}$ we may use the same construction as in the proof of Lemma 5.3.9 to get $\varepsilon_{1,2}, \varepsilon_{1,3} \in T$. Thus T contains $\mathcal{I}_{n-1}$ by Lemma 5.3.12.

In the case $\epsilon = \begin{pmatrix} 1 & 2 & 3 & 4 \\ 1 & 1 & 3 & 3 \end{pmatrix}$ we may use the same construction as in the proof of Lemma 5.3.14 to get $\varepsilon_{1,2}, \varepsilon_{3,4} \in T$. Thus T contains $\mathcal{I}_{n-1}$ by Lemma 5.3.13. $\qquad\square$

Lemma 5.3.16 *Let $\epsilon \in \mathcal{E}(\mathcal{T}_n)$ be of rank $(n - 1)$. Then $\sqrt{\epsilon} = G_\epsilon$.*

Proof. The inclusion $G_\epsilon \subset \sqrt{\epsilon}$ is obvious. Let now $\alpha \in \sqrt{\epsilon}$. Then $\alpha^m = \epsilon$ for some $m > 0$. From the inequalities

$$n > \text{rank}(\alpha) \geq \text{rank}(\epsilon) = n - 1$$

we have $\text{rank}(\alpha) = n - 1$. As $\text{im}(\alpha)$ is invariant with respect to α, and $\alpha^m = \epsilon$, we have $\text{stim}(\alpha) = \text{im}(\alpha)$. From Proposition 5.2.8 it now follows that the index of α is 1 and hence by Theorem 5.2.7 the semigroup $\langle \alpha \rangle$ is a group. Hence α is a group element and thus $\alpha \in G_\epsilon$. This proves the inclusion $\sqrt{\epsilon} \subset G_\epsilon$ and completes the proof. ☐

Theorem 5.3.17 *Let $n > 2$. A subsemigroup T of $\mathcal{T}_n$ is isolated if and only if it belongs to the following set:*

$$\{\mathcal{T}_n, \mathcal{S}_n, \mathcal{I}_{n-1}\} \cup \{G_{\varepsilon_{m,k}} \cup G_{\varepsilon_{k,m}} : m, k \in \mathbf{N}, m \neq k\} \cup$$
$$\cup \{\cup_{m \in M} G_{\varepsilon_{m,k}} : k \in \mathbf{N}, \varnothing \neq M \subset \mathbf{N} \backslash \{k\}\}.$$

Proof. As both $\mathcal{S}_n$ and $\mathcal{I}_{n-1}$ are subsemigroups, $\mathcal{S}_n \cap \mathcal{I}_{n-1} = \varnothing$ and $\mathcal{T}_n = \mathcal{S}_n \cup \mathcal{I}_{n-1}$, each of the semigroups $\mathcal{S}_n$, $\mathcal{I}_{n-1}$, and $\mathcal{T}_n$ is an isolated subsemigroup of $\mathcal{T}_n$.

Let T be an isolated subsemigroup of $\mathcal{T}_n$ such that $T \cap \mathcal{S}_n \neq \varnothing$. As $\mathcal{S}_n$ is a finite group, T must contain ε and hence $\mathcal{S}_n = \sqrt{\varepsilon} \subset T$ by Lemma 5.3.4(i). If at the same time $T \cap \mathcal{I}_{n-1} \neq \varnothing$, then Lemma 5.3.14 guarantees that T contains an idempotent of rank $(n-1)$. Thus T contains a generating system for $\mathcal{T}_n$ by Theorem 3.1.3, which implies that $T = \mathcal{T}_n$. If, on the other hand, $T \cap \mathcal{I}_{n-1} = \varnothing$, we have $T = \mathcal{S}_n$.

It is left to consider the case when $T \cap \mathcal{S}_n = \varnothing$ and $T \neq \mathcal{I}_{n-1}$. From Lemmas 5.3.14 and 5.3.15 it follows that in this case T contains only idempotents of rank $(n-1)$, that is, of the form $\varepsilon_{m,k}$.

Let $\mathcal{E}(T) = \{\varepsilon_{m_i,k_i} : i = 1, \dots, t\}$. If $\{m_i, k_i\} \cap \{m_j, k_j\} = \varnothing$ for some i, j, then $T \supset \mathcal{I}_{n-1}$ by Lemma 5.3.13, which is not possible. Hence for all i, j we have $\{m_i, k_i\} \cap \{m_j, k_j\} \neq \varnothing$. Moreover,

$$\varepsilon_{l,m} \varepsilon_{m,k} = \begin{pmatrix} k & l & m \\ l & l & l \end{pmatrix}$$

is an idempotent of rank $(n-2)$ as well. For $i \neq j$ the possibility $m_i = m_j$ is prohibited by Lemma 5.3.12. Hence we either have $k_i = k_j$, or $m_i = k_j$ and $m_j = k_i$. In the last case the only possibility is $\mathcal{E}(T) = \{\varepsilon_{m,k}, \varepsilon_{k,m}\}$ as for any other $\varepsilon_{m',k'}$ such that $\{m', k'\} \cap \{k, m\} \neq \varnothing$ we would have $k = k' = m$, which is impossible.

So, we have either $\mathcal{E}(T) = \{\varepsilon_{m,k}, \varepsilon_{k,m}\}$, or $\mathcal{E}(T) = \{\varepsilon_{m_i,k} : i = 1, \dots, t\}$. Because of Lemmas 5.3.4(ii) and 5.3.16 it remains to show that each of the sets

$$P = G_{\varepsilon_{m,k}} \cup G_{\varepsilon_{k,m}} \quad \text{and} \quad Q = G_{\varepsilon_{m_1,k}} \cup \cdots \cup G_{\varepsilon_{m_t,k}} \qquad (5.4)$$

is a subsemigroup of $\mathcal{T}_n$. In the first case of (5.4) for arbitrary $\alpha \in G_{\varepsilon_{m,k}}$ and $\beta \in G_{\varepsilon_{k,m}}$ we have $\rho_{\alpha\beta} = \rho_\beta = \rho_\alpha$ and $\operatorname{im}(\alpha\beta) = \operatorname{im}(\alpha)$. Hence $\alpha\beta \in G_{\varepsilon_{m,k}}$ by Theorem 4.5.1(iii) and because of the symmetry of the situation we conclude that $G_{\varepsilon_{m,k}} \cup G_{\varepsilon_{k,m}}$ is a semigroup.

In the second case of (5.4) we have

$$\operatorname{im}(\varepsilon_{m_1,k}) = \operatorname{im}(\varepsilon_{m_2,k}) = \cdots = \operatorname{im}(\varepsilon_{m_t,k}) = \mathbf{N}\backslash\{k\}.$$

For any $\alpha, \beta \in Q$ we have $\rho_{\alpha\beta} = \rho_\beta$ and $\operatorname{im}(\alpha\beta) = \mathbf{N}\backslash\{k\}$. By Theorem 4.5.1(iii) we have $\alpha\beta \in \mathcal{H}(\beta) \subset Q$. Hence Q is a subsemigroup as well. This completes the proof. $\square$

Remark 5.3.18 The only isolated subsemigroup of $\mathcal{T}_1$ is $\mathcal{T}_1$ itself. The semigroup $\mathcal{T}_2$ has five isolated subsemigroups: $\mathcal{T}_2$, $\mathcal{S}_2$, $\mathcal{I}_1$, $\{\varepsilon_{1,2}\}$, $\{\varepsilon_{2,1}\}$.

Corollary 5.3.19 *For $n > 1$ the only completely isolated subsemigroups of $\mathcal{T}_n$ are $\mathcal{T}_n$, $\mathcal{S}_n$, and $\mathcal{I}_{n-1}$.*

Proof. Since each completely isolated subsemigroup is isolated, we have just to check which of the subsemigroups, given by Theorem 5.3.17, are completely isolated. It is obvious that $\mathcal{T}_n$ is completely isolated. Further, both $\mathcal{S}_n$ and $\mathcal{I}_{n-1}$ are completely isolated as each of these semigroups is the complement of the other one.

Let T be an isolated subsemigroup of $\mathcal{T}_n$, different from $\mathcal{T}_n$, $\mathcal{S}_n$, and $\mathcal{I}_{n-1}$. Then from Theorem 5.3.17 we get that T consists only of group elements of rank $(n-1)$.

If $n = 2$, then neither $\{\varepsilon_{1,2}\}$ nor $\{\varepsilon_{2,1}\}$ is completely isolated. Indeed, assume $\{\varepsilon_{1,2}\}$ is completely isolated. Then $(1,2) \cdot \varepsilon_{2,1} = \varepsilon_{1,2} \in T$. As $(1,2)$ has rank 2, we have $(1,2) \notin T$, implying $\varepsilon_{2,1} \in T$ and hence $T = \mathcal{I}_{n-1}$, a contradiction. Analogously one shows that $\{\varepsilon_{2,1}\}$ is not completely isolated.

Assume $n > 2$. Then the complement $\overline{T} = \mathcal{T}_n \backslash T$ contains $\mathcal{S}_n$ and the element

$$\alpha = \begin{pmatrix} 1 & 2 & 3 & 4 & \cdots & n \\ 1 & 1 & 2 & 3 & \cdots & n-1 \end{pmatrix},$$

which is not a group element by Exercise 5.1.5. By Theorem 3.1.3 the set $\mathcal{S}_n \cup \{\alpha\}$ generates the whole $\mathcal{T}_n$ and hence $\overline{T}$ cannot be a subsemigroup. Hence T is not completely isolated by Exercise 5.3.1. $\square$

Let us now describe isolated and completely isolated subsemigroups of $\mathcal{IS}_n$. The elements of $\mathbf{N}$ which will be omitted in the notation of some transformation α will be assumed to be fixed points of α.

Lemma 5.3.20 *If an isolated subsemigroup $T \subset \mathcal{IS}_n$ contains some idempotent ϵ of rank at most $(n-2)$, then T contains at least two different idempotents of rank $(n-1)$.*

Proof. Let rank(ϵ) = $n - k$, $k > 1$, and $\overline{\text{dom}}(\epsilon) = \{a_1, a_2, \ldots, a_k\}$. Then the elements $\alpha = [a_1, a_2, \ldots, a_k]$ and $\beta = [a_k, a_2, \ldots, a_1]$ satisfy $\alpha^k = \beta^k = \epsilon$. Hence $\alpha, \beta \in T$ and thus $\beta\alpha, \alpha\beta \in T$. But both $\beta\alpha = [a_k]$ and $\alpha\beta = [a_1]$ are idempotents of rank $(n-1)$, and they are obviously different. $\qquad\square$

Lemma 5.3.21 *If an isolated subsemigroup $T \subset \mathcal{IS}_n$ contains two different idempotents of rank $(n-1)$, then T contains $\mathcal{I}_{n-1}$.*

Proof. Assume that T contains the idempotents $[x]$ and $[y]$, $x \neq y$. If $n = 2$, then T contains $[x][y]$ and thus contains all singular idempotents. This means that T contains $\mathcal{I}_{n-1}$ by Lemma 5.3.8.

Assume now that $n > 2$ and let $z \notin \{x, y\}$. Then $([x](y, z))^2 = [x]$ and hence $[x](y, z) \in T$. This means that T also contains the element

$$\alpha = [y] \cdot [x](y, z) \cdot [y] = [x][y][z].$$

Further, T contains both $\beta = [x, y, z]$ and $\gamma = [z, y, x]$ since $\beta^3 = \gamma^3 = \alpha$. Hence the idempotent $[z] = \gamma\beta$ is also contained in T. In particular, T contains all singular idempotents of rank $(n-1)$. As every singular idempotent $\epsilon = [a_1][a_2] \ldots [a_k]$ of $\mathcal{IS}_n$ is a product of singular idempotents $[a_i]$, $i = 1, \ldots, k$, of rank $(n-1)$, we obtain that T contains all singular idempotents. Hence T contains $\mathcal{I}_{n-1}$ by Lemma 5.3.8. $\qquad\square$

Theorem 5.3.22 *The only isolated subsemigroups of $\mathcal{IS}_n$ are $\mathcal{IS}_n$, $\mathcal{S}_n$, $\mathcal{I}_{n-1}$, and $G_{[x]}$, $x \in \mathbf{N}$.*

Proof. Let T be an isolated subsemigroup of $\mathcal{IS}_n$. Repeating the arguments of the first part of the proof of Theorem 5.3.17 one shows that either $T \in \{\mathcal{IS}_n, \mathcal{S}_n, \mathcal{I}_{n-1}\}$, or $T \subsetneq \mathcal{I}_{n-1}$.

If $T \subsetneq \mathcal{I}_{n-1}$, from Lemmas 5.3.20 and 5.3.21 we have that $\mathcal{E}(T) = \{[x]\}$ for some $x \in \mathbf{N}$. But then $T = \sqrt{[x]} = G_{[x]}$ by Lemma 5.3.4(ii). On the other hand, the semigroup $\sqrt{[x]} = G_{[x]}$ is obviously isolated. $\qquad\square$

Corollary 5.3.23 *The only completely isolated subsemigroups of $\mathcal{IS}_n$ are $\mathcal{IS}_n$, $\mathcal{S}_n$, and $\mathcal{I}_{n-1}$.*

Proof. Each completely isolated subsemigroup is isolated. So, we have just to go through the list given by Theorem 5.3.22. The semigroups $\mathcal{IS}_n$, $\mathcal{S}_n$, and $\mathcal{I}_{n-1}$ are obviously completely isolated. On the other hand, if $n > 1$, none of the semigroups $G_{[x]}$, $x \in \mathbf{N}$, is completely isolated. Indeed, let $y \neq x$. If $G_{[x]}$ were completely isolated, from $(x, y)[y](x, y) = [x]$ and $(x, y) \notin G_{[x]}$ it would follow $[y] \in G_{[x]}$, which is not the case. $\qquad\square$

Finally, let us describe all completely isolated subsemigroups of $\mathcal{PT}_n$. Note that $\mathcal{PT}_1 = \mathcal{IS}_1$. We shall need the following obvious result:

Lemma 5.3.24 *(i) Let S be a semigroup and $A < B < S$. If A is completely isolated in S, then A is completely isolated in B.*

(ii) Let S be a semigroup, $A < S$ completely isolated and $B < S$. Then $A \cap B$ is either empty or a completely isolated subsemigroup of B.

Theorem 5.3.25 *For $n > 1$ the semigroup $\mathcal{PT}_n$ has seven completely isolated subsemigroups, namely, $\mathcal{PT}_n$, $\mathcal{T}_n$, $\mathcal{S}_n$, $\mathcal{PT}_n \backslash \mathcal{S}_n$, $\mathcal{T}_n \backslash \mathcal{S}_n$, $\mathcal{PT}_n \backslash \mathcal{T}_n$, $\mathcal{PT}_n \backslash (\mathcal{T}_n \backslash \mathcal{S}_n)$.*

Proof. A product of two nontotal transformations is not total either. Hence the complement $\mathcal{PT}_n \backslash \mathcal{T}_n$ of the subsemigroup $\mathcal{T}_n$ is a subsemigroup itself. This means that both $\mathcal{PT}_n \backslash \mathcal{T}_n$ and $\mathcal{T}_n$ are completely isolated in $\mathcal{PT}_n$. Further, as both $\mathcal{S}_n$ and $\mathcal{I}_{n-1} = \mathcal{PT}_n \backslash \mathcal{S}_n$ are subsemigroups of $\mathcal{PT}_n$, they both are completely isolated by Exercise 5.3.1.

Let T be a completely isolated subsemigroup of $\mathcal{PT}_n$. The intersection $T \cap \mathcal{T}_n$ is either a empty or a completely isolated subsemigroup of $\mathcal{T}_n$ because of Lemma 5.3.24(ii). Hence $T \cap \mathcal{T}_n$ is either $\mathcal{T}_n$, or $\mathcal{S}_n$, or $\mathcal{T}_n \backslash \mathcal{S}_n$ by Corollary 5.3.19. The product of two noninvertible transformations of $\mathbf{N}$ is not invertible, and the product of two nontotal transformations is not total. Hence the subsemigroup $\mathcal{T}_n \backslash \mathcal{S}_n$ is completely isolated.

Assume that $T \cap (\mathcal{PT}_n \backslash \mathcal{T}_n) \neq \varnothing$. Then T contains some idempotent

$$\epsilon = \begin{pmatrix} A & \cdots & B & C \\ a & \cdots & b & \varnothing \end{pmatrix} \in \mathcal{PT}_n \backslash \mathcal{T}_n.$$

Let $x \in \text{im}(\epsilon)$ and $c \in C$. Set $\epsilon_x = (x,c)\epsilon(x,c)$. We claim that $\epsilon_x \in T$. Indeed, this is obvious if $(x,c) \in T$. If $(x,c) \notin T$, then $(x,c)\epsilon_x(x,c) = \epsilon \in T$ and hence $\epsilon_x \in T$ since T is completely isolated. Thus T contains the element $\epsilon' = \epsilon\epsilon_x$, which is an idempotent and $\text{im}(\epsilon') = \text{im}(\epsilon) \backslash \{x\}$. Continuing inductively we get that T contains the element $\mathbf{0}_n$. From $\mathbf{0}_n = [1][2]\dots[n]$ we have that $T \ni [k]$ for some $k \in \mathbf{N}$. By the same arguments as above we get that for each $l \neq k$ the semigroup T contains the element $(k,l)[k](k,l) = [l]$. In particular, T contains all $[k]$, $k \in \mathbf{N}$, and all their products, that is, all singular idempotents from $\mathcal{IS}_n$.

If $n = 2$, the above gives us that $\mathcal{PT}_2$ contains three minimal completely isolated subsemigroups: $\mathcal{S}_2$, $\mathcal{T}_2 \backslash \mathcal{S}_2$, and $\mathcal{PT}_2 \backslash \mathcal{T}_2 = \mathcal{IS}_2 \backslash \mathcal{S}_2$. Any union of these subsemigroups is again a semigroup, hence completely isolated by Proposition 5.3.3(ii).

Assume now that $n \geq 3$. We continue the consideration of the case $T \cap (\mathcal{PT}_n \backslash \mathcal{T}_n) \neq \varnothing$ and recall that we already know that T contains all singular idempotents from $\mathcal{IS}_n$. Consider the following elements in $\mathcal{PT}_n$:

$$\alpha = \begin{pmatrix} 1 & 2 & 3 \\ 1 & 1 & \varnothing \end{pmatrix}, \quad \beta = \begin{pmatrix} 1 & 2 & 3 \\ \varnothing & 3 & 3 \end{pmatrix}, \quad \alpha\beta = \begin{pmatrix} 1 & 2 & 3 \\ \varnothing & \varnothing & \varnothing \end{pmatrix}.$$

$\alpha\beta$ is a singular idempotent of $\mathcal{IS}_n$. Hence $\alpha \in T$, or $\beta \in T$. Without loss of generality we may assume $\alpha \in T$. As in the previous paragraph one shows that $\pi\alpha\tau \in T$ for all $\pi, \tau \in \mathcal{S}_n$, which implies that T contains all idempotents from $\mathcal{PT}_n \backslash \mathcal{T}_n$ of rank $(n-2)$, that is, all elements of the form $\varepsilon_{z,y}[x]$, where $x, y, z \in \mathbf{N}$ are arbitrary different elements.

Now let

$$\delta = \begin{pmatrix} A_1 & \cdots & A_k & B \\ a_1 & \cdots & a_k & \varnothing \end{pmatrix} \in \mathcal{PT}_n \backslash \mathcal{T}_n$$

be any idempotent, written such that all blocks A_i, $i = 1, \ldots, k$, consist of more than one element. For each $i = 1, \ldots, k$ consider the idempotent

$$\delta_i = \begin{pmatrix} A_i & B \\ a_i & \varnothing \end{pmatrix}.$$

Let $A_i = \{a_i, x_1, \ldots, x_k\}$ and $b \in B$. We have

$$\delta_i = \varepsilon_{\mathbf{N}\backslash B} \cdot \varepsilon_{a_i, x_1}[b] \cdot \varepsilon_{a_i, x_2}[b] \cdots \varepsilon_{a_i, x_k}[b]$$

and hence $\delta_i \in T$. At the same time $\delta = \delta_1 \delta_2 \cdots \delta_k$ and hence $\delta \in T$. Therefore T contains all idempotents from $\mathcal{PT}_n \backslash \mathcal{T}_n$ and thus T contains $\mathcal{PT}_n \backslash \mathcal{T}_n$ by Lemma 5.3.4(ii).

The above explanation gives us that for $n > 2$ the semigroup $\mathcal{PT}_n$ contains three minimal completely isolated subsemigroups: $\mathcal{S}_n$, $\mathcal{T}_n \backslash \mathcal{S}_n$, and $\mathcal{PT}_n \backslash \mathcal{T}_n$. Any union of these subsemigroups is again a semigroup, hence completely isolated by Proposition 5.3.3(ii). This completes the proof. $\square$

5.4   Addenda and Comments

5.4.1 For each semigroup S the Boolean $\mathcal{B}(S)$ has the following natural structure of a semigroup with respect to the multiplication: $A \cdot B = \{a \cdot b : a \in A, b \in B\}$, $A, B \subset S$ (it is very easy to check that this multiplication is associative). The semigroup $\mathcal{B}(S)$ is called the *global semigroup* or the *power semigroup* of S. Theorem 5.2.7 was proved by Frobenius in 1895 for the elements of the semigroup $\mathcal{B}(S)$, where S is a finite group (this restriction did not play an essential role in the proof). Since that time Theorem 5.2.7 has been rediscovered by many different authors.

5.4.2 Finite semigroups have the following property:

Theorem 5.4.1 *Let S be a finite semigroup. Then $\mathcal{D} = \mathcal{J}$.*

Proof. As we already know that $\mathcal{D} \subset \mathcal{J}$ (see Sect. 4.4), it is enough to show that $\mathcal{J} \subset \mathcal{D}$. Let $a\mathcal{J}b$, that is, $S^1 a S^1 = S^1 b S^1$. Then there exist $u, v, x, y \in S^1$ such that $uav = b$ and $xby = a$. Thus $(xu)a(vy) = a$ and even $(xu)^k a(vy)^k = a$ for all $k \in \mathbb{N}$. As S^1 is finite, there exist k and l such that $e = (xu)^k$ and $f = (vy)^l$ are idempotents. We have

$$a = (xu)^{kl} a(vy)^{kl} = eaf,$$

which implies $ea = a$ and $af = a$. From the first equality we have

$$(xu)^{k-1}x \cdot ua = a \quad \text{and} \quad u \cdot a = ua.$$

Hence $a\mathcal{L}ua$. Analogously one shows that $a\mathcal{R}av$, implying $ua\mathcal{R}uav$, that is, $ua\mathcal{R}b$. Thus $a\mathcal{D}b$.   □

5.4.3 A subsemigroup $T \subset S$ is called *maximal* provided that $T \neq S$ and for any subsemigroup $X \subset S$ the inclusion $T \subset X$ implies $T = X$, or $X = S$.

Obviously, if S is one of the semigroups $\mathcal{PT}_n$, $\mathcal{T}_n$, or $\mathcal{IS}_n$, and G is a subgroup of $\mathcal{S}_n$, then the set $\mathcal{I}_{n-1} \cup G$ is a subsemigroup of S.

Theorem 5.4.2 *Let S be one of the semigroups $\mathcal{T}_n$ or $\mathcal{IS}_n$. Then the only maximal subsemigroups of S are $\mathcal{S}_n \cup \mathcal{I}_{n-2}$ and $\mathcal{I}_{n-1} \cup G$, where G is a maximal subgroup of $\mathcal{S}_n$.*

Proof. Let T be a maximal subsemigroup of S. If $\mathcal{S}_n \not\subset T$, then $T \cup \mathcal{I}_{n-1}$ is a subsemigroup and $T \subset T \cup \mathcal{I}_{n-1} \subsetneq S$. Hence $T = T \cup \mathcal{I}_{n-1}$. On the other hand, $G = T \cap \mathcal{S}_n$ is a subsemigroup of $\mathcal{S}_n$ and hence is a subgroup since $\mathcal{S}_n$ is finite. Furthermore, G has to be a maximal subgroup for otherwise there would exist a subgroup G' of $\mathcal{S}_n$ such that $G \subsetneq G' \subsetneq \mathcal{S}_n$. In this case $G' \cup \mathcal{I}_{n-1}$ is a subsemigroup of S and $T \subsetneq G' \cup \mathcal{I}_{n-1} \subsetneq \mathcal{S}_n$, a contradiction.

If $\mathcal{S}_n \subset T$, the semigroup T cannot contain any element of rank $(n-1)$ for otherwise $T = S$ by Theorems 3.1.3 or 3.1.4. Hence $T \subset \mathcal{S}_n \cup \mathcal{I}_{n-2}$. On the other hand, $\mathcal{S}_n \cup \mathcal{I}_{n-2}$ is a subsemigroup of S and hence $T = \mathcal{S}_n \cup \mathcal{I}_{n-2}$.   □

Theorem 5.4.3 *The only maximal subsemigroups of $\mathcal{PT}_n$ are:*

(a) $\mathcal{I}_{n-1} \cup G$, where G is a maximal subgroup of $\mathcal{S}_n$;

(b) $\mathcal{PT}_n \backslash A$, where $A = \{\alpha \in \mathcal{T}_n : \text{rank}(\alpha) = n - 1\}$;

(c) $\mathcal{PT}_n \backslash B$, where $B = \{\alpha \in \mathcal{IS}_n : \text{rank}(\alpha) = n - 1\}$.

Proof. The proof is similar to that of Theorem 5.4.2. One has only to note that each element of rank $(n - 1)$ from $\mathcal{PT}_n$ is either a total transformation or a partial injection.   □

By the above theorems, the classification of maximal subsemigroups in $\mathcal{PT}_n$, $\mathcal{T}_n$, and $\mathcal{IS}_n$ reduces to the classification of maximal subgroups of the symmetric group $\mathcal{S}_n$. The latter problem is very difficult. The only known solution so far, which can be found in [LPS], is based on the so-called *classification of finite simple groups*. The correctness of the latter result is still questioned by many specialists.

5.4.4 From Theorem 5.4.2 it is easy to see that each maximal subsemigroup of $\mathcal{IS}_n$ is inverse. Thus as a bonus we get a description of all maximal inverse subsemigroups of $\mathcal{IS}_n$, see [X.Ya1]. Analogous questions can be asked for both $\mathcal{PT}_n$ and $\mathcal{T}_n$, especially taking into account that every $\alpha \in \mathcal{T}_n$ is contained in some inverse subsemigroup of $\mathcal{T}_n$, see [Sc5]. These questions were studied in particular in [Ni, Re2, KS, H.Ya].

5.4.5 Let $\alpha \in \mathcal{PT}_n$. By Proposition 5.2.8 and Exercises 5.2.9 and 5.2.10 the computation of the order of α can be easily reduced to the computation of the order of the permutation $\alpha|_{\mathrm{stim}(\alpha)}$. If $\pi = (a_1, \ldots, a_k) \cdots (b_1, \ldots, b_l)$, then the order of π is just the least common multiple (lcm) of the lengths $k, \ldots, l$ of the cycles. However, given some m to find the number of those $\sigma \in \mathcal{S}_n$ which have order m is a difficult problem. It reduces to the determination of all decompositions $n = n_1 + \cdots + n_k$ such that $\mathrm{lcm}(n_1, \ldots, n_k) = m$. Determination of the maximal possible order $P_{max}(\mathcal{S}_n)$ for elements in $\mathcal{S}_n$ is equivalent to finding

$$\max_{n_1 + \cdots + n_k = n} \mathrm{lcm}(n_1, \ldots, n_k),$$

which is also very difficult. Analogous problems for the semigroups $\mathcal{PT}_n$, $\mathcal{T}_n$, and $\mathcal{IS}_n$ are of the same level of difficulty.

5.4.6 On the other hand, the asymptotic behavior of $P_{\max}(\mathcal{S}_n)$ is much easier to describe. Already in 1903 E. Landau has proved in [La] the following formula:

$$\ln(P_{\max}(\mathcal{S}_n)) \sim \sqrt{n \ln(n)}, n \to \infty. \tag{5.5}$$

A very elementary proof of this statement can be found in [Mi]. Some additional comments about the asymptotic behavior of $P_{\max}(\mathcal{S}_n)$ can be found in [Sz]. From the formula (5.5) one easily obtains the following result about the asymptotic behavior of the orders of elements in the semigroups $\mathcal{T}_n$, $\mathcal{PT}_n$, and $\mathcal{IS}_n$.

Theorem 5.4.4

$$\ln(P_{\max}(\mathcal{T}_n)) \sim \ln(P_{\max}(\mathcal{PT}_n)) \sim \ln(P_{\max}(\mathcal{IS}_n)) \sim \sqrt{n \ln(n)}, n \to \infty.$$

5.4.7 From the proof of Theorem 2.4.3 it follows that each finite semigroup S can be considered as a subsemigroup of some $\mathcal{T}_n$, moreover if S is a monoid, then one can even assume that the identity element of S coincides with the identity transformation in $\mathcal{T}_n$. The group S^* becomes, under such identification, a subgroup of $\mathcal{S}_n$. As a consequence of this and Lemma 5.3.24(ii) we obtain:

Proposition 5.4.5 *If S is a finite semigroup, then both S^* and $S \backslash S^*$ are completely isolated subsemigroups of S.*

The statement of Proposition 5.4.5 is wrong in the general case as the group of units of an infinite semigroup does not have to be completely isolated (a bijective transformation can be a composition of two transformations, each of which is not bijective).

5.4.8 Let S be a semigroup. There is a natural partial order on the set of all $\mathcal{J}$-classes of S, given by the inclusion of the principal two-sided ideals:

$$\mathcal{J}(a) \leq \mathcal{J}(b) \quad \Leftrightarrow \quad S^1 a S^1 \subset S^1 b S^1.$$

If S is finite, then $\mathcal{J} = \mathcal{D}$ by Theorem 5.4.1 and hence the above partial order is also a partial order on the set of all $\mathcal{D}$-classes of S.

Theorem 5.4.6 *Let S be a finite monoid and assume that in the set of all $\mathcal{D}$-classes of S, different from $\mathcal{D}(1)$, there is the maximum $\mathcal{D}$-class D.*

(i) For any idempotent $e \in D$ we have $\sqrt{e} = G_e$.

(ii) Let $e_1, e_2, \ldots, e_k$ be a collection of idempotents from D which either belong to the same $\mathcal{R}$-class or to the same $\mathcal{L}$-class. Then $G_{e_1} \cup G_{e_2} \cup \cdots \cup G_{e_k}$ is an isolated subsemigroup of S.

We note that the semigroups $\mathcal{T}_n$, $\mathcal{PT}_n$, and $\mathcal{IS}_n$ satisfy all conditions of Theorem 5.4.6.

Proof. Consider S as a subsemigroup of some $\mathcal{T}_n$. Then $a\mathcal{R}b$ in S implies $a\mathcal{R}b$ in $\mathcal{T}_n$ by Proposition 4.4.2. Hence $a\mathcal{R}b$ implies $\text{im}(a) = \text{im}(b)$ by Theorem 4.2.1. Analogously, $a\mathcal{L}b$ implies $\rho_a = \rho_b$ by Theorem 4.2.4. In particular, all elements from the same $\mathcal{D}$-class of S have the same rank.

Obviously $\mathcal{D}(a^k) \leq \mathcal{D}(a)$. Hence for any idempotent $e \in D$ we have $\sqrt{e} \subset D$. Let $a \in \sqrt{e}$. From the equalities $a^m = e$, $\text{rank}(a) = \text{rank}(e)$, and the fact that $\text{im}(a)$ is invariant with respect to a, we get that $\text{stim}(a) = \text{im}(a)$. Hence, by Proposition 5.2.8, a has index one and thus is a group element. It follows that $\sqrt{e} \subset G_e$, which proves (i).

To prove (ii) it is enough to show that $G_{e_1} \cup G_{e_2} \cup \cdots \cup G_{e_k}$ is closed with respect to multiplication. Let $e = e_i$ and $f = e_j$ for $i \neq j$ and $a \in G_e$, $b \in G_f$. Assume first that all e_is belong to the same $\mathcal{R}$-class. We claim that in this case $ab \in G_f$. As S is a subsemigroup of $\mathcal{T}_n$, we have $\text{im}(a) = \text{im}(b)$ by Theorem 4.5.1(i). Since both a and b are group elements, different elements of $\text{im}(a) = \text{im}(b)$ must belong to different classes of the partitions ρ_a and ρ_b, respectively. Hence $\rho_{ab} = \rho_b$ and $\text{im}(ab) = \text{im}(a)$. This means that $ab \in G_f$ by Theorem 4.5.1(iii).

If all e_is belong to the same $\mathcal{L}$-class, then using analogous arguments one shows that $ab \in G_e$. This completes the proof. $\square$

5.4.9 The description of isolated semigroups for $\mathcal{PT}_n$ is even more technical than that for $\mathcal{T}_n$. Hence we just present here the result and an idea of the proof. To work out all the details of the proof is left to the reader.

Theorem 5.4.7 *A subsemigroup T of $\mathcal{PT}_n$ is isolated if and only if T belongs to the following list:*

(a) Completely isolated subsemigroups of $\mathcal{PT}_n$, given by Theorem 5.3.25.

(b) Isolated but not completely isolated subsemigroup of $\mathcal{T}_n \backslash \mathcal{S}_n$ (given by the list from Theorem 5.3.17 with $\{\mathcal{S}_n, \mathcal{T}_n, \mathcal{I}_{n-1}\}$ taken away).

(c) Semigroups $T(x, P)$, where $x \in \mathbf{N}$ and P is an isolated subsemigroup of $\mathcal{T}(\mathbf{N}\backslash\{x\})$, defined as follows:

$$T(x, P) = \{\alpha \in \mathcal{PT}_n : \overline{\mathrm{dom}}(\alpha) = \{x\},\ \alpha|_{\mathbf{N}\backslash\{x\}} \in P\}.$$

All possibilities for the semigroup P are given by Theorem 5.3.17.

(d) Semigroups $T(x, \mathcal{S}(\mathbf{N}\backslash\{x\})) \bigcup \bigcup_{m \in M} G_{\varepsilon_{m,x}}$, where $x \in \mathbf{N}$ and $\varnothing \neq M \subset \mathbf{N}\backslash\{x\}$.

Sketch of the Proof. By Theorem 5.3.25, the semigroup $\mathcal{PT}_n$ has three minimal completely isolated subsemigroups, namely, $\mathcal{S}_n$, $\mathcal{T}_n \backslash \mathcal{S}_n$, and $\mathcal{PT}_n \backslash \mathcal{T}_n$. If $T \subset \mathcal{PT}_n$ is an isolated subsemigroup and S is a minimal completely isolated subsemigroup of $\mathcal{PT}_n$, then $S \cap T$ is an isolated subsemigroup of S.

Let us first determine all possible intersections of T with the minimal completely isolated subsemigroups of $\mathcal{PT}_n$. The semigroup $T \cap \mathcal{S}_n$ is obviously either empty or equals $\mathcal{S}_n$. From Theorem 5.3.17 the semigroup $T \cap (\mathcal{T}_n \backslash \mathcal{S}_n)$ is either empty, or equals $\mathcal{T}_n \backslash \mathcal{S}_n$ or is given exactly by (b). The complicated part is to show that if $T \cap (\mathcal{PT}_n \backslash \mathcal{T}_n)$ is a proper subsemigroup of $\mathcal{PT}_n \backslash \mathcal{T}_n$, then it is given by (c).

So, let T be a proper isolated subsemigroup of $\mathcal{PT}_n \backslash \mathcal{T}_n$. First we claim that T contains an idempotent e such that $|\overline{\mathrm{dom}}(e)| = 1$. This can be proved analogously to the proof of Lemma 5.3.20. Then one shows that such T cannot contain $\mathbf{0}_n$. For this one shows that if T contains $\mathbf{0}_n$, then T contains all idempotents of $\mathcal{PT}_n \backslash \mathcal{T}_n$ and hence must coincide with the latter. This again can be done generalizing the arguments of Lemma 5.3.20. The next step is to show that T contains an idempotent e such that all blocks of ρ_e on which e is defined contain at most two elements. This is similar to the proof of Lemma 5.3.9. Generalizing the arguments of the proof of Lemma 5.3.21 one shows that all idempotents of T must satisfy $|\overline{\mathrm{dom}}(e)| = 1$. From this it follows easily that there should exist $x \in \mathbf{N}$ such that $\mathrm{dom}(\alpha) = \mathbf{N}\backslash\{x\}$ for all $\alpha \in T$. In particular, it follows that $x \notin \mathrm{im}(\alpha)$ for all $\alpha \in T$ and hence the set $\mathbf{N}\backslash\{x\}$ is invariant with respect to all elements from T. The restriction to $\mathbf{N}\backslash\{x\}$ defines a homomorphism from T to $\mathcal{T}(\mathbf{N}\backslash\{x\})$, the image of which is an isolated subsemigroup of $\mathcal{T}(\mathbf{N}\backslash\{x\})$, call it P. This gives $T = T(x, P)$ as in (c). On the other hand, it is easy to check that every $T = T(x, P)$ is isolated (this also follows from Exercise 5.5.13(a)).

It is left to see whether we can form a union of some of the semigroups given by (a), (b), and (c). A case-by-case analysis shows that the only possible unions are either the ones already described in (a) or the ones given by (d). The main argument of the analysis is that multiplying elements from different components of our union it is prohibited to get elements of smaller ranks. We leave the details to the reader.   □

5.4.10 There are of course many other natural classes of subsemigroups in the semigroups $\mathcal{T}_n$, $\mathcal{PT}_n$, and $\mathcal{IS}_n$, defined using various conditions. For instance, one can consider abelian subsemigroups, or nilpotent subsemigroups, or subsemigroups defined via some combinatorial conditions (for example, consisting of all α such that $\alpha(x) \leq x$ for all $x \in \mathbf{N}$, or consisting of all α such that $x \leq y$ implies $\alpha(x) \leq \alpha(y)$ for all $x, y \in \mathbf{N}$). Some of these classes are of independent interest and are intensively studied. In this book, we shall only briefly discuss nilpotent subsemigroups later on in Chap. 8.

5.5   Additional Exercises

5.5.1 Let G be a maximal subgroup of $\mathcal{T}_n$, or $\mathcal{IS}_n$. Show that G is a maximal subgroup of $\mathcal{PT}_n$.

5.5.2 Let S be a finite semigroup and $a \in S$. Prove that the following conditions are equivalent:

(a) a is a group element.

(b) $\langle a \rangle$ is a group.

(c) $a^k = a$ for some $k > 1$.

(d) a is regular and commutes with at least one of its inverses.

5.5.3 Let $\alpha \in \mathcal{PT}_n$ and assume that $\mathrm{rank}(\alpha^k) = \mathrm{rank}(\alpha^{k+1})$. Prove that

(a) $\mathrm{rank}(\alpha^{k+j}) = \mathrm{rank}(\alpha^k)$ for all $j > 0$

(b) $\mathrm{strank}(\alpha) = \mathrm{strank}(\alpha^k)$

(c) $\mathrm{im}(\alpha^{k+j}) = \mathrm{im}(\alpha^k)$ for all $j > 0$

(d) $\mathrm{stim}(\alpha) = \mathrm{stim}(\alpha^k)$

5.5.4 Let $\alpha \in \mathcal{PT}_n$. Assume that for some $m > 1$ the restriction $\alpha^m|_{\mathrm{stim}(\alpha)}$ is the permutation, which is inverse to the permutation $\alpha|_{\mathrm{stim}(\alpha)}$. Prove that the element α is a group element if and only if α and α^m form a pair of inverse elements.

5.5.5 Let S be a semigroup and $e \in \mathcal{E}(S)$. Show that eSe is the maximum submonoid of S, which has e as the identity element.

5.5.6 Let $\alpha \in \mathcal{PT}_n$ be a group element. Show that the restriction $\alpha|_{\mathrm{im}(\alpha)}$ is a permutation.

5.5.7 Let $\alpha \in \mathcal{PT}_n$. Show that α is a group element if and only if $\mathrm{im}(\alpha) = \mathrm{im}(\alpha^2)$.

5.5.8 (a) Let S be a cyclic semigroup, which is not a group. Show that S has a unique irreducible generating system.

(b) Show that for each $k \in \mathbb{N}$ there exists a finite cyclic group and an irreducible generating system in this group, containing exactly k elements.

5.5.9 Let $[p]_q = p(p-1)(p-2)\cdots(p-q+1)$, and let $P(n,m)$ denote the number of elements of order m in the symmetric group $\mathcal{S}_n$.

(a) Show that the number of elements of type (k,m) in the semigroup $\mathcal{T}_n$ equals
$$\sum_{n_1+\cdots+n_k+n_{k+1}=n} n_2^{n_1} n_3^{n_2} \cdots n_k^{n_{k-1}} n_{k+1}^{n_k} P(n_{k+1}, m).$$

(b) Show that the number of elements of type (k,m) in the semigroup $\mathcal{PT}_n$ equals
$$\sum_{n_1+\cdots+n_k+n_{k+1}\leq n} n_2^{n_1} n_3^{n_2} \cdots n_k^{n_{k-1}} n_{k+1}^{n_k} P(n_{k+1}, m).$$

(c) Show that the number of elements of type (k,m) in the semigroup $\mathcal{IS}_n$ equals
$$\sum_{n_1+\cdots+n_k+n_{k+1}=n} [n_2]_{n_1} [n_3]_{n_2} \cdots [n_k]_{n_{k-1}} P(n_{k+1}, m).$$

5.5.10 Prove that a (one-sided) ideal I of a semigroup S is an isolated subsemigroup of S if and only if I is semiprime.

5.5.11 Construct a semigroup S and a subsemigroup T such that $a^2 \in T$ implies $a \in T$ for any $a \in S$, but at the same time T is not an isolated subsemigroup of S.

5.5.12 Find the number of isolated subsemigroups in $\mathcal{T}_n$, $n > 2$.

5.5.13 (a) Let T be an isolated subsemigroup of S and X be an isolated subsemigroup of T. Show that X is an isolated subsemigroup of S.

(b) Let T be a completely isolated subsemigroup of S and X be a completely isolated subsemigroup of T. Is it true that X is a completely isolated subsemigroup of S?

5.5.14 Show that the intersection of completely isolated subsemigroups is not completely isolated in general.

Chapter 6

Other Relations on Semigroups

6.1 Congruences and Homomorphisms

Recall that a binary relation ρ on a semigroup S is called *left compatible* provided that $a\,\rho\,b$ implies $ca\,\rho\,cb$ for all $a, b, c \in S$; ρ is called *right compatible* provided that $a\,\rho\,b$ implies $ac\,\rho\,bc$ for all $a, b, c \in S$; and ρ is called *compatible* provided that it is both left and right compatible.

A left compatible equivalence relation is called a *left congruence*. A right compatible equivalence relation is called a *right congruence*. A compatible equivalence relation is called a *congruence*.

Example 6.1.1 (a) From Sect. 4.4 we know that Green's relation $\mathcal{L}$ is a right congruence and Green's relation $\mathcal{R}$ is a left congruence.

(b) Each semigroup S has two trivial congruences: the equality relation (the *identity* congruence ι_S) and the uniform congruence $\omega_S = S \times S$ consisting of just one equivalence class S.

If S is a set and ρ is an equivalence relation on S, there are several ways to show that some $a, b \in S$ belong to the same class of ρ. This can be written as follows: $a\,\rho\,b$, or $(a, b) \in \rho$, or $a \equiv b(\rho)$ or simply $a \equiv b$ if there is no confusion about which congruence we are considering. In what follows we may use any of the above notation.

Exercise 6.1.2 Prove that the intersection of an arbitrary family of congruences on S is a congruence on S.

Lemma 6.1.3 *Let S be a semigroup and ρ be a congruence on S. Let further K be an equivalence class of ρ. If K contains some ideal of S, then K is an ideal of S itself. In addition, at most one class of ρ can be an ideal.*

O. Ganyushkin, V. Mazorchuk, *Classical Finite Transformation Semigroups*, Algebra and Applications 9, DOI: 10.1007/978-1-84800-281-4_6,
© Springer-Verlag London Limited 2009

Proof. Let I be the ideal of S, contained in K. Fix some $b \in I$ and let $a \in K$ and $x \in S$ be arbitrary. As $a \equiv b$, we have $xa \equiv xb$ and $ax \equiv bx$. But $xb, bx \in I$ since I is an ideal. This implies $xa \in K$ and $ax \in K$.

Let I and J be two ideals of S. Then $IJ \subset I \cap J$, in particular, $I \cap J \neq \varnothing$. Hence I and J cannot be at the same time two different equivalence classes of an equivalence relation on S.   □

Let S be a semigroup and $I \subset S$ an ideal. Define the equivalence relation ρ_I on S as follows:

$$\rho_I = (I \times I) \cup \{(a,a) : a \in S \backslash I\}.$$

In other words, all elements from I form one equivalence class, and all other equivalence classes are trivial (each consists of a single element). The fact that I is an ideal guarantees that ρ_I is a congruence. Indeed, if $a \equiv b(\rho_I)$, then either $a = b$, or $a, b \in I$. In the last case $ac, bc, ca, cb \in I$ for any $c \in S$. The congruence ρ_I is called the *Rees congruence* on S with respect to the ideal I.

If ρ is a congruence on S and $a \in S$, the equivalence class of a in ρ is denoted by $\bar{a}_\rho$ or simply by $\bar{a}$ if the congruence ρ is clear from the context.

Let ρ be a congruence on a semigroup S. Consider the set S/ρ of all equivalence classes. For $a, b \in S$ set

$$\bar{a} \cdot \bar{b} := \overline{ab}. \tag{6.1}$$

Lemma 6.1.4 *The formula* (6.1) *defines a binary associative operation on* S/ρ. *In other words,* $(S/\rho, \cdot)$, *where* $\cdot$ *is defined by* (6.1), *is a semigroup.*

Proof. First we have to show that (6.1) defines a well-defined binary operation on S/ρ. Let $a' \in \bar{a}$ and $b' \in \bar{b}$. Since ρ is compatible we have $ab \equiv ab' \equiv a'b'$ and hence $\bar{a} \cdot \bar{b} = \bar{a'} \cdot \bar{b'}$, that is, the operation $\cdot$ is well-defined. The associativity of $\cdot$ follows from the associativity of the multiplication in S:

$$(\bar{a} \cdot \bar{b}) \cdot \bar{c} = \overline{ab} \cdot \bar{c} = \overline{abc} = \bar{a} \cdot \overline{bc} = \bar{a} \cdot (\bar{b} \cdot \bar{c}).$$

□

The semigroup $(S/\rho, \cdot)$ constructed above is called the *quotient* or the *factor* of S modulo the congruence ρ. If $\rho = \rho_I$, one usually simply writes S/I instead of S/ρ_I. The semigroup S/I is called the *Rees quotient* of S modulo the ideal I.

If S is a monoid with the identity element 1, then from (6.1) we get that $\bar{1}$ will be the identity element in S/ρ. The quotient S/ι_S is naturally identified with S. The quotient S/ω_S is the one-element semigroup.

Let now $(S, \cdot)$ and $(T, *)$ be two semigroups. A mapping $\varphi : S \to T$ is called a *homomorphism* provided that for all $a, b \in S$ we have

$$\varphi(a \cdot b) = \varphi(a) * \varphi(b). \tag{6.2}$$

The notion of an isomorphism from Sect. 2.4 is just a special case of the notion of a homomorphism. An *isomorphism* is just a bijective homomorphism. Other special cases of homomorphisms are (1) injective homomorphisms, which are called *monomorphisms* and (2) surjective homomorphisms, which are called *epimorphisms*. Monomorphisms are usually denoted by the symbol $\hookrightarrow$, and epimorphisms by the symbol $\twoheadrightarrow$.

A homomorphism $\varphi : S \to S$ is called an *endomorphism* of S. A bijective endomorphism is called an *automorphism*. The set of all endomorphisms of S is denoted by $\mathrm{End}(S)$ and the set of all automorphisms of S is denoted by $\mathrm{Aut}(S)$. Both these sets are not empty since they both contain the identity transformation on S.

Proposition 6.1.5 *The set* $\mathrm{End}(S)$ *is a monoid with respect to the composition of mappings.* $\mathrm{Aut}(S)$ *is the group of units in* $\mathrm{End}(S)$, *in particular,* $\mathrm{Aut}(S)$ *is a group.*

Proof. Let $\varphi, \psi \in \mathrm{End}(S)$. For any $a, b \in S$ we have

$$\varphi(\psi(ab)) = \varphi(\psi(a)\psi(b)) = \varphi(\psi(a))\varphi(\psi(b)).$$

Hence $\mathrm{End}(S)$ is closed with respect to the composition of mappings. Since this composition is associative (by Proposition 2.1.1), $\mathrm{End}(S)$ is a semigroup. The identity element of $\mathrm{End}(S)$ is the identity transformation. As the composition of bijections is a bijection, $\mathrm{Aut}(S)$ is a subsemigroup of $\mathrm{End}(S)$. Let $\varphi \in \mathrm{Aut}(S)$ and consider the inverse bijection φ^{-1}. Applying φ^{-1} to (6.2) for any $a, b \in S$ we have

$$ab = \varphi^{-1}(\varphi(a)\varphi(b)). \tag{6.3}$$

Take $x = \varphi(a)$ and $y = \varphi(b)$. Then $a = \varphi^{-1}(x)$ and $b = \varphi^{-1}(y)$ and (6.3) becomes

$$\varphi^{-1}(x)\varphi^{-1}(y) = \varphi^{-1}(xy)$$

and $x, y \in S$ are arbitrary since $a, b \in S$ were arbitrary and φ^{-1} is a bijection. This means that $\varphi^{-1} \in \mathrm{Aut}(S)$ and thus all elements of $\mathrm{Aut}(S)$ are invertible. On the other hand, each invertible transformation from $\mathrm{End}(S)$ is a bijection and hence belongs to $\mathrm{Aut}(S)$. This completes the proof. $\square$

If S is a semigroup and ρ is a congruence on S, we can define the *canonical projection* or the *canonical epimorphism* $\pi_\rho : S \twoheadrightarrow S/\rho$ via $a \mapsto \bar{a}$. The mapping π_ρ is surjective by definition and is a homomorphism by (6.1).

Let S and T be two semigroups and $\varphi : S \to T$ be a homomorphism. The equivalence relation

$$\mathrm{Ker}(\varphi) = \{(a, b) \in S \times S : \varphi(a) = \varphi(b)\}$$

is called the *kernel* of φ. This relation is a congruence on S. Indeed,

$$\varphi(a) = \varphi(b) \quad \Rightarrow \quad \varphi(ca) = \varphi(c)\varphi(a) = \varphi(c)\varphi(b) = \varphi(cb),$$

which shows that $\mathrm{Ker}(\varphi)$ is left compatible. That $\mathrm{Ker}(\varphi)$ is right compatible is proved analogously. On the other hand, if ρ is a congruence on S, then we have the homomorphism $\pi_\rho : S \twoheadrightarrow S/\rho$ and $\mathrm{Ker}(\pi_\rho) = \rho$. This means that each congruence is the kernel of some homomorphism.

If $\varphi : S \to T$ is an epimorphism, T is called a *homomorphic image* of S. It turns out that all homomorphic images of a semigroup can be characterized up to isomorphism.

Theorem 6.1.6 *Let $\varphi : S \to T$ be a homomorphism of semigroups, and π be the canonical projection $\pi : S \twoheadrightarrow S/\mathrm{Ker}(\varphi)$. Then the mapping $\psi : S/\mathrm{Ker}(\varphi) \to T$ defined via $\psi(\bar{a}) = \varphi(a)$ is a monomorphism and $\varphi = \psi\pi$, that is, the following diagram commutes:*

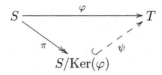

Moreover, if φ is an epimorphism, then ψ is an isomorphism.

Proof. First, we have to check that ψ is well defined. For $a' \in \bar{a}$ we have $\varphi(a') = \varphi(a)$ by the definition of $\mathrm{Ker}(\varphi)$. Hence $\psi(\bar{a'}) = \varphi(a') = \varphi(a) = \psi(\bar{a})$.

Now let us check that ψ is injective. Indeed, if $\psi(\bar{a}) = \psi(\bar{b})$, then $\varphi(a) = \varphi(b)$ by definition and hence $a \in \bar{b}$, that is, $\bar{a} = \bar{b}$.

The next step is to check that ψ is a homomorphism. Let $\bar{a}, \bar{b} \in S/\mathrm{Ker}(\varphi)$. Then

$$\psi(\bar{a} \cdot \bar{b}) = \psi(\overline{ab}) = \varphi(ab) = \varphi(a)\varphi(b) = \psi(\bar{a})\psi(\bar{b}).$$

Now for $a \in S$ we have $\psi(\pi(a)) = \psi(\bar{a}) = \varphi(a)$ and hence $\varphi = \psi\pi$.

Finally, if φ is surjective, then so is ψ since $\varphi = \psi\pi$. We already know that ψ is always injective. Hence ψ is bijective for surjective φ. This completes the proof. $\qquad\qquad\qquad\qquad\qquad\qquad\qquad\qquad\qquad\qquad\qquad\qquad\qquad\square$

From Theorem 6.1.6 it follows that, up to isomorphism, all homomorphic images of a semigroup S are exhausted by the quotients of S.

6.2   Congruences on Groups

Our main goal in the first part of this chapter is to describe all congruences on the semigroups $\mathcal{T}_n$, $\mathcal{PT}_n$, and $\mathcal{IS}_n$. This will be done in Sect. 6.3. In the present section, we briefly recall the description of all congruences on groups.

Lemma 6.2.1 *(i) Let S be a monoid and ρ be a congruence on S. Then the class $\bar{1}$ is a subsemigroup of S.*

(ii) If S is a group, then $\bar{1}$ is a group as well.

Proof. Let $a, b \in \bar{1}$. Since ρ is left compatible we have that $1 \equiv b$ implies $a \equiv ab$. Hence $ab \in \bar{1}$, proving (i).

Assume that S is a group and let $a \in \bar{1}$. Since ρ is left compatible we have that $1 \equiv a$ implies $a^{-1} \equiv a^{-1}a = 1$. Hence $a^{-1} \in \bar{1}$, proving (ii). $\square$

Lemma 6.2.2 *Let G be a group, ρ a congruence on G and $H = \bar{1}$. Then for any $g \in G$ and $h \in H$ we have $g^{-1}hg \in H$.*

Proof. Since ρ is left compatible, we have that $1 \equiv h$ implies $g^{-1} \equiv g^{-1}h$. Since ρ is right compatible, we have that $g^{-1} \equiv g^{-1}h$ implies $1 = g^{-1}g \equiv g^{-1}hg$. Hence $g^{-1}hg \in H$. $\square$

Recall that a subgroup H of a group G is called *normal* provided that $g^{-1}hg \in H$ for all $g \in G$ and $h \in H$. The fact that H is a normal subgroup of G is usually denoted by $H \triangleleft G$.

Let G be a group and H be a subgroup of G. For $g \in G$ the set $gH = \{gh : h \in H\}$ is called a *left coset* of G modulo H. The set of all left cosets of G modulo H is denoted by G/H. Given G and H as above define the binary relation ρ_H on G as follows: $(a, b) \in \rho_H$ if and only if $a \in bH$.

Lemma 6.2.3 *Let G be a group and H be a subgroup of G. Then ρ_H is an equivalence relation on G.*

Proof. H is a subgroup, in particular, $1 \in H$. Hence $a = a \cdot 1 \in aH$, which means that ρ_H is reflexive.

Let $(a, b) \in \rho_H$. Then $a = bh$ for some $h \in H$ by definition. Since H is a group, $h^{-1} \in H$ and hence $b = bhh^{-1} = ah^{-1}$, implying $b \in aH$. Hence $(b, a) \in \rho_H$ and ρ_H is symmetric.

Let $(a, b), (b, c) \in \rho_H$. Then $a = bh$ and $b = ch'$ for some $h, h' \in H$. Hence $a = ch'h$. Since H is a subgroup, we have $h'h \in H$ and thus $a \in cH$ implying $(a, c) \in \rho_H$. This means that ρ_H is transitive and completes the proof. $\square$

As an immediate corollary from Lemma 6.2.3 we obtain:

Corollary 6.2.4 *Let G be a group, H be a subgroup of G, and $a, b \in G$. Then either $aH = bH$ or $aH \cap bH = \varnothing$.*

Now we are ready to describe all possible congruences and quotients of a group.

Theorem 6.2.5 *Let G be a group.*

(i) If $H \lhd G$, then ρ_H is a congruence on G.

(ii) Every congruence on G has the form ρ_H for some normal subgroup $H \lhd G$.

(iii) If ρ is a congruence on G, then the semigroup G/ρ is, in fact, a group.

Proof. Let $H \lhd G$. We already know from Lemma 6.2.3 that ρ_H is an equivalence relation on G. Let $(a, b) \in \rho_H$ and $g \in G$. Then $a = bh$ for some $h \in H$ and multiplying with g from the left we get $ga = gbh$, that is, $(ga, gb) \in \rho_H$. Thus ρ_H is left compatible. On the other hand, multiplying with g from the right, we have $ag = bhg = bg(g^{-1}hg)$. As H is normal, we have $g^{-1}hg \in H$ and hence $(ag, bg) \in \rho_H$. Thus ρ_H is right compatible as well. This proves (i).

Let ρ be a congruence on G. Set $H = \overline{1}$. By Lemmas 6.2.1(ii) and 6.2.2, H is a normal subgroup of G. Let $a, b \in G$ be such that $(a, b) \in \rho$. Since ρ is left compatible we get $1 = a^{-1}a \equiv a^{-1}b$. Hence $a^{-1}b \in H$ and multiplying with a from the left we get $b \in aH$, that is, $(a, b) \in \rho_H$. This shows that $\rho \subset \rho_H$. On the other hand, let $(a, b) \in \rho_H$. Then $(1, a^{-1}b) \in \rho_H$ by the left stability of ρ_H. But then $a^{-1}b \in H = \overline{1}$ and hence $(1, a^{-1}b) \in \rho$. Multiplying the latter with a from the left we get $(a, b) \in \rho$ as ρ is left compatible. Hence $\rho_H \subset \rho$ and thus $\rho = \rho_H$, proving (ii).

We already know that G/ρ is a monoid with the identity element $\overline{1}$. To prove (iii) one just has to observe that $\overline{a} \cdot \overline{a^{-1}} = \overline{1}$ and hence all elements of G/ρ are invertible. This completes the proof. $\qquad\square$

6.3   Congruences on $\mathcal{T}_n$, $\mathcal{PT}_n$, and $\mathcal{IS}_n$

Let S denote one of the semigroups $\mathcal{T}_n$, $\mathcal{PT}_n$, or $\mathcal{IS}_n$.

From Theorem 4.5.1(iii) we have that some elements $\alpha, \beta \in S$ belong to the same $\mathcal{H}$-class of the $\mathcal{D}$-class D_k if and only if they have the following form:

$$\alpha = \begin{pmatrix} A_1 & A_2 & \cdots & A_k & A_{k+1} \\ a_1 & a_2 & \cdots & a_k & \varnothing \end{pmatrix},$$

$$\beta = \begin{pmatrix} A_1 & A_2 & \cdots & A_k & A_{k+1} \\ a_{\mu(1)} & a_{\mu(2)} & \cdots & a_{\mu(k)} & \varnothing \end{pmatrix}, \tag{6.4}$$

where the sets $A_1, \ldots, A_k, A_{k+1}$ are pairwise disjoint, $\mu \in \mathcal{S}_k$, and the set A_{k+1} may be empty. If $S = \mathcal{T}_n$, we always have $A_{k+1} = \varnothing$. If $S = \mathcal{IS}_n$, we have $|A_1| = \cdots = |A_k| = 1$. Sometimes we shall omit the set A_{k+1} in our notation.

For every $k = 1, 2, \ldots, n$ and for every normal subgroup $R \lhd \mathcal{S}_k$ we define an equivalence relation $\equiv_R$ on S as follows:

- If $\operatorname{rank}(\alpha) < k$, then $\alpha \equiv_R \beta$ if and only if $\operatorname{rank}(\beta) < k$.

- If $\operatorname{rank}(\alpha) > k$, then $\alpha \equiv_R \beta$ if and only if $\alpha = \beta$.

- If $\operatorname{rank}(\alpha) = k$, then $\alpha \equiv_R \beta$ if and only if $\alpha \mathcal{H} \beta$ and if α and β are given by (6.4), then $\mu \in R$.

Lemma 6.3.1 *The relation* $\equiv_R$ *is well defined, that is, it does not depend on the order of the blocks* $A_1, \ldots, A_{k+1}$ *in the presentation of* α *in the form* (6.4).

Proof. Assume that we permute the A_is in some way, say $B_1 = A_{\eta(1)}$, $B_2 = A_{\eta(2)}, \ldots, B_k = A_{\eta(k)}$. In this numeration, we have

$$\alpha = \begin{pmatrix} B_1 & B_2 & \cdots & B_k \\ b_1 & b_2 & \cdots & b_k \end{pmatrix}, \ \beta = \begin{pmatrix} B_1 & B_2 & \cdots & B_k \\ b_{\tau(1)} & b_{\tau(2)} & \cdots & b_{\tau(k)} \end{pmatrix},$$

where $b_i = a_{\eta(i)}$, $i = 1, \ldots, k$. Then for all i we have

$$b_{\tau(i)} = a_{\mu(\eta(i))} = b_{\eta^{-1}(\mu(\eta(i)))} = b_{(\eta^{-1}\mu\eta)(i)}.$$

Hence $\tau = \eta^{-1}\mu\eta$. As R is a normal subgroup of $\mathcal{S}_k$, we have $\tau \in R$ if and only if $\mu \in R$. The claim follows. $\qquad\square$

Lemma 6.3.2 *The relation* $\equiv_R$ *is a congruence on* S.

Proof. First we check that $\equiv_R$ is an equivalence relation on S. That $\equiv_R$ is reflexive is obvious. That R is symmetric and transitive for elements of rank $\neq k$ is also obvious. Let us show that R is symmetric and transitive for elements of rank k. Let

$$\alpha = \begin{pmatrix} A_1 & A_2 & \cdots & A_k \\ a_1 & a_2 & \cdots & a_k \end{pmatrix},$$

$$\beta = \begin{pmatrix} A_1 & A_2 & \cdots & A_k \\ a_{\mu(1)} & a_{\mu(2)} & \cdots & a_{\mu(k)} \end{pmatrix} = \begin{pmatrix} A_1 & A_2 & \cdots & A_k \\ b_1 & b_2 & \cdots & b_k \end{pmatrix},$$

$$\gamma = \begin{pmatrix} A_1 & A_2 & \cdots & A_k \\ b_{\tau(1)} & b_{\tau(2)} & \cdots & b_{\tau(k)} \end{pmatrix}.$$

Then

$$\alpha = \begin{pmatrix} A_1 & A_2 & \cdots & A_k \\ b_{\mu^{-1}(1)} & b_{\mu^{-1}(2)} & \cdots & b_{\mu^{-1}(k)} \end{pmatrix},$$

$$\gamma = \begin{pmatrix} A_1 & A_2 & \cdots & A_k \\ a_{\mu(\tau(1))} & a_{\mu(\tau(2))} & \cdots & a_{\mu(\tau(k))} \end{pmatrix}.$$

If $\alpha \equiv_R \beta$ and $\beta \equiv_R \gamma$, then $\mu, \tau \in R$ and hence both μ^{-1} and $\mu\tau$ are elements of R as well, since R is a group. Thus $\beta \equiv_R \alpha$ and $\alpha \equiv_R \gamma$, which means that $\equiv_R$ is indeed an equivalence relation.

Let us now show that $\equiv_R$ is both left and right compatible. If we have
$\mathrm{rank}(\alpha), \mathrm{rank}(\beta) < k$, then for any γ the ranks of the elements $\alpha\gamma$, $\beta\gamma$, $\gamma\alpha$,
and $\gamma\beta$ do not exceed $(k-1)$ as well (by Exercise 2.1.4(c)). Hence

$$\alpha\gamma \equiv_R \beta\gamma \text{ and } \gamma\alpha \equiv_R \gamma\beta. \tag{6.5}$$

If $\alpha = \beta$, then relations (6.5) are obvious. So, we are left with the case
$\mathrm{rank}(\alpha) = \mathrm{rank}(\beta) = k$. Let α and β be given by (6.4) and $\alpha \equiv_R \beta$. For any

$$\gamma = \begin{pmatrix} a_1 & a_2 & \cdots & a_k & \cdots \\ c_1 & c_2 & \cdots & c_k & \cdots \end{pmatrix}$$

we have

$$\gamma\alpha = \begin{pmatrix} A_1 & A_2 & \cdots & A_k \\ c_1 & c_2 & \cdots & c_k \end{pmatrix}, \; \gamma\beta = \begin{pmatrix} A_1 & A_2 & \cdots & A_k \\ c_{\mu(1)} & c_{\mu(2)} & \cdots & c_{\mu(k)} \end{pmatrix}.$$

If some of the symbols $c_1, c_2, \ldots, c_k$ are equal to $\varnothing$ or coincide, then we
have $\mathrm{rank}(\gamma\alpha), \mathrm{rank}(\gamma\beta) < k$ and hence $\gamma\alpha \equiv_R \gamma\beta$. In the other case we
have $\gamma\alpha \mathcal{H} \gamma\beta$. Hence we can consider $\gamma\alpha$ and $\gamma\beta$ as two elements, written in
the form (6.4). The corresponding permutation of indices is $\mu \in R$. Hence
$\gamma\alpha \equiv_R \gamma\beta$ again. This proves that $\equiv_R$ is left compatible.

To prove that $\equiv_R$ is right compatible set $B_i = \{x \in \mathbf{N} : \gamma(x) \in A_i\}$,
$i = 1, \ldots, k$. If $B_i = \varnothing$ for some i, then $\mathrm{rank}(\alpha\gamma), \mathrm{rank}(\beta\gamma) < k$ and hence
$\alpha\gamma \equiv_R \beta\gamma$. In the other case we have

$$\alpha\gamma = \begin{pmatrix} B_1 & B_2 & \cdots & B_k \\ c_1 & c_2 & \cdots & c_k \end{pmatrix}, \; \beta\gamma = \begin{pmatrix} B_1 & B_2 & \cdots & B_k \\ c_{\mu(1)} & c_{\mu(2)} & \cdots & c_{\mu(k)} \end{pmatrix}$$

and the same argument as in the previous paragraph shows that $\alpha\gamma \equiv_R \beta\gamma$
again. This proves that $\equiv_R$ is right compatible and completes the proof. $\square$

Remark 6.3.3 If $R = \mathcal{S}_1$, then $\equiv_R$ is the identity congruence on S.

Our ultimate goal in this section is to show that any congruence on S
is either the uniform congruence or has the form $\equiv_R$ for some R. To prove
this we will need a series of auxiliary lemmas. Recall from Sect. 2.3 that for
$a \in \mathbf{N}$ we denote by 0_a the element $0_a = \begin{pmatrix} \mathbf{N} \\ a \end{pmatrix} \in \mathcal{T}_n$. Recall also from
Sect. 2.7 that for $A \subset \mathbf{N}$ the element ε_A is the idempotent of $\mathcal{IS}_n$ such that
$\mathrm{dom}(\varepsilon_A) = A$.

Lemma 6.3.4 *Let ρ be a congruence on $\mathcal{T}_n$. If $\rho \neq \iota_S$, then the ideal $\mathcal{I}_1$ is
contained in some class of ρ.*

Proof. By assumption, there exist $\alpha \neq \beta \in \mathcal{T}_n$ such that $\alpha \equiv \beta$. This means
that for some $x \in \mathbf{N}$ we have $a = \alpha(x) \neq \beta(x) = b$. Take arbitrary $y, z \in \mathbf{N}$
and let

$$\gamma = \begin{pmatrix} a & b & \cdots \\ y & z & \cdots \end{pmatrix}.$$

Then $\gamma\alpha 0_x = 0_y$ and $\gamma\beta 0_x = 0_z$, implying $0_y \equiv 0_z$. Hence all elements from $\mathcal{I}_1$ belong to the same class of ρ. $\square$

Lemma 6.3.5 *Let ρ be a congruence on S, and $\alpha, \beta \in S$ be such that* rank$(\alpha) = k$, rank$(\beta) = m$, $k > m$, *and $\alpha \equiv \beta$. Then $\mathcal{I}_k \subset \bar{\alpha}$.*

Proof. First, we consider the case $S = \mathcal{PT}_n$ or $S = \mathcal{IS}_n$. As $\alpha \in \mathcal{I}_k$, it is enough to show that $\mathcal{I}_k \subset \bar{0}$. For this we show that $\mathcal{I}_l \subset \bar{0}$ for all $l \le k$ by induction on l. That $0 \in \bar{0}$ is obvious. Assume that $l < k$ and that $\mathcal{I}_l \subset \bar{0}$. Let us show that $\mathcal{I}_{l+1} \subset \bar{0}$ as well.

We have im$(\alpha)\backslash$im$(\beta) \ne \varnothing$ because of our assumptions. Take some $a_0 \in$ im$(\alpha)\backslash$im(β) and complete it to some $(l+1)$-element subset $A = \{a_0, a_1, \ldots, a_l\} \subset$ im(α). From $\alpha \equiv \beta$ we have $\varepsilon_A\alpha \equiv \varepsilon_A\beta$. However, im$(\varepsilon_A\beta) \subset \{a_1, \ldots, a_l\}$ and thus rank$(\varepsilon_A\beta) \le l$. This implies $\varepsilon_A\beta \in \bar{0}$ and thus $\varepsilon_A\alpha \in \bar{0}$ as well. But im$(\varepsilon_A\alpha) = A$ and thus rank$(\varepsilon_A\alpha) = l+1$. As $\bar{0}$ contains the ideal $\mathcal{I}_0$, by Lemma 6.1.3 the set $\bar{0}$ must be an ideal itself. Hence $\bar{0}$ contains the principal ideal $S\varepsilon_A\alpha S$, which coincides with $\mathcal{I}_{l+1}$ by Theorem 4.2.8.

Let now $S = \mathcal{T}_n$. By Lemma 6.3.4 one of the classes of ρ contains $\mathcal{I}_1$. Now the proof is analogous to the one given above. The only difference is that instead of the idempotent ε_A one should consider the following idempotent:

$$\varepsilon'_A = \begin{pmatrix} a_0 & X_1 & \cdots & X_l \\ a_0 & a_1 & \cdots & a_l \end{pmatrix},$$

where X_i are arbitrary such that $a_i \in X_i$, $i = 1, \ldots, l$. $\square$

If ρ is a congruence on S, different from the identity congruence, then by Lemmas 6.3.4 and 6.1.3 it always contains a unique congruence class which is an ideal of S. We denote this class by $\mathcal{I}_\rho$.

Lemma 6.3.6 *Let ρ be a congruence on S, different from the identity congruence. If $\alpha \equiv \beta$ and $\alpha \notin \mathcal{I}_\rho$, then $\alpha\mathcal{H}\beta$.*

Proof. Without loss of generality we may assume $n > 1$.

Assume that im$(\alpha) \ne$ im(β). Without loss of generality we may assume im$(\alpha)\backslash$im$(\beta) \ne \varnothing$. Let $a \in$ im$(\alpha)\backslash$im(β) and $b \in$ im$(\alpha)\backslash\{a\}$. Consider the idempotent ϵ, uniquely defined by the following conditions:

$$\epsilon(x) = \begin{cases} x, & x \ne a; \\ b, & x = a \text{ and } S = \mathcal{T}_n; \\ \varnothing, & x = a \text{ and } S \ne \mathcal{T}_n. \end{cases}$$

Then we have $\epsilon\alpha \equiv \epsilon\beta = \beta$. Hence $\alpha \equiv \epsilon\alpha$. On the other hand, im$(\epsilon\alpha) =$ im$(\alpha)\backslash\{a\}$ and hence rank$(\epsilon\alpha) <$ rank(α). Applying Lemma 6.3.5 we get $\alpha \in \mathcal{I}_\rho$, a contradiction. Hence im$(\alpha) =$ im(β).

Let $\text{im}(\alpha) = \{a_1, \ldots, a_k\}$. Then we may assume that

$$\alpha = \begin{pmatrix} A_1 & A_2 & \cdots & A_k & A_{k+1} \\ a_1 & a_2 & \cdots & a_k & \varnothing \end{pmatrix}, \quad \beta = \begin{pmatrix} B_1 & B_2 & \cdots & B_k & B_{k+1} \\ a_1 & a_2 & \cdots & a_k & \varnothing \end{pmatrix},$$

where the sets A_{k+1} and/or B_{k+1} may be empty. Assume that $\rho_\alpha \neq \rho_\beta$. Then either $A_{k+1} \neq B_{k+1}$, or $A_{k+1} = B_{k+1}$ (that is, $\text{dom}(\alpha) = \text{dom}(\beta)$) and there exist elements $x, y \in \text{dom}(\alpha)$, $x \neq y$, which belong to the same block of ρ_α but to different blocks of ρ_β.

Consider the second case first. In this case $k \geq 2$ and without loss of generality we may assume that there exist $b_1 \in B_1$ and $b_2 \in B_2$ such that $b_1, b_2 \in A_l$, $l < k + 1$. Consider the element

$$\gamma = \begin{pmatrix} a_1 & a_2 & \cdots & a_k & \cdots \\ b_1 & b_2 & \cdots & b_k & \cdots \end{pmatrix}.$$

From $\gamma\alpha \equiv \gamma\beta$ we have $(\gamma\alpha)^2 \equiv (\gamma\beta)^2$. But

$$(\gamma\alpha)^2 = \begin{pmatrix} A_1 & A_2 & A_3 & \cdots & A_k & A_{k+1} \\ b_l & b_l & c_3 & \cdots & c_k & \varnothing \end{pmatrix},$$

$$(\gamma\beta)^2 = \begin{pmatrix} B_1 & B_2 & B_3 & \cdots & B_k & B_{k+1} \\ b_1 & b_2 & b_3 & \cdots & b_k & \varnothing \end{pmatrix}.$$

Hence

$$\text{rank}((\gamma\alpha)^2) \leq k - 1 < k = \text{rank}((\gamma\beta)^2).$$

Applying Lemma 6.3.5 we get $\mathcal{I}_k \subset \mathcal{I}_\rho$, and hence $\alpha \in \mathcal{I}_\rho$ as $\text{rank}(\alpha) = k$, a contradiction. Therefore $\rho_\alpha = \rho_\beta$ in this case.

Consider now the case $A_{k+1} \neq B_{k+1}$. Without loss of generality we may assume that there exists $d_1 \in A_{k+1} \cap B_1$. Choose any $d_i \in B_i$, $i = 2, \ldots, k$, and consider the element

$$\delta = \begin{pmatrix} a_1 & a_2 & a_3 & \cdots & a_k & \cdots \\ d_1 & d_2 & d_3 & \cdots & d_k & \cdots \end{pmatrix}.$$

From $\delta\alpha \equiv \delta\beta$ we have $(\delta\alpha)^2 \equiv (\delta\beta)^2$. But

$$(\delta\alpha)^2 = \begin{pmatrix} A_1 & A_2 & A_3 & \cdots & A_k & A_{k+1} \\ \varnothing & s_2 & s_3 & \cdots & s_k & \varnothing \end{pmatrix},$$

$$(\delta\beta)^2 = \begin{pmatrix} B_1 & B_2 & B_3 & \cdots & B_k & B_{k+1} \\ d_1 & d_2 & d_3 & \cdots & d_k & \varnothing \end{pmatrix}.$$

Using the same arguments as in the previous paragraph, we conclude that $\alpha \in \mathcal{I}_\rho$, a contradiction. Therefore $\rho_\alpha = \rho_\beta$ in this case as well.

From Theorem 4.5.1(iii) it now follows that $\alpha \mathcal{H} \beta$. This completes the proof.   □

Lemma 6.3.7 *Let ρ be a congruence on S, and $\alpha \in S$ be an element of rank k such that there exists $\beta \neq \alpha$ with the property $\alpha \equiv \beta$. Then $\mathcal{I}_\rho \supset \mathcal{I}_{k-1}$.*

Proof. If $\alpha \in \mathcal{I}_\rho$, then $\mathcal{I}_\rho$ contains even $\mathcal{I}_k$ (the principal ideal, generated by α). So, we may now assume that $\alpha \notin \mathcal{I}_\rho$. By Lemma 6.3.6, we may assume that

$$\alpha = \left(\begin{array}{cccc} A_1 & A_2 & \cdots & A_k \\ a_1 & a_2 & \cdots & a_k \end{array} \right), \quad \beta = \left(\begin{array}{cccc} A_{\mu(1)} & A_{\mu(2)} & \cdots & A_{\mu(k)} \\ a_1 & a_2 & \cdots & a_k \end{array} \right),$$

where $\mu \in \mathcal{S}_k$ is different from the identity. In particular, $k > 1$. Let $k = 2$. If $S = \mathcal{T}_n$, then $\mathcal{I}_1 \subset \mathcal{I}_\rho$ follows from Lemma 6.3.4. If $S \neq \mathcal{T}_n$, then $\varepsilon_{\{a_1\}} \in S$ and since $\mu = (1, 2)$ we have

$$\gamma_1 = \varepsilon_{\{a_1\}} \alpha = \left(\begin{array}{c} A_1 \\ a_1 \end{array} \right) \neq \left(\begin{array}{c} A_2 \\ a_1 \end{array} \right) = \varepsilon_{\{a_1\}} \beta = \gamma_2.$$

Moreover, even $\mathrm{dom}(\gamma_1) = A_1 \neq A_2 = \mathrm{dom}(\gamma_2)$, that is, $\gamma_1 \notin \mathcal{H}(\gamma_2)$. However, $\gamma_1 \equiv \gamma_2$ and as $\mathrm{rank}(\gamma_1) = 1$, the inclusion $\mathcal{I}_1 \subset \mathcal{I}_\rho$ follows now from Lemma 6.3.6.

If $k > 2$, then, since μ is different from the identity transformation, it is easy to see that we can choose $t, t+1 \in \mathbf{N}$ such that $\{t, t+1\} \neq \{\mu(t), \mu(t+1)\}$. Consider first the case $S = \mathcal{T}_n$, or $S = \mathcal{PT}_n$. Take any element

$$\gamma = \left(\begin{array}{cccccccc} a_1 & \cdots & a_t & a_{t+1} & a_{t+2} & \cdots & a_k & \cdots \\ a_1 & \cdots & a_t & a_t & a_{t+2} & \cdots & a_k & \cdots \end{array} \right).$$

Then $\gamma \alpha \equiv \gamma \beta$. The partition $\rho_{\gamma\alpha}$ is obtained from ρ_α uniting the blocks A_t and A_{t+1}. On the other hand, the partition $\rho_{\gamma\beta}$ is obtained from ρ_β uniting the blocks $A_{\mu(t)}$ and $A_{\mu(t+1)}$. As $\{t, t+1\} \neq \{\mu(t), \mu(t+1)\}$, we get $A_t \cup A_{t+1} \neq A_{\mu(t)} \cup A_{\mu(t+1)}$. Hence $\rho_{\gamma\alpha} \neq \rho_{\gamma\beta}$. From Lemma 6.3.6 it thus follows that $\gamma\alpha \in \mathcal{I}_\rho$. As $\mathrm{rank}(\gamma\alpha) = k - 1$, it follows that $\mathcal{I}_\rho \supset \mathcal{I}_{k-1}$.

In the case $S = \mathcal{IS}_n$ the proof is similar, one has just to choose t such that $\mu(t) \neq t$ and consider the element

$$\gamma = \left(\begin{array}{cccccccc} a_1 & \cdots & a_{t-1} & a_t & a_{t+1} & \cdots & a_k & \cdots \\ a_1 & \cdots & a_{t-1} & \varnothing & a_{t+1} & \cdots & a_k & \cdots \end{array} \right).$$

$\square$

Lemma 6.3.8 *Let ρ be a congruence on S and $\mathcal{I}_\rho = \mathcal{I}_k$. If $\alpha, \beta \in S$ are such that $\mathrm{rank}(\alpha) > k + 1$ and $\mathrm{rank}(\beta) > k + 1$, then $\alpha \equiv \beta$ if and only if $\alpha = \beta$.*

Proof. By Lemma 6.3.7, the inequality $\alpha \neq \beta$ implies $\mathcal{I}_\rho \supset \mathcal{I}_{k+1}$. The claim follows. $\square$

Lemma 6.3.9 *Let ρ be a congruence on S, which is neither identity nor uniform. Assume that $\mathcal{I}_\rho = \mathcal{I}_{k-1}$. Then ρ has the form $\equiv_R$ for some normal subgroup R of $\mathcal{S}_k$.*

Proof. If $\mathcal{I}_\rho = \mathcal{I}_{k-1}$, all elements of rank at least $(k+1)$ form one-element equivalence classes by Lemma 6.3.8. Hence we have only to find out the structure of equivalence classes for elements of rank k. By Lemma 6.3.6, for such elements we have that $\gamma \equiv \delta$ implies $\gamma \mathcal{H} \delta$.

Let γ be an idempotent of rank k. Then $\mathcal{H}(\gamma)$ is a group with the identity element γ by Theorem 4.4.11. Moreover, $\mathcal{H}(\gamma) \cong \mathcal{S}_k$ by Theorem 5.1.4. The set $\mathcal{H}(\gamma) \times \mathcal{H}(\gamma)$ is a union of congruence classes of ρ by Lemma 6.3.6. This also induces a congruence on $\mathcal{H}(\gamma)$, call it ρ^γ. This means that the set $G(\gamma) = \{\delta \in \mathcal{H}(\gamma) : \delta \equiv \gamma\}$ is a normal subgroup of $\mathcal{H}(\gamma)$ and ρ^γ coincides with the partition of $\mathcal{H}(\gamma)$ into left cosets of $\mathcal{H}(\gamma)$ modulo $G(\gamma)$ by Theorem 6.2.5.

We shall need an explicit form of the above statement. Let the idempotent γ be of the following form:

$$\gamma = \begin{pmatrix} M_1 & M_2 & \cdots & M_k \\ m_1 & m_2 & \cdots & m_k \end{pmatrix}.$$

Then each element $\delta \in \mathcal{H}(\gamma)$ has the form

$$\delta = \begin{pmatrix} M_1 & M_2 & \cdots & M_k \\ m_{\mu(1)} & m_{\mu(2)} & \cdots & m_{\mu(k)} \end{pmatrix}.$$

And an isomorphism $\mathcal{H}(\gamma) \cong \mathcal{S}_k$ is given by using the mapping $\delta \mapsto \mu$, say. Under this isomorphism $G(\gamma)$ is mapped to some normal subgroup of $\mathcal{S}_k$, call it R.

Let us now show that $\rho = \equiv_R$. Let α and β be two $\mathcal{H}$-related elements of rank k in S. Assume that they are given by (6.4). Choose in each block A_i some element a_i' and consider the elements

$$\nu = \begin{pmatrix} M_1 & M_2 & \cdots & M_k \\ a_1' & a_2' & \cdots & a_k' \end{pmatrix}, \quad \pi = \begin{pmatrix} a_1 & \cdots & a_k & \cdots \\ m_1 & \cdots & m_k & \cdots \end{pmatrix},$$

$$\nu_1 = \begin{pmatrix} A_1 & A_2 & \cdots & A_k \\ m_1 & m_2 & \cdots & m_k \end{pmatrix}, \quad \pi_1 = \begin{pmatrix} m_1 & \cdots & m_k & \cdots \\ a_1 & \cdots & a_k & \cdots \end{pmatrix},$$

where $\pi, \pi_1 \in \mathcal{S}_n$. Then $\gamma = \pi \alpha \nu$, $\delta = \pi \beta \nu$, $\alpha = \pi_1 \gamma \nu_1$, $\beta = \pi_1 \delta \nu_1$. Hence $\alpha \equiv \beta$ if and only if $\gamma \equiv \delta$. From the previous paragraph we know that the latter is the case if and only if $\mu \in R$. Hence $\rho = \equiv_R$, completing the proof. $\square$

Now we can summarize the above results into the following:

Theorem 6.3.10 *Let S be one of the semigroups $\mathcal{PT}_n$, $\mathcal{T}_n$, or $\mathcal{IS}_n$.*

(i) *For each k, $1 \le k \le n$, and for each normal subgroup R of $\mathcal{S}_k$ the relation $\equiv_R$ is a congruence on S. The congruence $\equiv_R$ is uniform if and only if $n = 1$ and $S = \mathcal{T}_n$, and is the identity congruence if and only if $k = 1$.*

(ii) *Let ρ be a congruence on S, which is not uniform. Then there exists $k \in \{1, 2, \ldots, n\}$ and a normal subgroup R of S_k such that $\rho = \equiv_R$.*

Proof. The statement (i) is proved in Lemma 6.3.2. The statement (ii) follows from Lemmas 6.3.4 to 6.3.9. $\qquad\square$

6.4 Conjugate Elements

Let G be a group. An element $a \in G$ is said to be *conjugate* to an element $b \in G$ provided that there exists $c \in G$ such that $a = c^{-1}bc$.

Exercise 6.4.1 Show that the binary relation "*a* is conjugate to *b*" is an equivalence relation on G.

Conjugation plays a very important role in group theory. Hence it is very natural to study various generalizations of this notion for certain classes of semigroups, in the best case, for all semigroups. There are several ways to generalize this notion. The most direct one works for monoids and is defined as follows: Let S be a monoid and $G = S^*$ be the group of units of S. An element $a \in S$ is said to be *G-conjugate* to an element $b \in S$ provided that there exists $c \in S^*$ such that $a = c^{-1}bc$. Obviously, if S is a group, then the relations of the usual conjugation and of G-conjugation coincide.

Exercise 6.4.2 Show that the binary relation "*a* is *G*-conjugate to *b*" is an equivalence relation on S.

Recall that two graphs $\Gamma_1 = (V_1, E_1)$ and $\Gamma_2 = (V_2, E_2)$ are said to be *isomorphic* provided that there exists a bijection $\varphi : V_1 \to V_2$ such that for all $a, b \in V_1$ we have $(a, b) \in E_1$ if and only if $(\varphi(a), \varphi(b)) \in E_2$. In other words, two graphs are isomorphic if they can be obtained from each other via a renumeration of vertices.

Proposition 6.4.3 *Let S be one of the semigroups $\mathcal{PT}_n$, $\mathcal{IS}_n$, or $\mathcal{T}_n$. Two elements $\alpha, \beta \in S$ are G-conjugate if and only if the graphs Γ_α and Γ_β are isomorphic.*

Proof. In our case we have $S^* = \mathcal{S}_n$. Let $\alpha, \beta \in S$ and $\pi \in \mathcal{S}_n$ be such that $\alpha = \pi^{-1}\beta\pi$. For $x, y \in \mathbf{N}$ the arrow (x, y) is an arrow of the graph Γ_α, or Γ_β if and only if $y = \alpha(x)$, or $y = \beta(x)$, respectively. We first rewrite $\alpha = \pi^{-1}\beta\pi$ as follows: $\pi\alpha = \beta\pi$. Then for any $x \in \mathrm{dom}(\alpha)$ we have $\pi(\alpha(x)) = \beta(\pi(x))$, that is, the following diagram is commutative:

$$
\begin{array}{ccc}
x & \xrightarrow{\quad\pi\quad} & \pi(x) \\
{\scriptstyle\alpha}\Big\downarrow & & \Big\downarrow{\scriptstyle\beta} \\
\alpha(x) & \xrightarrow{\quad\pi\quad} & \pi(\alpha(x)) = \beta(\pi(x))
\end{array}
\qquad (6.6)
$$

The above diagram says that π maps arrows from Γ_α to arrows from Γ_β. Analogously one shows that the inverse permutation π^{-1} maps arrows from Γ_β to arrows from Γ_α. Hence π is an isomorphism from Γ_α to Γ_β.

The converse statement is obvious. If $\pi : \mathbf{N} \to \mathbf{N}$ is an isomorphism from Γ_α to Γ_β, then from (6.6) it follows that $\alpha = \pi^{-1}\beta\pi$. $\square$

A more interesting semigroup generalization of the notion of group conjugation is the following one: Let S be a semigroup. Elements $a, b \in S$ are said to be *primarily S-conjugate* provided that there exist $u, v \in S^1$ such that $a = uv$ and $b = vu$. The fact that a and b are primarily S-conjugate is denoted by $a \sim_{pS} b$. From the definition it is obvious that the relation $\sim_{pS}$ is reflexive and symmetric. However, it is not transitive in general.

Exercise 6.4.4 Let $\alpha = [1, 2, 3]$, $\beta = [1, 2][3]$, $\gamma = [1][2][3]$ be elements from $\mathcal{IS}_3$. Show that $\alpha \sim_{pS} \beta$ and $\beta \sim_{pS} \gamma$ but $\alpha \nsim_{pS} \gamma$.

Denote by $\sim_S$ the minimal transitive binary relation on S, which contains $\sim_{pS}$. Such minimal binary relation is usually called the *transitive closure* of the initial relations.

Exercise 6.4.5 Show that $\sim_S$ is an equivalence relation on S.

We will say that the elements $a, b \in S$ are *S-conjugate* provided that $a \sim_S b$. The notion of S-conjugation generalizes that of G-conjugation. Indeed, if $a, b \in S$ are G-conjugate and $c \in S^*$ is such that $a = c^{-1}bc$, then for $u = c^{-1}$ and $v = bc$ we have $a = uv$ and $vu = bcc^{-1} = b$. Hence $a \sim_S b$ (and even $a \sim_{pS} b$).

Proposition 6.4.6 *Let S be a semigroup and let $x, y \in S$. Then we have:*

(i) *If $x \sim_{pS} y$, then $x^i \sim_{pS} y^i$ for all $i \in \mathbb{N}$.*

(ii) *If $e = x^i$ and $f = y^j$ are idempotents and $x \sim_{pS} y$, then $e \sim_{pS} f$.*

Proof. Let $x = ab$ and $y = ba$ for some $a, b \in S$. Then $x^i = (ab)^i = a\left((ba)^{i-1}b\right)$ and $y^i = (ba)^i = \left((ba)^{i-1}b\right)a$ implying $x^i \sim_{pS} y^i$ and proving (i). To prove (ii) we observe that $e = x^i = x^{ij}$ and $f = y^j = y^{ij}$ and that $x^{ij} \sim_{pS} y^{ij}$ by (i). $\square$

Lemma 6.4.7 *Let S be a semigroup and e, f, g be three idempotents from S such that $e \sim_{pS} f$ and $e \sim_{pS} g$. Then $f \sim_{pS} g$.*

Proof. Let $x, y, u, v \in S$ be such that $e = xy$, $f = yx$, $e = uv$, $g = vu$. Since e is an idempotent we also have $e = e^2 = xyxy = xfy$ and analogously $f = yex$, $e = ugv$, $g = veu$. This implies $g = vxfyu$ and $f = yugvx$. Therefore $g = g^2 = (vxf)(yug)$ and $f = f^2 = (yug)(vxf)$ and thus $g \sim_{pS} f$. $\square$

An immediate corollary from Lemma 6.4.7 is the following:

Corollary 6.4.8 *Let S be a semigroup. Then the restriction of $\sim_{pS}$ to $\mathcal{E}(S)$ is an equivalence relation.*

Corollary 6.4.9 *Let S be a finite semigroup and $x, y \in S$ be such that $x \sim_S y$. Let $i, j \in \mathbb{N}$ be such that $e = x^i$ and $f = y^j$ are idempotents. Then $e \sim_{pS} f$.*

Proof. Let $x = x_1, x_2, \ldots, x_l = y$ be a sequence of elements from S such that $x_i \sim_{pS} x_{i+1}$ for $i = 1, \ldots, l-1$. Since S is finite, for every $i = 2, \ldots, l-1$ there exists $m_i \in \mathbb{N}$ such that $y_i = x_i^{m_i}$ is an idempotent. Let $y_1 = e$ and $y_l = f$. From Proposition 6.4.6(ii) we obtain $y_i \sim_{pS} y_{i+1}$ for all $i = 1, \ldots, l-1$. Applying Lemma 6.4.7 inductively, we get that $e \sim_{pS} y_i$ for all $i = 2, \ldots, l$. In particular, $e \sim_{pS} f$. □

Corollary 6.4.10 *Let S be a finite semigroup. Then the restrictions of the relation $\sim_{pS}$ and $\sim_S$ to $\mathcal{E}(S)$ coincide.*

Proof. For $e, f \in \mathcal{E}(S)$ that $e \sim_{pS} f$ implies $e \sim_S f$ is obvious. The converse implication is a special case of Corollary 6.4.9. □

Corollary 6.4.11 *Let S be a finite semigroup and $e, f \in \mathcal{E}(S)$. Then $e \sim_{pS} f$ if and only if $e\mathcal{D}f$.*

Proof. Let $u, v \in S^1$ be such that $uv = e$ and $vu = f$. Then $e = e^2 = uvuv = ufv$ and $f = f^2 = vuvu = veu$. Hence $S^1 e S^1 = S^1 f S^1$ and we have $e\mathcal{J}f$. By Theorem 5.4.1 we thus also have $e\mathcal{D}f$ since S is finite.

Conversely, if $e\mathcal{D}f$, we can take $a \in \mathcal{R}(e) \cap \mathcal{L}(f)$ and $a' \in \mathcal{R}(f) \cap \mathcal{L}(e)$ as in the proof of Theorem 4.7.5. For such a and a' we have $e = aa'$ and $f = a'a$ and hence $e \sim_{pS} f$. □

Let $\pi \in \mathcal{S}_n$. The *cyclic type* of π is the vector $\mathrm{ct}(\pi) = (l_1, l_2, \ldots, l_n)$, where l_i is the number of cycles of length i in the cyclic decomposition of π (or, equivalently, in the graph Γ_π). Obviously $1 \cdot l_1 + 2 \cdot l_2 + \cdots + n \cdot l_n = n$.

Lemma 6.4.12 *Let $\alpha, \beta \in \mathcal{PT}_n$ and $\alpha \sim_{pS} \beta$. Then we have $\mathrm{strank}(\alpha) = \mathrm{strank}(\beta)$ and $\mathrm{ct}(\alpha|_{\mathrm{stim}(\alpha)}) = \mathrm{ct}(\beta|_{\mathrm{stim}(\beta)})$.*

Proof. Let $\alpha = \mu\eta$ and $\beta = \eta\mu$. For each $a_1 \in \mathrm{stim}(\alpha)$ consider the cycle $(a_1, a_2, \ldots, a_k)$ of the permutation $\alpha|_{\mathrm{stim}(\alpha)}$, which contains a_1. We have

$$a_1 \overset{\eta}{\mapsto} b_1 \overset{\mu}{\mapsto} a_2 \overset{\eta}{\mapsto} b_2 \overset{\mu}{\mapsto} \ldots \overset{\mu}{\mapsto} a_{k-1} \overset{\eta}{\mapsto} b_{k-1} \overset{\mu}{\mapsto} a_k \overset{\eta}{\mapsto} b_k \overset{\mu}{\mapsto} a_1.$$

It follows that $(b_1, b_2, \ldots, b_k) = (\eta(a_1), \ldots, \eta(a_k))$ is a cycle for β. Obviously $b_1, \ldots, b_k \in \mathrm{stim}(\beta)$ and hence $(b_1, b_2, \ldots, b_k)$ is a cycle of the permutation $\beta|_{\mathrm{stim}(\beta)}$. Note that the lengths of the cycles $(a_1, a_2, \ldots, a_k)$ and $(b_1, b_2, \ldots, b_k)$ coincide. It follows that η and μ induce mutually inverse bijections between cycles of α and cycles of β, which preserve lengths of cycles. The claim follows. □

Theorem 6.4.13 *Let S denote one of the semigroups $\mathcal{PT}_n$, $\mathcal{T}_n$, or $\mathcal{IS}_n$, and $\alpha, \beta \in S$. Then $\alpha \sim_S \beta$ if and only if $\operatorname{strank}(\alpha) = \operatorname{strank}(\beta)$ and $\operatorname{ct}(\alpha|_{\operatorname{stim}(\alpha)}) = \operatorname{ct}(\beta|_{\operatorname{stim}(\beta)})$.*

Proof. The necessity follows from Lemma 6.4.12. Let us prove the sufficiency. For any $\gamma \in S$ define the following sets:

$$
\begin{aligned}
M_0'(\gamma) &= \operatorname{stim}(\gamma), \\
M_0''(\gamma) &= \operatorname{im}(\gamma)\backslash\operatorname{dom}(\gamma), \\
M_0(\gamma) &= M_0'(\gamma) \cup M_0''(\gamma), \\
M_1(\gamma) &= \{x \notin M_0(\gamma) : \alpha(x) \in M_0(\gamma)\}, \\
M_i(\gamma) &= \{x : \alpha(x) \in M_{i-1}(\gamma)\}, \ i \geq 2.
\end{aligned}
\tag{6.7}
$$

Note that $M_0''(\gamma) = \varnothing$ for $\gamma \in \mathcal{T}_n$. From Theorem 1.2.9, which describes the structure of connected components of Γ_γ it follows that the sets $M_0'(\gamma)$, $M_0''(\gamma)$, $M_1(\gamma)$, $M_2(\gamma),\ldots$ are pairwise disjoint and that $\gamma(M_i(\gamma)) \subset M_{i-1}(\gamma)$ for all $i > 0$. Obviously $M_i(\gamma) = \varnothing$ for some i implies $M_j(\gamma) = \varnothing$ for all $j > i$. Of course it makes sense to consider only those sets from (6.7), which are nonempty, say these are the sets $M_0(\gamma),\ldots, M_k(\gamma)$. Then we have the following picture:

$M_0''(\gamma)$

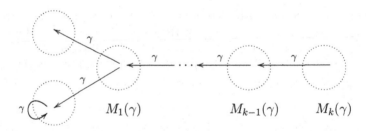

$M_1(\gamma)$ $M_{k-1}(\gamma)$ $M_k(\gamma)$

$M_0'(\gamma)$

Now let us apply the above to our element α. If $S = \mathcal{PT}_n$, or $S = \mathcal{IS}_n$ and $k > 0$, or if $S = \mathcal{T}_n$ and $k > 1$, we consider an idempotent ϵ, which satisfies the following conditions:

- $\operatorname{im}(\epsilon) = M_0(\alpha) \cup M_1(\alpha) \cup \cdots \cup M_{k-1}(\alpha)$;

- $\operatorname{dom}(\epsilon) = \operatorname{im}(\epsilon)$ if $S = \mathcal{PT}_n$, or $S = \mathcal{IS}_n$; and $\epsilon(M_k(\alpha)) \subset M_{k-1}(\alpha)$ if $S = \mathcal{T}_n$.

In the case $S = \mathcal{PT}_n$, or $S = \mathcal{IS}_n$ the idempotent ϵ is uniquely determined, while in the case $S = \mathcal{T}_n$ we might have a choice. For such ϵ we obviously have $\alpha = \epsilon\alpha$. On the other hand, for the element $\alpha_1 = \alpha\epsilon$ the nonempty part of the sequence (6.7) will be one step shorter. Further, for all $i = 0,\ldots,k-2$

we obviously have $M_i(\alpha_1) = M_i(\alpha)$. If $S = \mathcal{PT}_n$, or $S = \mathcal{IS}_n$, we also have $M_{k-1}(\alpha_1) = M_{k-1}(\alpha)$ if $k > 1$ and $M_0(\alpha_1) = M_0'(\alpha)$ if $k = 1$. If $S = \mathcal{T}_n$, it is easy to see that $M_{k-1}(\alpha_1) = M_{k-1}(\alpha) \cup M_k(\alpha)$. Further, $\mathrm{strank}(\alpha_1) = \mathrm{strank}(\alpha)$ and $\alpha_1|_{\mathrm{stim}(\alpha)} = \alpha_1|_{\mathrm{stim}(\alpha_1)}$.

Now we can apply the same procedure to α_1, obtaining the element α_2, and so on. We obtain the sequence $\alpha_0 = \alpha$, $\alpha_1, \ldots$, such that $\alpha_i \sim_{pS} \alpha_{i+1}$ for all i. Now we have to consider two different cases.

Case 1. $S = \mathcal{PT}_n$, or $S = \mathcal{IS}_n$. Let $\alpha' = \alpha_k$ if $k > 0$, and $\alpha' = \alpha$ in the opposite case. Then $M_0(\alpha') = M_0'(\alpha') = \mathrm{stim}(\alpha) = \mathrm{im}(\alpha)$. Analogously we obtain the element β'. We have $\alpha \sim_S \alpha'$, $\beta \sim_S \beta'$ and using the assumptions of our theorem we also have

$$\begin{aligned}
\alpha|_{\mathrm{stim}(\alpha)} &= \alpha'|_{\mathrm{stim}(\alpha)} = (a_1, \ldots, a_p)(b_1, \ldots, b_q) \ldots (c_1, \ldots, c_r), \\
\beta|_{\mathrm{stim}(\beta)} &= \beta'|_{\mathrm{stim}(\beta)} = (a_1', \ldots, a_p')(b_1', \ldots, b_q') \ldots (c_1', \ldots, c_r').
\end{aligned} \tag{6.8}$$

Consider the elements

$$\begin{aligned}
\mu &= \begin{pmatrix} a_1 & \cdots & a_p & b_1 & \cdots & b_q & \cdots & c_1 & \cdots & c_r \\ a_1' & \cdots & a_p' & b_1' & \cdots & b_q' & \cdots & c_1' & \cdots & c_r' \end{pmatrix}, \\
\eta &= \begin{pmatrix} a_1' & \cdots & a_p' & b_1' & \cdots & b_q' & \cdots & c_1' & \cdots & c_r' \\ a_1 & \cdots & a_p & b_1 & \cdots & b_q & \cdots & c_1 & \cdots & c_r \end{pmatrix},
\end{aligned} \tag{6.9}$$

where $\mathrm{dom}(\mu) = \mathrm{dom}(\alpha') = \mathrm{im}(\alpha')$ and $\mathrm{dom}(\eta) = \mathrm{dom}(\beta') = \mathrm{im}(\beta')$. We have $\alpha' = \eta\beta' \cdot \mu$ and $\beta' = \mu \cdot \eta\beta'$. Hence $\alpha' \sim_{pS} \beta'$ and thus $\alpha \sim_S \beta$.

Case 2. $S = \mathcal{T}_n$. Let $\alpha' = \alpha_{k-1}$ if $k > 1$, and $\alpha' = \alpha$ in the opposite case. For this element the sequence (6.7) has at most two nonempty sets: $M_0(\alpha') = M_0'(\alpha')$ and possibly $M_1(\alpha') = \mathbf{N} \backslash M_0(\alpha')$. Analogously we get the element β'. As above, we have $\mathrm{stim}(\alpha) = \mathrm{stim}(\alpha') = \mathrm{im}(\alpha')$, $\mathrm{stim}(\beta) = \mathrm{stim}(\beta') = \mathrm{stim}(\beta')$, $\alpha \sim_S \alpha'$, $\beta \sim_S \beta'$ and by the assumptions of the theorem we also have the decompositions (6.8). Consider some permutation $\tilde{\mu}$, which coincides with the element μ from (6.9) on $\mathrm{im}(\alpha')$. Let $\tilde{\eta} = \tilde{\mu}^{-1}$. Then $\tilde{\eta}$ coincides with the element η from (6.9) on $\mathrm{im}(\beta')$ and for the element $\delta = \tilde{\mu}\alpha' \cdot \tilde{\eta}$ we have $\alpha' \sim_{pS} \delta$ as $\alpha' = \tilde{\eta} \cdot \tilde{\mu}\alpha'$. Moreover, we also have $\mathrm{stim}(\delta) = \mathrm{im}(\beta')$ and $\delta|_{\mathrm{im}(\beta')} = \beta'|_{\mathrm{im}(\beta')}$.

Set $\pi = \beta'|_{\mathrm{im}(\beta')}$. If $\mathrm{im}(\beta') = \{d_1, d_2, \ldots, d_m\}$, then β' can be written as follows:

$$\beta' = \begin{pmatrix} D_1 & \cdots & D_m & d_1 & \cdots & d_m \\ d_1 & \cdots & d_m & \pi(d_1) & \cdots & \pi(d_m) \end{pmatrix}$$

(some D_is may be empty). At the same time we can write δ as follows:

$$\delta = \begin{pmatrix} F_1 & \cdots & F_m & d_1 & \cdots & d_m \\ d_1 & \cdots & d_m & \pi(d_1) & \cdots & \pi(d_m) \end{pmatrix}.$$

For the element

$$\tau = \begin{pmatrix} F_1 & \cdots & F_m & d_1 & \cdots & d_m \\ \pi^{-1}(d_1) & \cdots & \pi^{-1}(d_m) & d_1 & \cdots & d_m \end{pmatrix}$$

we then have $\tau\beta' = \beta'$ and $\beta'\tau = \delta$. Hence $\beta' \sim_{pS} \delta$, which implies $\alpha \sim_S \beta$. This completes the proof. $\qquad\square$

Corollary 6.4.14 *Let* $\alpha, \beta \in \mathcal{S}_n$. *Then* α *and* β *are conjugate if and only if* $\mathrm{ct}(\alpha) = \mathrm{ct}(\beta)$.

Proof. If α and β are conjugate, then $\alpha = \pi^{-1} \cdot \beta\pi$ for some $\pi \in \mathcal{S}_n$. As $\beta = \beta\pi \cdot \pi^{-1}$, we get $\alpha \sim_{pS} \beta$ if we consider α and β as elements of $\mathcal{PT}_n$. Hence $\mathrm{ct}(\alpha) = \mathrm{ct}(\beta)$ by Theorem 6.4.13. Conversely, if $\mathrm{ct}(\alpha) = \mathrm{ct}(\beta)$, choosing μ and η as in (6.9) we get $\mu \in \mathcal{S}_n$, $\eta = \mu^{-1}$ and $\alpha = \eta\beta\mu$. Hence α and β are conjugate. $\qquad\square$

6.5 Addenda and Comments

6.5.1 Congruences on $\mathcal{T}_n$ were described by Mal'cev in [Ma1]. This description was used by Liber in [Lib] to describe congruences on $\mathcal{IS}_n$. Congruences on $\mathcal{PT}_n$ were described by Sutov in [Sut2].

6.5.2 A subsemigroup T of $\mathcal{PT}_n$ is said to be $\mathcal{S}_n$-*normal* provided that for all $\alpha \in T$ and $\sigma \in \mathcal{S}_n$ we have $\sigma^{-1}\alpha\sigma \in T$. For example, the semigroups $\mathcal{PT}_n, \mathcal{T}_n$ and $\mathcal{IS}_n$ are $\mathcal{S}_n$-normal subsemigroups of $\mathcal{PT}_n$. Moreover, any ideal in each of these semigroups is an $\mathcal{S}_n$-normal subsemigroup of $\mathcal{PT}_n$ as well.

Levi has shown in [Le] that for $n > 4$ all nontrivial congruences on almost all $\mathcal{S}_n$-normal subsemigroups have the form $\equiv_R$. The results of Mal'cev, Liber, and Sutov are special cases of this general result of Levi.

6.5.3 Theorem 6.1.6 is true for a very general class of universal algebras of the same type. The proof of this very general result is technically slightly more difficult as one has to consider the cases of an arbitrary collection of operations with different numbers of arguments.

6.5.4 Let Sem denote the category, whose objects are semigroups, and morphisms are all possible homomorphisms between semigroups. Categorical monomorphisms in Sem are exactly the injective homomorphisms of semigroups, while the non-surjective embedding $(\mathbb{N}, +) \hookrightarrow (\mathbb{Z}, +)$ is a categorical epimorphism in Sem. It is a good exercise to prove this.

6.5.5 Normal subgroups of the symmetric group $\mathcal{S}_k$ are very well known. If $n \neq 4$, then the only normal subgroups of $\mathcal{S}_k$ are: the group $\mathcal{S}_k$ itself, the subgroup $\mathcal{A}_k$ of all even permutation (also known as the *alternating group*), and the subgroup $\mathcal{E}_k$ consisting of the identity transformation. For $n \neq 4$ this was proved by Galois in about 1830. A short and easy argument for this can be found for example in [KM]. In addition to $\mathcal{S}_4$, $\mathcal{A}_4$, and $\mathcal{E}_4$, the group $\mathcal{S}_4$ has one more normal subgroup, namely, the so-called *Klein 4-group* $\mathcal{V}_4 = \{\varepsilon, (12)(34), (13)(24), (14)(23)\}$. We note that $\mathcal{S}_1 = \mathcal{A}_1 = \mathcal{E}_1$ and $\mathcal{A}_2 = \mathcal{E}_2$. Hence from Theorem 6.3.10 we have the following corollary:

Corollary 6.5.1 *(i) For $n > 1$ all congruences on each of the semigroups $\mathcal{PT}_n$, $\mathcal{T}_n$ and $\mathcal{IS}_n$ form the following chain:*

$$\iota_S \; =\equiv_{S_1} \subsetneqq \equiv_{\mathcal{E}_2} \subsetneqq \equiv_{S_2} \subsetneqq \equiv_{\mathcal{E}_3} \subsetneqq \equiv_{A_3} \subsetneqq \equiv_{S_3} \subsetneqq \equiv_{\mathcal{E}_4} \subsetneqq \equiv_{V_4} \subsetneqq \equiv_{A_4} \subsetneqq \equiv_{S_4} \subsetneqq$$
$$\subsetneqq \equiv_{\mathcal{E}_5} \subsetneqq \equiv_{A_5} \subsetneqq \equiv_{S_5} \subsetneqq \cdots \subsetneqq \equiv_{\mathcal{E}_n} \subsetneqq \equiv_{A_n} \subsetneqq \equiv_{S_n} \subsetneqq \omega_S.$$

(ii) Each congruence on $\mathcal{IS}_n$, or $\mathcal{T}_n$ is a restriction of some congruence on $\mathcal{PT}_n$. Moreover, for $n > 1$ different congruences on $\mathcal{PT}_n$ restrict to different congruences on both $\mathcal{IS}_n$ and $\mathcal{T}_n$.

6.5.6 The criterion for S-conjugacy of two elements in $\mathcal{IS}_n$ given in Theorem 6.4.13 was obtained in [GK1]. For the semigroups $\mathcal{PT}_n$ and $\mathcal{T}_n$ an analogous result is obtained in [KuMa1]. Several abstract results about S-conjugacy can be found in [Ku2, KuMa1, KuMa3, KuMa4].

6.5.7 With each $\alpha \in \mathcal{PT}_n$ one associates the binary relation

$$\Phi_\alpha = \{(x,y) \; : \; x \in \mathbf{N}, \; y = \alpha(x)\}$$

on $\mathbf{N}$. This can be used to define the *natural partial order* on $\mathcal{PT}_n$:

$$\alpha \preceq \beta \quad \text{if and only if} \quad \Phi_\alpha \subset \Phi_\beta.$$

Note that the restriction of the natural partial order to $\mathcal{T}_n$ coincides with the equality relation.

It is easy to see that the natural partial order is a compatible relation. The restriction of the natural partial order to $\mathcal{IS}_n$ has a transparent algebraic interpretation:

Lemma 6.5.2 *Let $\alpha, \beta \in \mathcal{IS}_n$. Then $\alpha \preceq \beta$ if and only if $\alpha\beta^{-1} = \alpha\alpha^{-1}$.*

It is hence natural and interesting to try to understand the structure of all compatible partial orders on the semigroups $\mathcal{PT}_n$, $\mathcal{T}_n$, and $\mathcal{IS}_n$. It turns out that the only compatible partial order on $\mathcal{T}_n$ is the equality relation Δ (see [Mo2]). For $\mathcal{PT}_n$ and $\mathcal{IS}_n$ the answer is much more interesting.

Theorem 6.5.3 ([Mo1]) *(i) For each collection of numbers as follows: $0 \leq k_1 < k_2 < \cdots < k_m \leq l_m < \cdots < l_1 \leq n$, $m \geq 1$, $k_m < n$, the relation*

$$\Psi_{k_1,\ldots,k_m,l_m,\ldots,l_1} = (\Delta \cup (\mathcal{I}_{k_1} \times \mathcal{I}_{l_1}) \cup \cdots \cup (\mathcal{I}_{k_m} \times \mathcal{I}_{l_m})) \cap \preceq \quad (6.10)$$

is a compatible partial order on $\mathcal{PT}_n$. Moreover, different collections give different compatible partial orders.

(ii) Each compatible partial order on the semigroup $\mathcal{PT}_n$ has either the form $\Psi_{k_1,\ldots,k_m,l_m,\ldots,l_1}$ or is transposed to some $\Psi_{k_1,\ldots,k_m,l_m,\ldots,l_1}$.

Theorem 6.5.4 ([Mo1])   *(i) For each collection of numbers as follows:*
$0 \leq k_1 < k_2 < \cdots < k_m < l_m < \cdots < l_1 \leq n$, $m \geq 1$, *the rela-*
tion $\Psi_{k_1,\ldots,k_m,l_m,\ldots,l_1}$ *as in* (6.10) *is a compatible partial order on* $\mathcal{IS}_n$.
Moreover, different collections give different compatible partial orders.

(ii) *Each compatible partial order on* $\mathcal{IS}_n$ *different from* Δ *has either the*
form $\Psi_{k_1,\ldots,k_m,l_m,\ldots,l_1}$ *or is transposed to some* $\Psi_{k_1,\ldots,k_m,l_m,\ldots,l_1}$.

Corollary 6.5.5 ([Mo1])   *(i) There are exactly* $2^{n+2}-5$ *compatible partial*
orders on $\mathcal{PT}_n$.

(ii) *There are exactly* $2^{n+1}-1$ *compatible partial orders on* $\mathcal{IS}_n$.

6.6   Additional Exercises

6.6.1 Prove that the alternating group $\mathcal{A}_n$ of all even permutations is a
normal subgroup of $\mathcal{S}_n$.

6.6.2 Prove that the Klein group $\mathcal{V}_4$ is normal in $\mathcal{S}_4$.

6.6.3 Describe all left compatible and all right compatible equivalence relations on a group.

6.6.4 Let ρ be a congruence on a semigroup S. Let $\varphi : S \twoheadrightarrow S/\rho$ be an
epimorphism such that $\mathrm{Ker}(\varphi) = \rho$. Is it possible to claim that φ always
coincides with the canonical epimorphism?

6.6.5 ([Mo2]) Prove that the only compatible partial order on $\mathcal{T}_n$ is Δ.

6.6.6 Show that the set of all maximal elements of $\mathcal{PT}_n$ with respect to the
natural partial order coincides with $\mathcal{T}_n$.

6.6.7 Prove that for the semigroup $\mathcal{IS}_n$ the following conditions are equivalent:

(a) $\alpha \preceq \beta$

(b) $\alpha^{-1} \preceq \beta^{-1}$

(c) $\beta\alpha^{-1} = \alpha\alpha^{-1}$

(d) $\alpha^{-1}\beta = \alpha^{-1}\alpha$

(e) $\alpha\beta^{-1} = \alpha\alpha^{-1}$

(f) $\beta^{-1}\alpha = \alpha^{-1}\alpha$

(g) $\alpha\beta^{-1}\alpha = \alpha$

(h) $\alpha^{-1}\beta\alpha^{-1} = \alpha^{-1}$

6.6.8 Prove Corollary 6.5.5.

Chapter 7

Endomorphisms

7.1 Automorphisms of $\mathcal{T}_n$, $\mathcal{PT}_n$, and $\mathcal{IS}_n$

Recall from Sect. 6.1 that an *automorphism* of a semigroup S is a bijective mapping $\varphi : S \to S$ such that $\varphi(x \cdot y) = \varphi(x) \cdot \varphi(y)$ for all $x, y \in S$.

Proposition 7.1.1 *Let S be a monoid. For each $a \in S^*$ the mapping Λ_a: $x \mapsto a^{-1}xa$, $x \in S$, is an automorphism of S.*

Proof. For $x, y \in S$ we have

$$\Lambda_a(xy) = a^{-1}xya = a^{-1}xa \cdot a^{-1}ya = \Lambda_a(x)\Lambda_a(y)$$

and hence Λ_a is an endomorphism of S. Further for all $x \in S$ we have

$$\Lambda_{a^{-1}}\Lambda_a(x) = aa^{-1}xaa^{-1} = x, \ \Lambda_a\Lambda_{a^{-1}}(x) = a^{-1}axa^{-1}a = x.$$

Hence Λ_a is a bijection with the inverse $\Lambda_{a^{-1}}$. The claim follows. $\square$

Automorphisms of the form Λ_a, $a \in S^*$, are called *inner* automorphisms of S. The set of all inner automorphisms of S is denoted by $\mathrm{Inn}(S)$.

Proposition 7.1.2 *Let S be a monoid.*

(i) The mapping $a \mapsto \Lambda_{a^{-1}}$, $a \in S^*$, is an epimorphism from S^* to $\mathrm{Inn}(S)$.

(ii) $\mathrm{Inn}(S)$ is a normal subgroup of $\mathrm{Aut}(S)$.

Proof. The mapping $a \mapsto \Lambda_{a^{-1}}$, $a \in S^*$, is obviously surjective. Let $a, b \in S^*$. For any $x \in S$ we have

$$\Lambda_{(ab)^{-1}}(x) = abx(ab)^{-1} = abxb^{-1}a^{-1} = \Lambda_{a^{-1}}(bxb^{-1}) = \Lambda_{a^{-1}}\Lambda_{b^{-1}}(x).$$

Hence $\Lambda_{a^{-1}}\Lambda_{b^{-1}} = \Lambda_{(ab)^{-1}}$. This proves (i).

In the proof of Proposition 7.1.1 we have shown that $\Lambda_a^{-1} = \Lambda_{a^{-1}}$ for all $a \in S^*$. Together with the above proof of (i) we hence have that $\mathrm{Inn}(S)$

O. Ganyushkin, V. Mazorchuk, *Classical Finite Transformation Semigroups*, Algebra and Applications 9, DOI: 10.1007/978-1-84800-281-4_7, © Springer-Verlag London Limited 2009

is closed with respect to both composition of maps and taking the inverse map. This means that $\text{Inn}(S)$ is a subgroup of $\text{Aut}(S)$.

Let now $a \in S^*$, $\psi \in \text{Aut}(S)$ and $x \in S$. Then

$$\psi^{-1}\Lambda_a\psi(x) = \psi^{-1}(a^{-1}\psi(x)a) = \psi^{-1}(a^{-1})x\psi^{-1}(a) = \Lambda_{\psi^{-1}(a)}(x).$$

This means that $\psi^{-1}\Lambda_a\psi = \Lambda_{\psi^{-1}(a)} \in \text{Inn}(S)$ and completes the proof of (ii).
$\square$

Theorem 7.1.3 *Let S be one of the semigroups $\mathcal{T}_n$, $\mathcal{PT}_n$, or $\mathcal{IS}_n$. Then* $\text{Inn}(S) = \text{Aut}(S) \cong \mathcal{S}_n$.

Proof. We start our proof with the following lemma:

Lemma 7.1.4 *Let σ and τ be different elements of $\mathcal{S}_n$. Then the elements Λ_σ and Λ_τ of $\text{Inn}(S)$ are also different.*

Proof. As $\sigma \neq \tau$, there exists $x \in \mathbf{N}$ such that $y = \sigma(x) \neq \tau(x) = z$. If $S = \mathcal{T}_n$ or $S = \mathcal{PT}_n$, then $\Lambda_\sigma(0_x) = 0_y \neq 0_z = \Lambda_\tau(0_x)$. Hence $\Lambda_\sigma \neq \Lambda_\tau$. If $S = \mathcal{IS}_n$, then $\Lambda_\sigma(\varepsilon_{\{x\}}) = \varepsilon_{\{y\}} \neq \varepsilon_{\{z\}} = \Lambda_\tau(\varepsilon_{\{x\}})$. Hence $\Lambda_\sigma \neq \Lambda_\tau$ again. $\square$

Lemma 7.1.4 says that the epimorphism which was described in Proposition 7.1.2(i) is injective. In particular, this mapping induces an isomorphism $\text{Inn}(S) \cong S^* \cong \mathcal{S}_n$. To complete the proof we only need to show that any automorphism of S is inner. Since we already have $n!$ different (inner) automorphisms of S, it is in fact enough to show that $|\text{Aut}(S)| \leq n!$. We shall do it via a case-by-case analysis.

Let $\psi \in \text{Aut}(S)$. Since S is finite and all two-sided ideals of S form a chain (by Theorem 4.3.1), ψ induces the identity mapping on the set of all two-sided ideals. In particular, for all $i = 0, 1, \ldots, n$ we have that the restriction map $\psi|_{\mathcal{I}_i} : \mathcal{I}_i \to \mathcal{I}_i$ is bijective.

Case 1. $S = \mathcal{T}_n$. In this case we have $\mathcal{I}_1 = \{0_a : a \in \mathbf{N}\}$. Since the restriction of ψ to $\mathcal{I}_1$ is a bijection, we can define the permutation $\overline{\psi} \in \mathcal{S}_n$ by the following rule: $\psi(0_a) = 0_{\overline{\psi}(a)}$, $a \in \mathbf{N}$. The mapping $\psi \mapsto \overline{\psi}$ is a homomorphism of groups. Indeed,

$$0_{\overline{\eta\psi}(a)} = \eta\psi(0_a) = \eta(\psi(0_a)) = \eta(0_{\overline{\psi}(a)}) = 0_{\overline{\eta}\,\overline{\psi}(a)}$$

for all $a \in \mathbf{N}$ and hence $\overline{\eta\psi} = \overline{\eta}\,\overline{\psi}$. Since $|\mathcal{S}_n| = n!$, to complete the proof it is enough to show that the kernel of the mapping $\psi \mapsto \overline{\psi}$ is trivial.

In other words, let $\psi \in \text{Aut}(\mathcal{T}_n)$ be such that $\psi(0_a) = 0_a$ for all $a \in \mathbf{N}$. We have to show that $\psi(\alpha) = \alpha$ for all $\alpha \in \mathcal{T}_n$. For $\alpha \in \mathcal{T}_n$ set

$$M(\alpha) = \{(a, b) \in \mathbf{N} \times \mathbf{N} : \alpha \cdot 0_b = 0_a\}.$$

From the definition it is clear that $(a, b) \in M(\alpha)$ if and only if $\alpha(b) = a$. Take now some $\alpha \in \mathcal{T}_n$ and let $\psi(\alpha) = \beta$. Applying ψ to the equality $\alpha 0_b = 0_a$

we get that this equality is equivalent to the equality $\beta 0_b = 0_a$. This means that $M(\alpha) = M(\beta)$, in particular, $\alpha(x) = \beta(x)$ for all $x \in \mathbf{N}$. Therefore $\beta = \alpha$. Hence $\psi(\alpha) = \alpha$ for all $\alpha \in \mathcal{T}_n$ and the proof in the case $S = \mathcal{T}_n$ is completed.

Case 2. $S = \mathcal{IS}_n$. The set $X = \{\varepsilon_{\{a\}} : a \in \mathbf{N}\}$ is the set of all idempotents of rank one, that is, the set of all idempotents in $\mathcal{I}_1 \backslash \mathcal{I}_0$. As an automorphism, ψ sends idempotents to idempotents. Since the restriction of ψ to both $\mathcal{I}_1$ and $\mathcal{I}_0$ is a bijection, ψ induces a permutation on X, in particular, we can define the permutation $\overline{\psi} \in \mathcal{S}_n$ by the following rule: $\psi(\varepsilon_{\{a\}}) = \varepsilon_{\{\overline{\psi}(a)\}}$, $a \in \mathbf{N}$. As in the previous case one checks that the mapping $\psi \mapsto \overline{\psi}$ is a homomorphism of groups. To complete the proof it is enough to show that the kernel of this mapping is trivial.

In other words, let $\psi \in \mathrm{Aut}(\mathcal{IS}_n)$ be such that $\psi(\varepsilon_{\{a\}}) = \varepsilon_{\{a\}}$ for all $a \in \mathbf{N}$. We have to show that $\psi(\alpha) = \alpha$ for all $\alpha \in \mathcal{IS}_n$. For $\alpha \in \mathcal{IS}_n$ set

$$M(\alpha) = \{(a,b) \in \mathbf{N} \times \mathbf{N} : \varepsilon_{\{a\}} \alpha \varepsilon_{\{b\}} \neq \mathbf{0}\}.$$

From the definition it is clear that $(a,b) \in M(\alpha)$ if and only if $\alpha(b) = a$. Now take some $\alpha \in \mathcal{IS}_n$ and let $\psi(\alpha) = \beta$. Applying ψ to the inequality $\varepsilon_{\{a\}} \alpha \varepsilon_{\{b\}} \neq \mathbf{0}$ we get the inequality $\varepsilon_{\{a\}} \beta \varepsilon_{\{b\}} \neq \mathbf{0}$. This means that $M(\alpha) = M(\beta)$, and thus $\alpha = \beta$. Hence $\psi(\alpha) = \alpha$ for all $\alpha \in \mathcal{IS}_n$ and the proof in the case $S = \mathcal{IS}_n$ is completed as well.

Case 3. $S = \mathcal{PT}_n$. We could apply exactly the same arguments as in the proof given in Case 2 provided that we could prove that ψ preserves the set X. To prove this we will need the following:

Lemma 7.1.5 *For* $\alpha \in \mathcal{PT}_n$ *we have*

$$|\mathrm{dom}(\alpha)| = |\mathrm{dom}(\psi(\alpha))| \quad and \quad |\mathrm{im}(\alpha)| = |\mathrm{im}(\psi(\alpha))|.$$

Proof. Obviously $\alpha\beta = \mathbf{0}$ if and only if $\mathrm{dom}(\alpha) \cap \mathrm{im}(\beta) = \varnothing$, that is, if $\mathrm{im}(\beta) \subset \overline{\mathrm{dom}(\alpha)}$. Hence the right annihilator $\mathrm{Ann}_r(\alpha) = \{\beta : \alpha\beta = \mathbf{0}\}$ of the element α contains exactly $(|\overline{\mathrm{dom}}(\alpha)| + 1)^n$ elements. Moreover, the equality $\psi(\mathbf{0}) = \mathbf{0}$ implies that $\alpha\beta = 0$ if and only if $\psi(\alpha)\psi(\beta) = \mathbf{0}$. Hence $\psi(\mathrm{Ann}_r(\alpha)) = \mathrm{Ann}_r(\psi(\alpha))$. As any automorphism is a bijection, we have $|\mathrm{Ann}_r(\alpha)| = |\mathrm{Ann}_r(\psi(\alpha))|$ and hence $|\mathrm{dom}(\alpha)| = |\mathrm{dom}(\psi(\alpha))|$.

Analogously, we have $\beta\alpha = \mathbf{0}$ if and only if $\mathrm{im}(\alpha) \cap \mathrm{dom}(\beta) = \varnothing$, that is, if $\mathrm{dom}(\beta) \subset \overline{\mathrm{im}}(\alpha)$. Hence the left annihilator $\mathrm{Ann}_l(\alpha) = \{\beta : \beta\alpha = \mathbf{0}\}$ of the element α contains exactly $(n+1)^{n-|\mathrm{im}(\alpha)|}$ elements. Analogously to the previous paragraph one shows that $\psi(\mathrm{Ann}_l(\alpha)) = \mathrm{Ann}_l(\psi(\alpha))$ and $|\mathrm{Ann}_l(\alpha)| = |\mathrm{Ann}_l(\psi(\alpha))|$. Thus $|\mathrm{im}(\alpha)| = |\mathrm{im}(\psi(\alpha))|$. □

From Lemma 7.1.5 it follows that for any $a \in \mathbf{N}$ the element $\alpha = \psi(\varepsilon_{\{a\}})$ is an idempotent such that $|\mathrm{dom}(\alpha)| = 1$ and $|\mathrm{im}(\alpha)| = 1$. Hence $\alpha \in X$. This completes the proof. □

7.2 Endomorphisms of Small Ranks

Recall from Sect. 6.1 that an *endomorphism* of a semigroup S is a mapping
$\varphi : S \to S$ such that $\varphi(x \cdot y) = \varphi(x) \cdot \varphi(y)$ for all $x, y \in S$. Our main goal for
the rest of this chapter is to describe all endomorphisms of the semigroups
$\mathcal{T}_n$, $\mathcal{IS}_n$, and $\mathcal{PT}_n$. The final classification will be done basically using the
case-by-case analysis. If S is a semigroup and $\varphi \in \mathrm{End}(S)$, the number $|\varphi(S)|$
is called the *rank* of φ. We start with constructing some endomorphisms of
small ranks.

Let S denote one of the semigroups $\mathcal{T}_n$, $\mathcal{IS}_n$, or $\mathcal{PT}_n$. Let $\epsilon \in S$ be an
idempotent. Define the mapping $\Phi_\epsilon : S \to S$ via $\Phi_\epsilon(\alpha) = \epsilon$ for all $\alpha \in S$.
The following statement is obvious:

Lemma 7.2.1 *(i) For any idempotent ϵ the mapping Φ_ϵ is an endomorphism of S of rank one.*

(ii) If $\epsilon \neq \epsilon'$, then $\Phi_\epsilon \neq \Phi_{\epsilon'}$.

*(iii) Every endomorphism of S or rank one has the form Φ_ϵ for some idem-
potent $\epsilon \in S$.*

Let $\epsilon, \delta \in S$ be different idempotents such that $\epsilon\delta = \delta\epsilon = \delta$. Define the
mapping $\Psi_{\epsilon,\delta} : S \to S$ as follows:

$$\Psi_{\epsilon,\delta}(\alpha) = \begin{cases} \epsilon, & \alpha \in \mathcal{S}_n; \\ \delta, & \text{otherwise.} \end{cases}$$

Lemma 7.2.2 *(i) For any pair ϵ, δ of different idempotents of S satisfying
$\epsilon\delta = \delta\epsilon = \delta$ the mapping $\Psi_{\epsilon,\delta}$ is an endomorphism of S of rank two.*

(ii) If the pairs ϵ, δ and ϵ', δ' are different, then $\Psi_{\epsilon,\delta} \neq \Psi_{\epsilon',\delta'}$.

*(iii) Every endomorphism of S or rank two has the form $\Phi_{\epsilon,\delta}$ for some
unequal idempotents $\epsilon, \delta \in S$ satisfying $\epsilon\delta = \delta\epsilon = \delta$.*

Proof. The statement (i) is proved by a direct calculation and the statement
(ii) is obvious. To prove (iii) we should assume that $S \neq \mathcal{T}_1$ as $|\mathcal{T}_1| = 1$.
Let $\varphi \in \mathrm{End}(S)$ be an endomorphism of rank two. Then the congruence
$\mathrm{Ker}(\varphi)$ contains exactly two congruence classes. Going through the list of all
congruences on S given by Theorem 6.3.10, we see that the only congruence
on S with exactly two congruence classes is $\equiv_{\mathcal{S}_n}$. The congruence classes
are S^* and $\mathcal{I}_{n-1}$. As the representatives of the two congruence classes of
this congruence we can take for example the identity transformation ε_n and
any (left) zero element γ of S. These are idempotents and obviously satisfy
$\varepsilon_n\gamma = \gamma\varepsilon_n = \gamma$. Since they belong to different congruence classes, we have
$\epsilon = \varphi(\varepsilon_n) \neq \varphi(\gamma) = \delta$. Moreover, $\epsilon\delta = \delta\epsilon = \delta$. By definition, in this notation
we have $\varphi = \Phi_{\epsilon,\delta}$. This completes the proof of (iii). $\square$

Let $\epsilon, \delta \in S$ be different idempotents such that $\epsilon\delta = \delta\epsilon = \delta$. Let further $\tau \in G_\epsilon$ be an element of order two (i.e., $\tau^2 = \epsilon$, $\tau \neq \epsilon$) such that $\tau\delta = \delta\tau = \delta$. Define the mapping $\Theta^\tau_{\epsilon,\delta} : S \to S$ as follows:

$$\Theta^\tau_{\epsilon,\delta}(\alpha) = \begin{cases} \epsilon, & \alpha \in \mathcal{A}_n; \\ \tau, & \alpha \in \mathcal{S}_n \backslash \mathcal{A}_n; \\ \delta, & \text{otherwise.} \end{cases}$$

Lemma 7.2.3 *(i) For any triple ϵ, δ, τ as above the mapping $\Theta^\tau_{\epsilon,\delta}$ is an endomorphism of S of rank three.*

(ii) If the triples ϵ, δ, τ and $\epsilon', \delta', \tau'$ are different, then $\Theta^\tau_{\epsilon,\delta} \neq \Theta^{\tau'}_{\epsilon',\delta'}$.

(iii) Every endomorphism of S or rank three has the form $\Theta^\tau_{\epsilon,\delta}$ for some ϵ, δ, and τ as above.

Proof. The statement (i) is proved by a direct calculation and the statement (ii) is obvious. To prove (iii) we should assume $n > 1$ as $|\mathcal{PT}_1| = 2$. Let $\varphi \in \text{End}(S)$ be an endomorphism of rank three. Then the congruence $\text{Ker}(\varphi)$ contains exactly three congruence classes. Going through the list of all congruences on S given by Theorem 6.3.10, we see that the only congruence on S with exactly three congruence classes is $\equiv_{\mathcal{A}_n}$, $n > 1$. The congruence classes are $\mathcal{A}_n$, $\mathcal{S}_n \backslash \mathcal{A}_n$, and $\mathcal{I}_{n-1}$. As the representatives of the three congruence classes of this congruence we can take for example the identity transformation ε_n, the transposition $(1, 2)$ and any (left) zero element γ of S. These elements obviously satisfy $\varepsilon_n\gamma = \gamma\varepsilon_n = \gamma$, $\gamma(1, 2) = \gamma$, and $(1, 2) \in G_{\varepsilon_n}$. Further, either $(1, 2)\gamma = \gamma$ or $(1, 2)\gamma$ is another left zero, and hence belongs to $\mathcal{I}_{n-1}$. Since our representatives belong to different congruence classes, we have $\epsilon = \varphi(\varepsilon_n) \neq \varphi(\gamma) = \delta$ and $\varphi((1, 2)) = \tau$ is an element of order two in G_ϵ. Further, from the above we also get that $\epsilon\delta = \delta\epsilon = \delta$ and $\tau\delta = \delta\tau = \delta$. By definition, in this notation we have $\varphi = \Theta^\tau_{\epsilon,\delta}$. This proves (iii). □

7.3 Exceptional Endomorphism

Let $\varphi \in \text{Aut}(\mathcal{S}_n)$ and $S = \mathcal{IS}_n$ or $S = \mathcal{PT}_n$. Define the mapping $\Omega_\varphi : S \to S$ as follows:

$$\Omega_\varphi(\alpha) = \begin{cases} \varphi(\alpha), & \alpha \in \mathcal{S}_n; \\ \mathbf{0}, & \text{otherwise.} \end{cases}$$

Lemma 7.3.1 *Ω_φ is an endomorphism of both $\mathcal{IS}_n$ and $\mathcal{PT}_n$.*

Proof. Follows by a direct calculation. □

To proceed we need some additional notation. Let $\alpha \in \mathcal{PT}_n$ be an element of rank $(n-1)$. We would like to associate with α some element $\hat{\alpha}$ of rank one. There are two possibilities. The first one is $\mathrm{dom}(\alpha) \neq \mathbf{N}$. Then we define $\hat{\alpha}$ as the unique element of rank one such that

$$\mathrm{dom}(\hat{\alpha}) = \overline{\mathrm{dom}(\alpha)}, \quad \mathrm{im}(\hat{\alpha}) = \mathbf{N} \backslash \mathrm{im}(\alpha).$$

Here is an example of such α and $\hat{\alpha}$ for $n = 5$:

$$\alpha = \begin{pmatrix} 1 & 2 & 3 & 4 & 5 \\ 2 & 5 & \varnothing & 1 & 3 \end{pmatrix}, \quad \hat{\alpha} = \begin{pmatrix} 1 & 2 & 3 & 4 & 5 \\ \varnothing & \varnothing & 4 & \varnothing & \varnothing \end{pmatrix}.$$

The second possibility is that there exists a unique pair $a, b \in \mathbf{N}$ such that $\alpha(a) = \alpha(b)$. Then we define $\hat{\alpha}$ as the unique element of rank one such that

$$\mathrm{dom}(\hat{\alpha}) = \{a, b\}, \quad \mathrm{im}(\hat{\alpha}) = \mathbf{N} \backslash \mathrm{im}(\alpha).$$

Here is an example of such α and $\hat{\alpha}$ for $n = 5$:

$$\alpha = \begin{pmatrix} 1 & 2 & 3 & 4 & 5 \\ 2 & 5 & 2 & 1 & 3 \end{pmatrix}, \quad \hat{\alpha} = \begin{pmatrix} 1 & 2 & 3 & 4 & 5 \\ 4 & \varnothing & 4 & \varnothing & \varnothing \end{pmatrix}.$$

Define the mapping $\Xi : \mathcal{PT}_n \to \mathcal{PT}_n$, $n > 1$, as follows:

$$\Xi(\alpha) = \begin{cases} \alpha, & \mathrm{rank}(\alpha) = n; \\ \hat{\alpha}, & \mathrm{rank}(\alpha) = n - 1; \\ \mathbf{0}, & \text{otherwise.} \end{cases}$$

Note that Ξ maps $\mathcal{IS}_n$ to $\mathcal{IS}_n$. Abusing notation we denote the induced mapping on $\mathcal{IS}_n$ also by Ξ.

Lemma 7.3.2 *Let $S = \mathcal{IS}_n$ or $S = \mathcal{PT}_n$, $n > 1$.*

(i) Ξ is an endomorphism of S.

(ii) If $\varphi, \psi \in \mathrm{Aut}(\mathcal{PT}_n)$ and $\varphi \neq \psi$, then $\varphi \Xi \neq \psi \Xi$.

Proof. To prove (i) we have to check that

$$\Xi(\alpha\beta) = \Xi(\alpha)\Xi(\beta) \tag{7.1}$$

for all $\alpha, \beta \in S$. By the definition of Ξ this is obvious if both α and β are permutations or if one of them has rank at most $(n-2)$. There are three cases left.

   Case 1. $\alpha \in \mathcal{S}_n$, $\mathrm{rank}(\beta) = n-1$. In this case $\mathrm{rank}(\alpha\beta) = n-1$ and hence (7.1) reduces to the equality $\alpha\hat{\beta} = \widehat{\alpha\beta}$. Since α is a permutation, we have $\mathrm{dom}(\alpha\beta) = \mathrm{dom}(\beta)$ and $\mathrm{im}(\alpha\beta) = \alpha(\mathrm{im}(\beta))$. Hence the equality $\alpha\hat{\beta} = \widehat{\alpha\beta}$ follows from the definition of Ξ.

Case 2. $\mathrm{rank}(\alpha) = n-1$, $\beta \in S_n$. In this case $\mathrm{rank}(\alpha\beta) = n-1$ and hence (7.1) reduces to the equality $\hat{\alpha}\beta = \widehat{\alpha\beta}$. Since β is a permutation, we have $\mathrm{dom}(\alpha\beta) = \beta^{-1}(\mathrm{dom}(\alpha))$ and $\mathrm{im}(\alpha\beta) = \mathrm{im}(\alpha)$. Hence the equality $\hat{\alpha}\beta = \widehat{\alpha\beta}$ follows from the definition of Ξ.

Case 3. $\mathrm{rank}(\alpha) = n-1$, $\mathrm{rank}(\beta) = n-1$. Here we have two possibilities. The first one is that $\mathrm{rank}(\alpha\beta) < n - 1$. This is possible if $\mathrm{im}(\beta)$ contains two different a and b such that $\alpha(a) = \alpha(b)$ or if $\mathrm{im}(\beta) \cap \overline{\mathrm{dom}(\alpha)} \neq \varnothing$. In both cases the definition of Ξ gives $\hat{\alpha}\hat{\beta} = \mathbf{0}$. The second possibility is that $\mathrm{rank}(\alpha\beta) = n - 1$. This is possible if and only if $\mathrm{im}(\beta) \subset \mathrm{dom}(\alpha)$ and α is injective on $\mathrm{im}(\beta)$. In particular, from the definition of Ξ we have that the unique element in $\mathbf{N}\backslash\mathrm{im}(\beta)$ belongs to $\mathrm{dom}(\hat{\alpha})$, that is, $\mathrm{im}(\hat{\beta}) \subset \mathrm{dom}(\hat{\alpha})$. We thus get $\mathrm{dom}(\widehat{\alpha\beta}) = \mathrm{dom}(\hat{\beta})$ and $\mathrm{im}(\widehat{\alpha\beta}) = \mathrm{im}(\hat{\alpha})$, which implies $\hat{\alpha}\hat{\beta} = \widehat{\alpha\beta}$ since both elements in the last equality have rank one. This proves (i). To prove (ii) one can just observe that $\varphi\Xi$ and $\psi\Xi$ act differently on idempotents of rank $(n-1)$. $\qquad\square$

Now let $n = 4$. In this case we have an additional normal subgroup of S_4, namely, V_4. The cosets of S_4 modulo V_4 are the following sets:

$$X_1 = \left\{ \begin{pmatrix} 1 & 2 & 3 & 4 \\ 1 & 2 & 3 & 4 \end{pmatrix}, \begin{pmatrix} 1 & 2 & 3 & 4 \\ 2 & 1 & 4 & 3 \end{pmatrix}, \begin{pmatrix} 1 & 2 & 3 & 4 \\ 3 & 4 & 1 & 2 \end{pmatrix}, \begin{pmatrix} 1 & 2 & 3 & 4 \\ 4 & 3 & 2 & 1 \end{pmatrix} \right\},$$

$$X_2 = \left\{ \begin{pmatrix} 1 & 2 & 3 & 4 \\ 2 & 1 & 3 & 4 \end{pmatrix}, \begin{pmatrix} 1 & 2 & 3 & 4 \\ 1 & 2 & 4 & 3 \end{pmatrix}, \begin{pmatrix} 1 & 2 & 3 & 4 \\ 4 & 3 & 1 & 2 \end{pmatrix}, \begin{pmatrix} 1 & 2 & 3 & 4 \\ 3 & 4 & 2 & 1 \end{pmatrix} \right\},$$

$$X_3 = \left\{ \begin{pmatrix} 1 & 2 & 3 & 4 \\ 1 & 3 & 2 & 4 \end{pmatrix}, \begin{pmatrix} 1 & 2 & 3 & 4 \\ 2 & 4 & 1 & 3 \end{pmatrix}, \begin{pmatrix} 1 & 2 & 3 & 4 \\ 3 & 1 & 4 & 2 \end{pmatrix}, \begin{pmatrix} 1 & 2 & 3 & 4 \\ 4 & 2 & 3 & 1 \end{pmatrix} \right\},$$

$$X_4 = \left\{ \begin{pmatrix} 1 & 2 & 3 & 4 \\ 3 & 2 & 1 & 4 \end{pmatrix}, \begin{pmatrix} 1 & 2 & 3 & 4 \\ 4 & 1 & 2 & 3 \end{pmatrix}, \begin{pmatrix} 1 & 2 & 3 & 4 \\ 1 & 4 & 3 & 2 \end{pmatrix}, \begin{pmatrix} 1 & 2 & 3 & 4 \\ 2 & 3 & 4 & 1 \end{pmatrix} \right\},$$

$$X_5 = \left\{ \begin{pmatrix} 1 & 2 & 3 & 4 \\ 2 & 3 & 1 & 4 \end{pmatrix}, \begin{pmatrix} 1 & 2 & 3 & 4 \\ 1 & 4 & 2 & 3 \end{pmatrix}, \begin{pmatrix} 1 & 2 & 3 & 4 \\ 4 & 1 & 3 & 2 \end{pmatrix}, \begin{pmatrix} 1 & 2 & 3 & 4 \\ 3 & 2 & 4 & 1 \end{pmatrix} \right\},$$

$$X_6 = \left\{ \begin{pmatrix} 1 & 2 & 3 & 4 \\ 3 & 1 & 2 & 4 \end{pmatrix}, \begin{pmatrix} 1 & 2 & 3 & 4 \\ 4 & 2 & 1 & 3 \end{pmatrix}, \begin{pmatrix} 1 & 2 & 3 & 4 \\ 1 & 3 & 4 & 2 \end{pmatrix}, \begin{pmatrix} 1 & 2 & 3 & 4 \\ 2 & 4 & 3 & 1 \end{pmatrix} \right\}.$$

We see that each of these cosets contains a unique element α satisfying $\alpha(4) = 4$ (the first element of each coset). Mapping the whole coset to this element defines the endomorphism $\psi : S_4 \to S_4$ with kernel V_4. Define the following transformations of $\mathcal{PT}_4$:

$$\Upsilon_1(\alpha) = \begin{cases} \psi(\alpha), & \alpha \in S_4; \\ \mathbf{0}, & \text{otherwise;} \end{cases} \qquad \Upsilon_2(\alpha) = \begin{cases} \psi(\alpha), & \alpha \in S_4; \\ \varepsilon_{\{4\}}, & \text{otherwise;} \end{cases}$$

$$\Upsilon_3(\alpha) = \begin{cases} \psi(\alpha), & \alpha \in S_4; \\ \mathbf{0}_4, & \text{otherwise;} \end{cases} \qquad \Upsilon_4(\alpha) = \begin{cases} \psi(\alpha)\varepsilon_{\{1,2,3\}}, & \alpha \in S_4; \\ \mathbf{0}, & \text{otherwise;} \end{cases}$$

$$\Upsilon_5(\alpha) = \begin{cases} \psi(\alpha)\varepsilon_{\{1,4\}}, & \alpha \in S_4; \\ \mathbf{0}, & \text{otherwise;} \end{cases} \quad \Upsilon_6(\alpha) = \begin{cases} \psi(\alpha)\varepsilon_{\{2,4\}}, & \alpha \in S_4; \\ \mathbf{0}, & \text{otherwise;} \end{cases}$$

$$\Upsilon_7(\alpha) = \begin{cases} \psi(\alpha)\varepsilon_{\{3,4\}}, & \alpha \in S_4; \\ \mathbf{0}, & \text{otherwise.} \end{cases}$$

Note that Υ_1, Υ_2, and Υ_4 preserve $\mathcal{IS}_4$ and Υ_3 preserves $\mathcal{T}_4$. Abusing notation we shall denote the induced mappings on $\mathcal{IS}_4$ and $\mathcal{T}_4$ by the same symbols.

Exercise 7.3.3 Let $i \in \{1, 2, \ldots, 7\}$ and $\varphi, \psi \in \mathrm{Aut}(\mathcal{PT}_4)$ be two different automorphisms. Show that the restrictions of $\varphi\Upsilon_i$ and $\psi\Upsilon_i$ to S_4 are different.

Lemma 7.3.4 *(i) $\Upsilon_i \in \mathrm{End}(\mathcal{PT}_4)$ for all $i = 1, \ldots, 7$.*

(ii) If $\varphi, \psi \in \mathrm{Aut}(\mathcal{PT}_4)$ and $\varphi \neq \psi$, then $\varphi\Upsilon_i \neq \psi\Upsilon_i$ for all $i = 1, \ldots, 7$.

(iii) The statement (ii) holds as well for the restrictions to $\mathcal{IS}_4$ and $\mathcal{T}_4$ in appropriate cases.

Proof. We prove the statement (i) for Υ_1. All other cases are similar. We have to check that

$$\Upsilon_1(\alpha\beta) = \Upsilon_1(\alpha)\Upsilon_1(\beta) \tag{7.2}$$

for all $\alpha, \beta \in \mathcal{PT}_4$. If both α and β are permutations, (7.2) follows from the fact that φ is a homomorphism. If both α and β are not permutations, (7.2) follows from the fact that $\mathbf{0}$ is an idempotent. If exactly one of α and β is a permutation, then $\alpha\beta$ is not a permutation and hence both sides of (7.2) are equal to $\mathbf{0}$. This proves (i).

Finally, (ii) and (iii) follow from Exercise 7.3.3. $\square$

7.4 Classification of Endomorphisms

Theorem 7.4.1 *(i) Let $n \neq 4$. Then each endomorphism of $\mathcal{T}_n$ has one of the following forms:*

(a) Λ_π, where $\pi \in S_n$.

(b) Φ_ϵ, where $\epsilon \in \mathcal{T}_n$ is an idempotent.

(c) $\Psi_{\epsilon,\delta}$, where $\epsilon \neq \delta \in \mathcal{T}_n$ are idempotents such that $\epsilon\delta = \delta\epsilon = \delta$.

(d) $\Theta^\tau_{\epsilon,\delta}$, where $\epsilon \neq \delta \in \mathcal{T}_n$ are idempotents such that $\epsilon\delta = \delta\epsilon = \delta$ and $\tau \in G_\epsilon$ is an element of order two such that $\tau\delta = \delta\tau = \delta$.

(ii) In addition to the above endomorphisms, the semigroup $\mathcal{T}_4$ has also endomorphisms $\Lambda_\pi\Upsilon_3$, $\pi \in S_n$.

Proof. We start with some auxiliary statements:

Lemma 7.4.2 *For $n > 1$ the semigroup $\mathcal{T}_n$ does not contain any element α such that $\pi\alpha = \alpha$ for all $\pi \in \mathcal{S}_n$.*

Proof. Let $\alpha \in \mathcal{T}_n$, $x \in \mathbf{N}$ and $y = \alpha(x)$. Let further $z \in \mathbf{N}$ be such that $z \neq y$. Then $(y, z)\alpha(x) = z \neq y = \alpha(x)$ and hence $(y, z)\alpha \neq \alpha$. $\square$

Lemma 7.4.3 *Let S denote one of the semigroups $\mathcal{T}_n$, $\mathcal{IS}_n$, or $\mathcal{PT}_n$. Let $\varphi \in \mathrm{End}(S)$. Assume that all elements of $\mathcal{S}_n$ form one-element congruence classes of $\mathrm{Ker}(\varphi)$. Then $\varphi(\mathcal{S}_n) \subset \mathcal{S}_n$ and $\varphi|_{\mathcal{S}_n} \in \mathrm{Aut}(\mathcal{S}_n)$.*

Proof. By assumptions, $\varphi(\mathcal{S}_n) \cong \mathcal{S}_n$ is a subgroup of S. It is of course then contained in some maximal subgroup. But, by Theorem 5.1.4, all maximal subgroups of S are isomorphic to $\mathcal{S}_m$, $m \leq n$, and there is only one maximal subgroup isomorphic to $\mathcal{S}_n$, namely, $\mathcal{S}_n$ itself. Hence φ maps $\mathcal{S}_n$ to $\mathcal{S}_n$ and thus induces an automorphism of $\mathcal{S}_n$. $\square$

Lemma 7.4.4 *Let $\varphi \in \mathrm{End}(\mathcal{T}_n)$. Then either $\mathrm{Ker}(\varphi) = \iota_{\mathcal{T}_n}$ or $\mathrm{Ker}(\varphi) = \omega_{\mathcal{T}_n}$ or $\mathrm{Ker}(\varphi) = \equiv_R$, where R is a normal subgroup of $\mathcal{S}_n$.*

Proof. The statement is obvious for $n = 1$ so we assume $n > 1$.

Assume that $\mathrm{Ker}(\varphi)$ has the form $\equiv_R$, where R is a normal subgroup of $\mathcal{S}_k$ and $k < n$. From the definition of $\equiv_R$ (see Sect. 6.3) it follows that all elements of $\mathcal{S}_n$ form one-element congruence classes. From Lemma 7.4.3 we have that φ maps $\mathcal{S}_n$ to $\mathcal{S}_n$ and induces an automorphism of $\mathcal{S}_n$.

If $k = 1$, then $\equiv_R$ coincides with $\iota_{\mathcal{T}_n}$. If $k > 1$, then by the definition of $\equiv_R$ the ideal $\mathcal{I}_{k-1}$ forms one congruence class and hence is mapped by φ to some idempotent, say γ. But $\mathcal{I}_{k-1}$ is an ideal and hence satisfies $\alpha\beta \in \mathcal{I}_{k-1}$ for all $\alpha \in \mathcal{S}_n$ and $\beta \in \mathcal{I}_{k-1}$. Applying φ yields $\varphi(\alpha)\gamma = \gamma$. Since the restriction of φ to $\mathcal{S}_n$ is bijective by the previous paragraph, we have $\pi\gamma = \gamma$ for all $\pi \in \mathcal{S}_n$. But $\mathcal{T}_n$, $n > 1$, does not contain any γ with such property by Lemma 7.4.2. This means that the case $1 < k < n$ is impossible, which completes the proof. $\square$

Let $n \neq 4$ and $\varphi \in \mathrm{End}(\mathcal{T}_n)$. If $\mathrm{Ker}(\varphi)$ is a uniform congruence, then φ has rank one and thus coincides with some Φ_ϵ by Lemma 7.2.1(iii). If $\mathrm{Ker}(\varphi)$ is the identity congruence, then φ is an automorphism and thus coincides with some Λ_π by Theorem 7.1.3. In all other cases $\mathrm{Ker}(\varphi) = \equiv_R$, where R is $\mathcal{E}_n$, $\mathcal{A}_n$, or $\mathcal{S}_n$ by Lemma 7.4.4. If $R = \mathcal{S}_n$, then φ has rank two and thus coincides with some $\Psi_{\epsilon,\delta}$ by Lemma 7.2.2(iii). If $R = \mathcal{A}_n$, then φ has rank three and thus coincides with some $\Theta^\tau_{\epsilon,\delta}$ by Lemma 7.2.3(iii).

For $n \neq 4$ it remains to consider the case when $R = \mathcal{E}_n$. In this case all elements of $\mathcal{S}_n$ form separate congruence classes. Hence Lemma 7.4.3 shows that φ induces an automorphism of $\mathcal{S}_n$. The ideal $\mathcal{I}_{n-1}$ is a congruence class and should be sent by φ to some idempotent γ. As $\varphi|_{\mathcal{S}_n} \in \mathrm{Aut}(\mathcal{S}_n)$, as in the proof of Lemma 7.4.4 we obtain $\pi\gamma = \gamma$ for all $\pi \in \mathcal{S}_n$. But such element γ does not exist by Lemma 7.4.2. Hence the case when $R = \mathcal{E}_n$ is not possible.

Let now $n = 4$. In this case the only difference with the arguments above is the fact that R can be equal to $\mathcal{V}_4$. For such R we have $\varphi(\mathcal{S}_4) \cong \mathcal{S}_3$, which may be a subgroup of either $\mathcal{S}_4$ or some $G_\gamma \cong \mathcal{S}_3$, where γ is an idempotent of rank 3 (see Theorem 5.1.4).

Assume first that $\varphi(\mathcal{S}_4) \subset \mathcal{S}_4$. Then $\varphi(\mathcal{S}_4) \cong \mathcal{S}_3$ is a subgroup of $\mathcal{S}_4$.

Exercise 7.4.5 Show that $\mathcal{S}_4$ contains exactly four subgroups, isomorphic to $\mathcal{S}_3$, namely, the subgroups $H_i = \{\alpha \in \mathcal{S}_4 : \alpha(i) = i\}$, $i = 1, 2, 3, 4$.

From Exercise 7.4.5, $\varphi(\mathcal{S}_4) = H_i$ for some $i \in \{1, 2, 3, 4\}$. Consider now the endomorphism

$$\varphi' = \begin{cases} \varphi, & i = 4; \\ \Lambda_{(i,4)}\varphi, & i \neq 4. \end{cases} \tag{7.3}$$

We have $\varphi'(\mathcal{S}_4) = H_4$, in particular, φ' induces an automorphism of $H_4 \cong \mathcal{S}_3$ via restriction.

Exercise 7.4.6 Show that $\mathrm{Aut}(\mathcal{S}_3) = \mathrm{Inn}(\mathcal{S}_3) \cong \mathcal{S}_3$.

Taking into account (7.3) and using Exercise 7.4.6 we obtain that there exists $\pi \in \mathcal{S}_4$ such that $\Lambda_\pi\varphi$ restricts to the identity map on H_4. The ideal $\mathcal{I}_3$ is a congruence class and thus should be mapped by $\Lambda_\pi\varphi$ to some nonin-vertible idempotent α of $\mathcal{T}_4$. As in the proof of Lemma 7.4.4 we get $\beta\alpha = \alpha$ for all $\beta \in H_4$, in particular, for $\beta = (1, 2, 3)$. Hence for any $x \in \mathrm{im}(\alpha)$ we should have $\beta(x) = x$, which means that $\mathrm{im}(\alpha) = \{4\}$, and thus $\alpha = 0_4$. Therefore $\Lambda_\pi\varphi = \Upsilon_3$ and hence $\varphi = \Lambda_{\pi^{-1}}\Upsilon_3$.

Finally, assume that $\varphi(\mathcal{S}_4) = G_\gamma$, where γ has rank 3. Let $\delta = \varphi(\mathcal{I}_3)$. Then δ is an idempotent and the fact that $\mathcal{I}_3$ is an ideal implies that $\delta\alpha = \alpha\delta = \delta$ for all $\alpha \in G_\gamma$. From $\gamma\delta = \delta$ we have $\mathrm{im}(\delta) \subset \mathrm{im}(\gamma)$. In the proof of Theorem 4.7.4 it was shown that the restriction to $\mathrm{im}(\gamma)$ de-fines an isomorphism from G_γ to $\mathcal{S}(\mathrm{im}(\gamma))$. Take any $\alpha \in G_\gamma$, whose image in $\mathcal{S}(\mathrm{im}(\gamma))$ under this isomorphism does not have any fixed point. Then for any $x \in \mathbf{N}$ we have $\alpha(\delta(x)) \neq \delta(x)$ since $\delta(x) \in \mathrm{im}(\gamma)$. This means that the equality $\alpha\delta = \delta$ is not possible, a contradiction. Hence the case $\varphi(\mathcal{S}_4) = G_\gamma$, where γ has rank 3, is not possible. This completes the proof. $\square$

Theorem 7.4.7   (i) Let $n \neq 4$. Then each endomorphism of $\mathcal{IS}_n$ has one of the following forms:

    (a) Λ_π, where $\pi \in \mathcal{S}_n$.

    (b) Φ_ϵ, where $\epsilon \in \mathcal{IS}_n$ is an idempotent.

    (c) $\Psi_{\epsilon,\delta}$, where $\epsilon \neq \delta \in \mathcal{IS}_n$ are idempotents such that $\epsilon\delta = \delta\epsilon = \delta$.

    (d) $\Theta_{\epsilon,\delta}^\tau$, where $\epsilon \neq \delta \in \mathcal{IS}_n$ are idempotents such that $\epsilon\delta = \delta\epsilon = \delta$ and $\tau \in G_\epsilon$ is an element of order two such that $\tau\delta = \delta\tau = \delta$.

(e) Ω_ψ, $\psi \in \mathrm{Aut}(\mathcal{S}_n)$.

(f) $\Lambda_\pi\Xi$, $\pi \in \mathcal{S}_n$.

(ii) *In addition to the above endomorphisms, the semigroup $\mathcal{IS}_4$ also has endomorphisms $\Lambda_\pi \Upsilon_i$, $i = 1, 2, 4$, $\pi \in \mathcal{S}_n$.*

Proof. The statement is obvious for $n = 1$ so until the end of the proof we assume $n > 1$. Let $\varphi \in \mathrm{End}(\mathcal{IS}_n)$. Assume first that all elements of $\mathcal{S}_n$ form separate congruence classes of $\mathrm{Ker}(\varphi)$. By Lemma 7.4.3, φ preserves $\mathcal{S}_n$ and induces an automorphism on it. Consider first the case $n \neq 6$. Then this automorphism is inner (see 7.6.2) and hence there exists $\pi \in \mathcal{S}_n$ such that $\varphi' = \Lambda_\pi\varphi$ induces on $\mathcal{S}_n$ the identity mapping. Set $\epsilon = \varepsilon_{\mathbf{N}\setminus\{1\}}$. The element $\varphi'(\epsilon)$ must be an idempotent, say $\varphi'(\epsilon) = \varepsilon_B$ for some $B \subset \mathbf{N}$. To determine B more precisely we shall need an auxiliary notion and statement.

For $\alpha \in \mathcal{IS}_n$ the set $\{\pi \in \mathcal{S}_n : \alpha\pi = \pi\alpha\}$ is called the *centralizer of α* in $\mathcal{S}_n$ and is denoted by $\mathcal{C}_{\mathcal{S}_n}(\alpha)$.

Lemma 7.4.8 *For $A \subset \mathbf{N}$ we have*

$$\mathcal{C}_{\mathcal{S}_n}(\varepsilon_A) = \{\pi \in \mathcal{S}_n : \pi(A) = A\}.$$

In particular, $|\mathcal{C}_{\mathcal{S}_n}(\varepsilon_A)| = |A|! \cdot (n - |A|)!$.

Proof. Since π is invertible, $\pi\varepsilon_A = \varepsilon_A\pi$ is equivalent to $\varepsilon_A = \pi\varepsilon_A\pi^{-1}$. Obviously $\pi\varepsilon_A\pi^{-1} = \varepsilon_{\pi(A)}$. Hence $\varepsilon_A = \pi\varepsilon_A\pi^{-1}$ is equivalent to $\varepsilon_A = \varepsilon_{\pi(A)}$, that is, $A = \pi(A)$. This proves the first claim and the second claim follows immediately from the first one. $\qquad\square$

Applying φ' to $\pi\epsilon = \epsilon\pi$ we get $\pi\varepsilon_B = \varepsilon_B\pi$ as $\varphi'(\pi) = \pi$ for all $\pi \in \mathcal{S}_n$. This implies that $\mathcal{C}_{\mathcal{S}_n}(\epsilon) \subset \mathcal{C}_{\mathcal{S}_n}(\varepsilon_B)$. Using Lemma 7.4.8 we obtain that

$$\{\pi \in \mathcal{S}_n : \pi(\mathbf{N}\setminus\{1\}) = \mathbf{N}\setminus\{1\}\} \subset \{\pi \in \mathcal{S}_n : \pi(B) = B\}. \quad (7.4)$$

This gives the following possibilities for B: $B = \mathbf{N}\setminus\{1\}$, $B = \{1\}$ or $B = \mathbf{N}$.

Lemma 7.4.9 *Let S, T be semigroups and $A \subset S$ be a generating system of S. Let $f : A \to T$ be any mapping. Then there exists at most one homomorphism $F : S \to T$ such that $F|_A = f$.*

Proof. As $S = \langle A \rangle$, each $x \in S$ can be written in the form $x = a_1a_2\cdots a_k$ for some $a_i \in A$. Then

$$F(x) = F(a_1a_2\cdots a_k) = F(a_1)F(a_2)\cdots F(a_k) = f(a_1)f(a_2)\cdots f(a_k),$$

which means that $F(x)$ is uniquely determined by f. $\qquad\square$

By Theorem 3.1.4, the semigroup $\mathcal{IS}_n$ is generated by $\mathcal{S}_n$ and ϵ. By Lemma 7.4.9 this means that any endomorphism of $\mathcal{IS}_n$ is uniquely determined by its values on $\mathcal{S}_n$ and ϵ. In the situation we have $\varphi'(\pi) = \pi$ for all $\pi \in \mathcal{S}_n$ and we have three possibilities for $\varphi'(\epsilon)$, namely, $\varphi'(\epsilon) = \epsilon$, $\varphi'(\epsilon) = \varepsilon_{\{1\}}$ and $\varphi'(\epsilon) = \mathbf{0}$. The first possibility is realized by the identity automorphism, the second one is realized by the endomorphism Ξ, and the third one is realized by the endomorphism Ω_θ, where θ is the identity automorphism of $\mathcal{S}_n$. This means that the original automorphism $\varphi = \Lambda_{\pi^{-1}}\varphi'$ is of the form (ia), (if), or (ie).

Let now $n = 6$. In addition to the cases considered above we have to consider the case when φ induces on $\mathcal{S}_n$ an automorphism, which is not inner. Such automorphisms are known explicitly, see 7.6.2. In particular, each such automorphism maps transpositions from the subgroup

$$C_{\mathcal{S}_6}(\epsilon) = \{\pi \in \mathcal{S}_6 : \pi(1) = 1\} < \mathcal{S}_6$$

to permutations, each of which is a product of three different commuting transpositions. It is easy to see that this implies that the only subset of $\mathbf{N}$, invariant under all such permutations, is the set $\mathbf{N}$ itself. Hence Lemma 7.4.8 implies that the only possibility for $\varphi(\epsilon)$ in this case is $\mathbf{0}$. This means that φ is of type (ie).

Now we have to consider all cases when φ does not induce an automorphism of $\mathcal{S}_n$. If $\mathrm{Ker}(\varphi) = \omega_{\mathcal{IS}_n}$, then φ is of type (ib) by Lemma 7.2.1(iii). In all other cases from Theorem 6.3.10 we have $\mathrm{Ker}(\varphi) = \equiv_R$, where R is a nontrivial normal subgroup of $\mathcal{S}_n$. If $R = \mathcal{S}_n$, then φ is of type (ic) by Lemma 7.2.2(iii). If $R = \mathcal{A}_n$, then φ is of type (id) by Lemma 7.2.3(iii). It remains to consider the case $R = \mathcal{V}_4$. We leave it to the reader to verify that in this case φ is of type (ii). $\square$

Theorem 7.4.10   *(i) Let $n \neq 4$. Then each endomorphism of $\mathcal{PT}_n$ has one of the following forms:*

(a) Λ_π, where $\pi \in \mathcal{S}_n$

(b) Φ_ϵ, where $\epsilon \in \mathcal{PT}_n$ is an idempotent

(c) $\Psi_{\epsilon,\delta}$, where $\epsilon \neq \delta \in \mathcal{PT}_n$ are idempotents such that $\epsilon\delta = \delta\epsilon = \delta$

(d) $\Theta^\tau_{\epsilon,\delta}$, where $\epsilon \neq \delta \in \mathcal{PT}_n$ are idempotents such that $\epsilon\delta = \delta\epsilon = \delta$ and $\tau \in G_\epsilon$ is an element of order two such that $\tau\delta = \delta\tau = \delta$

(e) Ω_φ, $\varphi \in \mathrm{Aut}(\mathcal{S}_n)$

(f) $\Lambda_\pi\Xi$, $\pi \in \mathcal{S}_n$

(ii) In addition to the above endomorphisms, the semigroup $\mathcal{PT}_4$ also has endomorphisms $\Lambda_\pi\Upsilon_i$, $i = 1, 2, \ldots, 7$, $\pi \in \mathcal{S}_n$.

Exercise 7.4.11 Prove Theorem 7.4.10.

7.5 Combinatorics of Endomorphisms

Corollary 7.5.1 *(i) The semigroup T_1 has one endomorphism.*

(ii) The semigroup T_4 has 345 endomorphisms.

(iii) The semigroup T_n, $n \neq 1, 4$, has

$$n! \left(1 + \sum_{m=1}^{n} \sum_{r=0}^{\lfloor \frac{m-1}{2} \rfloor} \sum_{k=1}^{m-2r} \frac{m^{n-m} \cdot k^{m-r-k}}{2^r \cdot (n-m)! \cdot (m-2r-k)! \cdot k! \cdot r!} \right) \quad (7.5)$$

endomorphisms.

Proof. The statement (i) is obvious. The statement (ii) follows from the formula of (iii) and the note that T_4 has 24 additional endomorphisms described in Theorem 7.4.1(ii). Hence we have to prove (iii).

The semigroup T_n has $n!$ automorphisms by Theorem 7.1.3. By Theorem 7.4.1, if we forget about the additional endomorphisms T_4, we have only to count the number of endomorphisms of T_n of ranks 1, 2, and 3.

By Lemma 7.2.1(iii) the endomorphisms of rank 1 correspond bijectively to idempotents of T_n. Hence we have $\sum_{m=1}^{n} \binom{n}{m} m^{n-m}$ endomorphisms of rank 1 by Corollary 2.7.4.

By Lemma 7.2.2(iii) the endomorphisms of rank 2 correspond to pairs of idempotents (ϵ, δ) of T_n satisfying the condition $\epsilon\delta = \delta\epsilon = \delta$.

Lemma 7.5.2 *Let $\epsilon, \delta \in T_n$ be two idempotents. Then the following conditions are equivalent:*

(a) $\epsilon\delta = \delta\epsilon = \delta$

(b) $\operatorname{im}(\delta) \subset \operatorname{im}(\epsilon)$ and $\rho_\delta \preceq \rho_\epsilon$

Proof. If $\epsilon\delta = \delta$, then $\operatorname{im}(\delta) \subset \operatorname{im}(\epsilon)$ by Exercise 2.1.4(b). If $\delta\epsilon = \delta$, then $\rho_\delta \preceq \rho_\epsilon$ by Theorem 4.2.4. Hence (a) implies (b). Conversely, if $\operatorname{im}(\delta) \subset \operatorname{im}(\epsilon)$ and $\rho_\delta \preceq \rho_\epsilon$, then a direct calculation shows that $\epsilon\delta = \delta\epsilon = \delta$. Hence (b) implies (a). $\square$

Now we can use Lemma 7.5.2 to count the number of all pairs (ϵ, δ) of different idempotents of T_n satisfying the condition $\epsilon\delta = \delta\epsilon = \delta$. Let $m = \operatorname{rank}(\epsilon)$ and $k = \operatorname{rank}(\delta)$. Since $\epsilon \neq \delta$ and $\operatorname{im}(\delta) \subset \operatorname{im}(\epsilon)$, we have $1 \leq k < m \leq n$. The image of ϵ can be chosen in $\binom{n}{m}$ different ways. By Lemma 7.5.2, the image of δ should be chosen inside the $\operatorname{im}(\epsilon)$ and this can be done in $\binom{m}{k}$ different ways. As ϵ is an idempotent, $\epsilon(x) = x$ for all $x \in \operatorname{im}(\epsilon)$. To define ϵ on other elements, we should map each $x \in \mathbf{N}\backslash\operatorname{im}(\epsilon)$ to some element of $\operatorname{im}(\epsilon)$. This can be done in m^{n-m} different ways. For $x \in \operatorname{im}(\epsilon)\backslash\operatorname{im}(\delta)$ let $A_x = \{y \in \mathbf{N} : \epsilon(y) = x\}$. From Lemma 7.5.2, to define δ we should map each A_x, $x \in \operatorname{im}(\epsilon)\backslash\operatorname{im}(\delta)$ to some element of $\operatorname{im}(\delta)$. This can be done in k^{m-k} different ways. Using the multiplication rule and then adding everything up for all possible m and k we get

$$\sum_{m=2}^{n} \sum_{k=1}^{m-1} \binom{n}{m} \binom{m}{k} m^{n-m} k^{m-k} \tag{7.6}$$

endomorphisms of rank 2.

By Lemma 7.2.3(iii) the endomorphisms of rank 3 correspond to triples (ϵ, δ, τ) of $\mathcal{T}_n$, where ϵ and δ are different idempotents of $\mathcal{T}_n$ satisfying the condition $\epsilon\delta = \delta\epsilon = \delta$, and $\tau \in G_\epsilon$ is an element of order 2 such that $\tau\delta = \delta\tau = \delta$. The restrictions on ϵ and δ are given by Lemma 7.5.2.

Lemma 7.5.3 *Let $\epsilon, \delta \in \mathcal{T}_n$ be two idempotents such that $\epsilon\delta = \delta\epsilon = \delta$. Let further $\tau \in G_\epsilon$. Then the following conditions are equivalent:*

(a) $\tau\delta = \delta\tau = \delta$

(b) $\tau(x) = x$ for all $x \in \mathrm{im}(\delta)$; and $\delta(x) = \delta(y)$ for all $x, y \in \mathrm{im}(\epsilon)$ such that $x \neq y$ and $\tau(x) = y$

Proof. Assume (a). Let $x \in \mathrm{im}(\delta)$. Then $x = \delta(x) = \tau\delta(x) = \tau(x)$. Let $x, y \in \mathrm{im}(\epsilon)$ be such that $x \neq y$ and $\tau(x) = y$. Then $\delta(x) = \delta\tau(x) = \delta(y)$. Hence we have (b).

Assume (b) and let $x \in \mathbf{N}$. As $\tau(x) = x$ for all $x \in \mathrm{im}(\delta)$, we have $\tau\delta = \delta$. By assumptions we have $\tau\epsilon = \tau$ and $\delta\epsilon = \delta$. Hence the equality $\delta\tau = \delta$ is equivalent to $\delta\tau\epsilon = \delta\epsilon$, which means that the equality $\delta\tau = \delta$ should be checked only for $x \in \mathrm{im}(\epsilon)$. For those $x \in \mathrm{im}(\epsilon)$ for which $\tau(x) = x$, the equality $\delta\tau(x) = \delta(x)$ is obvious. If $\tau(x) = y \neq x$ for some $x, y \in \mathrm{im}(\epsilon)$, then $\delta\tau(x) = \delta(x)$ follows from the second condition of (b). Thus $\delta\tau = \delta$, implying (a). This completes the proof. $\square$

An element of order 2 in the symmetric group $\mathcal{S}_m$ is a product of r commuting transpositions, where $0 < r \leq \lfloor \frac{m}{2} \rfloor$. There are $\binom{m}{2}$ ways to choose the first transposition, $\binom{m-2}{2}$ ways to choose the second transposition, and so on. As the order of transpositions is not important, for each r as above we have that $\mathcal{S}_m$ contains exactly

$$\frac{\binom{m}{2} \cdot \binom{m-2}{2} \cdot \cdots \cdot \binom{m-2r+2}{2}}{r!} = \frac{m!}{(m-2r)! \cdot r! \cdot 2^r} \tag{7.7}$$

elements of order 2, which are products of r commuting transpositions.

Let us now count the number of triples (ϵ, δ, τ) as above. The rank m of ϵ can have values $m = 2, \ldots, n$. The image of ϵ can be chosen in $\binom{n}{m}$ different ways and to define ϵ completely we should assign some element from $\mathrm{im}(\epsilon)$ to each element outside this image. This can be done in m^{n-m} different ways. Now we choose τ. This is an element of order 2 in $G_\epsilon \cong \mathcal{S}_m$ and hence it can be a product of r commuting transpositions, where $0 < r \leq \lfloor \frac{m}{2} \rfloor$. However, $\tau(x) = x$ for all $x \in \mathrm{im}(\delta) \neq \emptyset$. Hence $r \neq \frac{m}{2}$, that is, $0 < r \leq \lfloor \frac{m-1}{2} \rfloor$. For each such r by (7.7) we have $\frac{m!}{(m-2r)!r!2^r}$ ways to choose τ.

After we have chosen τ let k be the rank of δ. We have $1 \leq k \leq m - 2r$. There are $\binom{m-2r}{k}$ different ways to choose $\mathrm{im}(\delta)$. For $x \in \mathrm{im}(\epsilon)$ set $A_x = \{y \in \mathbf{N} : \epsilon(y) = x\}$. As $\rho_\delta \preceq \rho_\epsilon$ by Lemma 7.5.2(b) and $\delta(x) = \delta(y)$ for all $x, y \in \mathrm{im}(\epsilon)$ such that $x \neq y$ and $\tau(x) = y$ by Lemma 7.5.3(b), to define δ we should assign to each A_x, $x \in \mathrm{im}(\epsilon) \backslash \mathrm{im}(\delta)$, an element of $\mathrm{im}(\delta)$, such that the same element is assigned to A_x and A_y as soon as $\tau(x) = y$. Such assignment can be done in k^{m-r-k} different ways. Adding everything up, we get

$$\sum_{m=2}^{n} \sum_{r=1}^{\lfloor \frac{m-1}{2} \rfloor} \sum_{k=1}^{m-2r} \binom{n}{m}\binom{m-2r}{k} \frac{m! \cdot m^{n-m} \cdot k^{m-r-k}}{(m-2r)! \cdot r! \cdot 2^r} \qquad (7.8)$$

endomorphisms of order 3. If for $r = 0$ in the formula (7.8) we allow k to vary from 1 to $(m-1)$, we will obtain the formula (7.6). Further, the formula (7.6) would count all endomorphisms of rank one for $\mathcal{T}_n$ as well if we allow $k = m$ and the value 1 for m. Taking this into account, adding $n!$ automorphisms, and rewriting binomial coefficients in terms of factorials gives the formula (7.5). $\qquad \square$

Corollary 7.5.4 *(i) The semigroup $\mathcal{IS}_1$ has three endomorphisms.*

(ii) The semigroup $\mathcal{IS}_2$ has 14 endomorphisms.

(iii) The semigroup $\mathcal{IS}_4$ has 282 endomorphisms.

(iv) The semigroup $\mathcal{IS}_6$ has 5,244 endomorphisms.

(v) The semigroup $\mathcal{IS}_n$, $n \neq 1, 2, 4, 6$, has

$$n! \left(3 + \sum_{m=0}^{n} \sum_{r=0}^{\lfloor \frac{m}{2} \rfloor} \frac{2^{m-3r}}{(n-m)! \cdot (m-2r)! \cdot r!} \right) \qquad (7.9)$$

endomorphisms.

Proof. The semigroup $\mathcal{IS}_1$ has the trivial automorphism and two endomorphisms of type Theorem 7.4.7(ib). This proves (i). The statements (iii) and (iv) follow from the general formula (v) taking into account 72 additional endomorphisms of $\mathcal{IS}_4$ given by Theorem 7.4.7(ii) and 6! additional endomorphisms of $\mathcal{IS}_6$ corresponding to the outer automorphisms of $\mathcal{S}_6$ in Theorem 7.4.7(ie). The statement (ii) follows from (v) taking into account that for $n = 2$ the only endomorphism of type Theorem 7.4.7(ie) has rank 3 (and hence is of type Theorem 7.4.7(id) as well), and all endomorphisms of type Theorem 7.4.7(if) are in fact automorphisms. So, we have to prove only the statement (v).

For $n > 1$ the semigroup $\mathcal{IS}_n$ has $n!$ automorphisms by Theorem 7.1.3, $n!$ endomorphisms of type (ie), which correspond to inner automorphisms of $\mathcal{S}_n$, and $n!$ endomorphisms of type (if). So again we are left to count the number of endomorphisms of ranks 1, 2, and 3.

Let us count endomorphisms of rank at most three. By Lemmas 7.2.1, 7.2.2, and 7.2.3 these are given by triples (ϵ, δ, τ) where ϵ and δ are idempotents satisfying $\epsilon\delta = \delta\epsilon = \delta$ (not necessarily different) and $\tau \in G_\epsilon$ is an element of order at most 2 satisfying $\tau\delta = \delta\tau = \delta$. Let $m = \operatorname{rank}(\epsilon)$. Then we have $\binom{n}{m}$ choices for $\operatorname{im}(\epsilon) = A$. Assume now that τ is a product of exactly r transpositions, $0 \le r \le \lfloor \frac{m}{2} \rfloor$. Then by (7.7) we have exactly $\frac{m!}{(m-2r)!r!2^r}$ possibilities to choose τ. To proceed we need to know the restrictions on δ imposed by the condition $\tau\delta = \delta\tau = \delta$.

Lemma 7.5.5 *For $\alpha \in \mathcal{IS}_n$ and $B \subset \mathbf{N}$ we have $\alpha\varepsilon_B = \varepsilon_B\alpha = \varepsilon_B$ if and only if $\alpha(x) = x$ for all $x \in B$.*

Proof. We have $\varepsilon_B(x) = x$ for all $x \in B$ by definition. Hence already $\alpha\varepsilon_B = \varepsilon_B$ implies $\alpha(x) = x$ for all $x \in B$. On the other hand, if $\alpha(x) = x$ for all $x \in B$, then $\alpha\varepsilon_B = \varepsilon_B$ is obvious, which also reduces $\varepsilon_B\alpha = \varepsilon_B$ to $\varepsilon_B\alpha\varepsilon_B = \varepsilon_B$, that is, $\varepsilon_B\alpha = \varepsilon_B$ should be checked only for $x \in B$. However, for such x the equality $\varepsilon_B\alpha = \varepsilon_B$ follows immediately from $\alpha(x) = x$. □

Lemma 7.5.5 says that $\operatorname{im}(\delta)$ can be an arbitrary set of fixed points for τ. By construction τ has $(m - 2r)$ fixed points. Hence we have 2^{m-2r} ways to choose δ. Summing everything up gives

$$\sum_{m=0}^{n} \sum_{r=0}^{\lfloor \frac{m}{2} \rfloor} \binom{n}{m} \frac{m! \cdot 2^{m-2r}}{(m-2r)! \cdot r! \cdot 2^r}$$

endomorphisms of rank at most 3. The claim follows.   □

Corollary 7.5.6   *(i) The semigroup $\mathcal{PT}_1$ has three endomorphisms.*

(ii) The semigroup $\mathcal{PT}_2$ has 18 endomorphisms.

(iii) The semigroup $\mathcal{PT}_4$ has 1,374 endomorphisms.

(iv) The semigroup $\mathcal{PT}_6$ has 170,772 endomorphisms.

(v) The semigroup $\mathcal{PT}_n$, $n \ne 1, 2, 4, 6$, has

$$n! \left(3 + \sum_{m=0}^{n} \sum_{r=0}^{\lfloor \frac{m}{2} \rfloor} \sum_{k=0}^{m-2r} \frac{(m+1)^{n-m} \cdot (k+1)^{m-r-k}}{2^r \cdot (n-m)! \cdot (m-2r-k)! \cdot r! \cdot k!} \right) \qquad (7.10)$$

endomorphisms.

Exercise 7.5.7 Prove Corollary 7.5.6.

7.6 Addenda and Comments

7.6.1 Automorphisms of $\mathcal{T}_n$ were described by Schreier in [Sch]. A different argument was later proposed by Mal'cev in [Ma1]. Automorphisms of $\mathcal{IS}_n$ and $\mathcal{PT}_n$ were described by Liber and Sutov in [Lib] and [Sut2], respectively.

7.6.2 For $n \neq 2, 6$ all automorphisms of $\mathcal{S}_n$ are inner and $\mathrm{Aut}(\mathcal{S}_n) \cong \mathcal{S}_n$. The semigroup $\mathcal{S}_2$ has only the trivial automorphism (which is of course inner). In contrast to $\mathcal{S}_2$, for the semigroup $\mathcal{S}_6$ we have $|\mathrm{Aut}(\mathcal{S}_6)/\mathrm{Inn}(\mathcal{S}_6)| = 2$. A noninner automorphism of $\mathcal{S}_6$ was first constructed in [Hoe]. The reason that all automorphisms of $\mathcal{S}_n$, $n \neq 6$, are inner is purely numerical. Any automorphism maps conjugate elements to conjugate elements and preserves the order of an element. Let X be the set of all transpositions in $\mathcal{S}_n$. This is a conjugacy class of elements of order two. A straightforward combinatorial computation shows that for $n \neq 6$ all other conjugacy classes of elements of order two have cardinalities different from $|X|$. Hence for $n \neq 6$ any automorphism maps transpositions to transpositions. From here it is fairly straightforward to deduce that any automorphism is inner. Details can be found for example in [KM]. Each outer (that is, not inner) automorphism of $\mathcal{S}_6$ maps a transposition to a product of three commuting transpositions.

7.6.3 Endomorphisms of $\mathcal{IS}_n$, $\mathcal{T}_n$, and $\mathcal{PT}_n$ were described and their numbers were counted in [ST1], [ST2], and [ST3], respectively. Although the generic statement is correct, there are some typos in the exceptional cases of [ST1] and [ST3]. In particular, for $n = 6$ the authors missed the endomorphisms of the form Ω_φ, where φ is an outer automorphism of $\mathcal{S}_6$.

7.6.4 Here is the table for $|\mathrm{End}(S)|$, where $S = \mathcal{IS}_n$, $\mathcal{T}_n$, and $\mathcal{PT}_n$ for small n:

n	1	2	3	4	5	6	7	8
$|\mathrm{End}(\mathcal{IS}_n)|$	3	14	54	282	918	5,244	25,560	168,828
$|\mathrm{End}(\mathcal{T}_n)|$	1	7	40	345	3,226	38,503	529,614	8,219,025
$|\mathrm{End}(\mathcal{PT}_n)|$	3	18	138	1,374	13,178	170,772	2,507,690	41,387,036

7.6.5 The problem to describe all homomorphisms (monomorphisms, epimorphisms) from S to T, where $S, T \in \{\mathcal{IS}_n, \mathcal{T}_n, \mathcal{PT}_n : n \geq 1\}$, is open in general.

7.6.6 In [ST1], [ST2], and [ST3] for $S = \mathcal{IS}_n$, $\mathcal{T}_n$, and $\mathcal{PT}_n$ Schein and Teclezghi asked whether $\frac{|\mathrm{End}(S)|}{|S|} \to 0$, $n \to \infty$. More generally they ask to find an estimate for the asymptotic of $|\mathrm{End}(S)|$. For the semigroup $\mathcal{IS}_n$ both questions were answered in [JM]:

Theorem 7.6.1 ([JM]) *(i)* $|\mathrm{End}(\mathcal{IS}_n)| \sim 3n!$, $n \to \infty$.

(ii) $\frac{|\mathrm{End}(\mathcal{IS}_n)|}{|\mathcal{IS}_n|} \to 0$, $n \to \infty$.

Proof. Set

$$X_n = n! \cdot \sum_{m=0}^{n} \sum_{r=0}^{\lfloor \frac{m}{2} \rfloor} \frac{2^{m-3r}}{(n-m)! \cdot (m-2r)! \cdot r!}.$$

Because of Corollary 7.5.4(v) to prove (i) it is enough to show that $X_n/n! \to 0$, $n \to \infty$. We have

$$X_n = \sum_{m=0}^{n} \sum_{r=0}^{\lfloor \frac{m}{2} \rfloor} \binom{n}{m} \binom{m}{2r} \binom{2r}{r} \cdot 2^{m-3r} \cdot r! \tag{7.11}$$

All binomial coefficients in (7.11) are smaller than 2^n and $r \le \lfloor n/2 \rfloor$. Hence

$$X_n \le 8^n \cdot \lfloor n/2 \rfloor! \cdot \sum_{m=0}^{n} \sum_{r=0}^{\lfloor \frac{m}{2} \rfloor} 2^{m-3r} \le 16^n \cdot (n/2)! \cdot n^2. \tag{7.12}$$

Using the Stirling formula we have

$$\frac{16^n \cdot (n/2)! \cdot n^2}{n!} \sim \frac{16^n \cdot n^2 \cdot n^{n/2} \cdot e^n \cdot \sqrt{\pi n}}{2^{n/2} \cdot e^{n/2} \cdot n^n \cdot \sqrt{2\pi n}} =$$

$$= \frac{1}{\sqrt{2}} \cdot e^{n \ln 16 + 2 \ln n + \frac{n}{2} \ln n + n - \frac{n}{2} \ln 2 - \frac{n}{2} - n \ln n}. \tag{7.13}$$

Since the exponent is $-\frac{1}{2} n \ln n + O(n)$, we obtain that the right-hand side of (7.13) approaches 0 for large n. This proves (i).

By Theorem 2.5.1 we have $|\mathcal{IS}_n| \ge \binom{n}{n-1}^2 (n-1)! = n \cdot n!$. Hence, using (i), we have

$$0 \le \frac{|\text{End}(\mathcal{IS}_n)|}{|\mathcal{IS}_n|} \sim \frac{3 \cdot n!}{|\mathcal{IS}_n|} \le \frac{3 \cdot n!}{n \cdot n!} \to 0, n \to \infty.$$

The claim (ii) follows. □

For $S = \mathcal{T}_n$ and $\mathcal{PT}_n$ the corresponding problems are still open.

7.7   Additional Exercises

7.7.1 Show that the only element $\alpha \in \mathcal{PT}_n$ satisfying the condition $\pi\alpha = \alpha$ for all $\pi \in \mathcal{S}_n$ is the element $\alpha = \mathbf{0}$.

7.7.2 Determine all elements $\alpha \in \mathcal{PT}_n$ which satisfy the condition $\alpha\pi = \alpha$ for all $\pi \in \mathcal{S}_n$.

7.7.3 Let $A \subset \mathbf{N}$. Show that $\alpha \in \mathcal{IS}_n$ satisfies the condition $\alpha\varepsilon_A = \varepsilon_A\alpha$ if and only if both A and $\mathbf{N}\backslash A$ are invariant with respect to α. Use this to show that the centralizer of ε_A in $\mathcal{IS}_n$ can be identified with $\mathcal{IS}(A) \times \mathcal{IS}(\mathbf{N}\backslash A)$.

7.7.4 Determine $|\{\alpha \in \mathcal{IS}_3 : \alpha\varepsilon_{\{1\}} = \varepsilon_{\{1\}}\alpha\}|$.

7.7.5 If S is a semigroup, then the *center* of S is the set

$$\mathbf{Z}(S) = \{\alpha \in S : \alpha\beta = \beta\alpha \text{ for all } \beta \in S\}.$$

Determine $\mathbf{Z}(S)$, where $S = \mathcal{IS}_n$, $\mathcal{PT}_n$, and $\mathcal{T}_n$.

7.7.6 Show that $\mathrm{Aut}(\mathcal{I}_1) = \mathcal{S}_n$ for the ideal $\mathcal{I}_1$ of $\mathcal{IS}_n$.

7.7.7 ([ST1, ST2, ST3]) An endomorphism $\varphi \in \mathrm{End}(S)$ is called a *retraction* provided that $\varphi \cdot \varphi = \varphi$. Classify all retractions of $\mathcal{IS}_n$, $\mathcal{T}_n$, and $\mathcal{PT}_n$.

7.7.8 Classify all endomorphisms and all retractions of $\mathcal{S}_n$.

7.7.9 Let S be a finite semigroup such that $\mathrm{End}(S) = \mathrm{Aut}(S)$. Prove that $|S| = 1$.

7.7.10 Let $S = (\mathbb{Q}_{>0}, +)$ be the semigroup of all positive rational numbers with respect to addition. Show that $\mathrm{End}(S) = \mathrm{Aut}(S) \cong (\mathbb{Q}_{>0}, \cdot)$.

7.7.11 Classify all endomorphisms of the semigroup $(\mathbb{N}, +)$ of all positive integers with respect to addition.

7.7.12 Classify all endomorphisms of the finite cyclic semigroup S generated by an element of type (k, m).

7.7.13 Let $\epsilon \in \mathcal{PT}_n$ be an idempotent of rank one with domain D and image $\{x\}$. Show that $\pi \in \mathcal{S}_n$ commutes with ϵ if and only if $\pi(x) = x$ and $\pi(D) = D$. Use this to show that the number of permutations commuting with ϵ equals $(|D| - 1)! \cdot (n - |D|)!$.

Chapter 8

Nilpotent Subsemigroups

8.1 Nilpotent Subsemigroups and Partial Orders

A semigroup S with the zero element 0 is said to be *nilpotent* provided that there exists $k \in \mathbb{N}$ such that $S^k = \{0\}$, that is, $a_1 \cdot a_2 \cdots a_k = 0$ for all $a_1, \ldots, a_k \in S$. If S is nilpotent, then the minimal $k \in \mathbb{N}$ such that $S^k = \{0\}$ is called the *nilpotency degree* or *nilpotency class* of S and is denoted by $\mathrm{nd}(S)$.

Remark 8.1.1 Note that here we slightly abuse the notation S^1. In the present chapter, we will use this notation to denote the first power of S, that is, the semigroup S itself, and not the semigroup S with the adjoint identity element, as we did before.

Proposition 8.1.2 *Let S be a finite semigroup with the zero element 0. Then the following conditions are equivalent:*

(a) S is nilpotent

(b) Every element $a \in S$ is nilpotent

Proof. If S is nilpotent, $\mathrm{nd}(S) = k$ and $a \in S$, then $a^k = 0$. Hence a is nilpotent. This proves the implication (a)⇒(b).

Conversely assume that each element of S is nilpotent. Let $a_1, a_2, \ldots, a_k$ be arbitrary elements of S such that $a_1 a_2 \cdots a_k \neq 0$. For $i = 1, \ldots, k$ set $b_i = a_1 a_2 \cdots a_i$ and note that $b_i \neq 0$. Assume that $b_i = b_j$ for some $i < j$ and let $x = a_{i+1} a_{i+2} \cdots a_j$. Then for all $m \in \mathbf{N}$ we have

$$b_i x^m = (b_i x) x^{m-1} = (b_j) x^{m-1} = b_i x^{m-1} = \cdots = b_i. \tag{8.1}$$

Hence (8.1) implies that $x^m \neq 0$ for all m. This contradicts (b). Hence all elements b_i, $i = 1, \ldots, k$, are different and nonzero and thus $k < |S|$. In particular, $S^{|S|} = \{0\}$ and thus S is nilpotent. This proves the implication (b)⇒(a). ∎

O. Ganyushkin, V. Mazorchuk, *Classical Finite Transformation Semigroups*, Algebra and Applications 9, DOI: 10.1007/978-1-84800-281-4_8, © Springer-Verlag London Limited 2009

Corollary 8.1.3 *Let S be a finite nilpotent semigroup. Then* $\mathrm{nd}(S) \leq |S|$.

Exercise 8.1.4 Let S be the Rees quotient of the semigroup $(\mathbb{Q}_{>0}, +)$ of all positive rational numbers with respect to the addition modulo the ideal, consisting of all rational number greater than 1. Show that each element of S is nilpotent and that S is not nilpotent.

The aim of the present chapter is to study nilpotent subsemigroups of the semigroups $\mathcal{T}_n$, $\mathcal{PT}_n$, and $\mathcal{IS}_n$. Note that only $\mathcal{PT}_n$, and $\mathcal{IS}_n$ contain a zero element, so the notion of nilpotent elements and nilpotent subsemigroups is really natural only for these two semigroups. In this natural case, one just considers those nilpotent subsemigroups of $\mathcal{PT}_n$ or $\mathcal{IS}_n$, which share the zero element with the original semigroup. Of course one can also consider nilpotent subsemigroups in which the zero element is different from the zero element of the original semigroup. In such interpretation, it also makes sense to study nilpotent subsemigroups of $\mathcal{T}_n$. However, it turns out that this study for $\mathcal{T}_n$, $\mathcal{PT}_n$, and $\mathcal{IS}_n$ can be basically reduced to the study of nilpotent subsemigroups of $\mathcal{PT}_n$ or $\mathcal{IS}_n$ with the global zero element. We shall explain this in the Addenda. Until then we restrict our attention to the semigroups $\mathcal{PT}_n$ and $\mathcal{IS}_n$ and consider only those nilpotent subsemigroups of these semigroups whose zero element is the global zero element **0**.

For a semigroup S with the zero element 0 denote by $\mathrm{Nil}(S)$ the set of all nilpotent subsemigroups of S with the zero element 0. The set $\mathrm{Nil}(S)$ is nonempty (it contains the nilpotent subsemigroup $\{0\}$) and is partially ordered with respect to inclusions. The subsemigroup $\{0\}$ is the minimum element of $\mathrm{Nil}(S)$. If the semigroup S itself is nilpotent, then S is the maximum element of S. In the case when S is not nilpotent, the set $\mathrm{Nil}(S)$ may contain many different maximal elements. They are called the *maximal nilpotent subsemigroups* of S.

For $k \in \mathbb{N}$ denote also by $\mathrm{Nil}_k(S)$ the subset of $\mathrm{Nil}(S)$ consisting of all nilpotent subsemigroups of nilpotency degree k. The set $\mathrm{Nil}(S)$ decomposes into a disjoint union of $\mathrm{Nil}_k(S)$, $k \in \mathbb{N}$. Note that $\mathrm{Nil}_1(S) = \{\{0\}\}$. For a given k the set $\mathrm{Nil}_k(S)$ may be empty. For example, $\mathrm{Nil}_k(S)$ is certainly empty for all k big enough if S is finite. If it is not empty, then it inherits a partial order with respect to inclusions from $\mathrm{Nil}(S)$. In general, the set $\mathrm{Nil}_k(S)$ may have many minimal and many maximal elements.

Our main goal for this chapter is to classify all maximal elements in all $\mathrm{Nil}_k(S)$, where $S = \mathcal{PT}_n$ or $\mathcal{IS}_n$. The combinatorial tool we shall use for this is the set $\mathfrak{O}_n$ of all antireflexive partial orders on the set $\mathbf{N}$. The set $\mathfrak{O}_n$ itself is partially ordered with respect to inclusions.

The inclusion $\mathcal{IS}_n \hookrightarrow \mathcal{PT}_n$ induces inclusions $\mathrm{Nil}(\mathcal{IS}_n) \hookrightarrow \mathrm{Nil}(\mathcal{PT}_n)$ and $\mathrm{Nil}_k(\mathcal{IS}_n) \hookrightarrow \mathrm{Nil}_k(\mathcal{PT}_n)$ for all $k \in \mathbb{N}$. Hence we can identify $\mathrm{Nil}(\mathcal{IS}_n)$ and $\mathrm{Nil}_k(\mathcal{IS}_n)$ with the corresponding images of the above inclusions.

For $T \in \mathrm{Nil}(\mathcal{PT}_n)$ define the binary relation τ_T on $\mathbf{N}$ as follows:

$$x\tau_T y \text{ if and only if there exists} \sigma \in T \text{such that } \sigma(y) = x.$$

Lemma 8.1.5 $\tau_T \in \mathfrak{O}_n$.

Proof. Let $x, y, z \in \mathbf{N}$ be such that $x\tau_T y$ and $y\tau_T z$. Then there exist $\alpha, \beta \in T$ such that $\alpha(y) = x$ and $\beta(z) = y$. This gives $(\alpha\beta)(z) = \alpha(\beta(z)) = \alpha(y) = x$. Hence $x\tau_T z$ and the relation τ_T is transitive.

Assume that $x\tau_T x$ for some $x \in \mathbf{N}$. Then there exists $\alpha \in T$ such that $\alpha(x) = x$. But this means that $\alpha^k(x) = x$ for all $k \in \mathbf{N}$. Hence α is not nilpotent, contradicting the nilpotency of T. Hence τ_T is antireflexive and thus belongs to $\mathfrak{O}_n$. $\square$

By Lemma 8.1.5, we have the mapping $T \mapsto \tau_T$ from the set $\mathrm{Nil}(\mathcal{PT}_n)$ to $\mathfrak{O}_n$. It restricts to a mapping from $\mathrm{Nil}(\mathcal{IS}_n)$ to $\mathfrak{O}_n$.

For each $\tau \in \mathfrak{O}_n$ define the subsemigroups $N_\tau \in \mathcal{PT}_n$ and $N'_\tau \in \mathcal{IS}_n$ as follows:

$$N_\tau = \{\alpha \in \mathcal{PT}_n : \alpha(x)\tau x \text{ for all} x \in \mathrm{dom}(\alpha)\}, \quad N'_\tau = N_\tau \cap \mathcal{IS}_n.$$

Lemma 8.1.6 (i) $N_\tau \in \mathrm{Nil}(\mathcal{PT}_n)$.

(ii) $N'_\tau \in \mathrm{Nil}(\mathcal{IS}_n)$.

Proof. The statement (ii) follows from (i). To prove (i) we consider any $\alpha_i \in N_\tau$, $i = 1, \ldots, n$. Let $\alpha = \alpha_1\alpha_2 \cdots \alpha_n$ and assume that $x \in \mathrm{dom}(\alpha)$. For $i = 1, \ldots, n$ set $\beta_i = \alpha_i\alpha_{i+1} \cdots \alpha_n$. Then $x \in \mathrm{dom}(\beta_i)$ for all $i = 1, \ldots, n$ and we set $x_i = \beta_i(x)$ for all such i. Set $x_{n+1} = x$. Then we have $x_i = \alpha_i(x_{i+1})$ for all $i = 1, \ldots, n$. Hence $x_i\tau x_{i+1}$ for all $i = 1, \ldots, n$. As τ is an antireflexive partial order, we get that all elements $x_1, x_2, \ldots, x_{n+1}$ must be different, contradicting $|\mathbf{N}| = n$. Hence $\mathrm{dom}(\alpha) = \varnothing$ and thus $\alpha = \mathbf{0}$. This implies that $N_\tau^n = \{\mathbf{0}\}$ and therefore $N_\tau \in \mathrm{Nil}(\mathcal{PT}_n)$. $\square$

By Lemma 8.1.6 we have the mapping $\tau \mapsto N_\tau$ from $\mathfrak{O}_n$ to $\mathrm{Nil}(\mathcal{PT}_n)$ and the mapping $\tau \mapsto N'_\tau$ from $\mathfrak{O}_n$ to $\mathrm{Nil}(\mathcal{IS}_n)$. The crucial properties of the mappings we have constructed are:

Proposition 8.1.7 (i) *If* $S, T \in \mathrm{Nil}(\mathcal{PT}_n)$ *and* $S \subset T$, *then* $\tau_S \subset \tau_T$.

(ii) *If* $\tau, \sigma \in \mathfrak{O}_n$ *and* $\tau \subset \sigma$, *then* $N_\tau \subset N_\sigma$ *and* $N'_\tau \subset N'_\sigma$.

(iii) *For any* $\sigma \in \mathfrak{O}_n$ *we have* $\tau_{N_\sigma} = \sigma$ *and* $\tau_{N'_\sigma} = \sigma$.

Proof. The statements (i) and (ii) are obvious from the definitions, so we have to prove only the statement (iii). Let $x, y \in \mathbf{N}$ be such that $x\sigma y$. Consider the element $\alpha \in \mathcal{IS}_n$ of rank one, defined as follows: $\mathrm{dom}(\alpha) = \{y\}$ and $\alpha(y) = x$. Then $\alpha \in N'_\sigma \subset N_\sigma$. From the definition it follows that $y\tau_{N'_\sigma} x$ and hence $\sigma \subset \tau_{N'_\sigma}$ and $\sigma \subset \tau_{N_\sigma}$.

Conversely, let $x, y \in \mathbf{N}$ be such that $x \tau_{N_\sigma} y$ (for N'_σ the arguments are similar). Then there exists $\alpha \in N_\sigma$ such that $\alpha(y) = x$. But the latter implies $x \sigma y$ by the definition of N_σ. Hence $\tau_{N_\sigma} \subset \sigma$ and the statement (iii) follows. $\square$

Let $\tau \in \mathfrak{O}_n$. The maximal possible $k \in \mathbf{N}$ for which there exist elements $x_1, \ldots, x_k \in \mathbf{N}$ such that $x_i \tau x_{i+1}$ for all $i = 1, \ldots, k-1$ is called the *height* of τ. Thus any linear order on $\mathbf{N}$ has height n, while the empty partial order on $\mathbf{N}$ has height 1. For $k \in \mathbb{N}$ we denote by $\mathfrak{O}_n^{(k)}$ the subset of $\mathfrak{O}_n$ consisting of all partial orders of height k. Obviously, we have the following decomposition into a disjoint union of subsets:

$$\mathfrak{O}_n = \bigcup_{k=1}^{n} \mathfrak{O}_n^{(k)}.$$

Lemma 8.1.8   *(i) If $T \in \mathrm{Nil}_k(\mathcal{PT}_n)$, then $\tau_T \in \mathfrak{O}_n^{(k)}$.*

(ii) If $\tau \in \mathfrak{O}_n^{(k)}$, then $N_\tau \in \mathrm{Nil}_k(\mathcal{PT}_n)$ and $N'_\tau \in \mathrm{Nil}_k(\mathcal{IS}_n)$.

Proof. Let $\alpha_1, \ldots, \alpha_{k-1} \in T$ be such that $\alpha = \alpha_1 \alpha_2 \cdots \alpha_{k-1} \neq \mathbf{0}$. Fix some $x \in \mathrm{dom}(\alpha)$ and for $i = 1, \ldots, k-1$ set $x_i = \alpha_i \alpha_{i+1} \cdots \alpha_{k-1}(x)$. Then under the convention $x_k = x$ we have $x_i \tau_T x_{i+1}$ for all $i = 1, \ldots, k-1$ and hence the nilpotency degree of T does not exceed the height of τ_T.

Let x_i, $i = 1, \ldots, m$, be elements of $\mathbf{N}$ such that $x_i \tau_T x_{i+1}$ for all $i = 1, \ldots, m-1$. By the definition of τ_T there exist $\alpha_i \in T$, $i = 1, \ldots, m-1$, such that $\alpha_i(x_{i+1}) = x_i$. Then for $\alpha = \alpha_1 \alpha_2 \cdots \alpha_m$ we have $x_m \in \mathrm{dom}(\alpha)$ and hence $\alpha \neq \mathbf{0}$. This implies that the height of τ_T does not exceed $\mathrm{nd}(T)$. This proves (i).

The statement (ii) follows from the statement (i) and the assertion of Proposition 8.1.7(iii). $\square$

Corollary 8.1.9

$$\mathrm{Nil}(\mathcal{PT}_n) = \bigcup_{k=1}^{n} \mathrm{Nil}_k(\mathcal{PT}_n), \quad \mathrm{Nil}(\mathcal{IS}_n) = \bigcup_{k=1}^{n} \mathrm{Nil}_k(\mathcal{IS}_n).$$

Proof. Follows from Lemma 8.1.8(i) since, obviously, the maximal possible height of a partial order on $\mathbf{N}$ is n. $\square$

8.2   Classification of Maximal Nilpotent Subsemigroups

Let $k \in \mathbb{N}$, $1 \leq k \leq n$, and $\mathbf{m} = (M_1, M_2, \ldots, M_k)$ be an ordered partition of $\mathbf{N}$ into a disjoint union of nonempty subsets, that is, $\mathbf{N} = M_1 \cup M_2 \cup \cdots \cup M_k$;

$M_i \neq \varnothing$, $i = 1, \ldots, k$; and $M_i \cap M_j = \varnothing$ if $i \neq j$. Define the partial order $\tau_{\mathbf{m}}$ on $\mathbf{N}$ in the following way:

$$x \tau_{\mathbf{m}} y \quad \text{if and only if } x \in M_i, y \in M_j, \text{ and } i < j.$$

Example 8.2.1 For $\mathbf{m} = \{\{1,2\}, \{3,4\}, \{5,6,7\}\}$ the Hasse diagram of the partial order $\tau_{\mathbf{m}}$ is as follows:

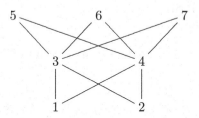

Proposition 8.2.2 *(i) For $\mathbf{m}$ as above the order $\tau_{\mathbf{m}}$ is a maximal element in $\mathfrak{O}_n^{(k)}$.*

(ii) Every maximal element in $\mathfrak{O}_n^{(k)}$ has the form $\tau_{\mathbf{m}}$ for some $\mathbf{m}$ as above.

(iii) If $\mathbf{m} \neq \mathbf{n}$, then $\tau_{\mathbf{m}} \neq \tau_{\mathbf{n}}$.

Proof. First we note that $\tau_{\mathbf{m}} \in \mathfrak{O}_n^{(k)}$. Indeed, if $x_1, \ldots, x_m$ are such that $x_i \tau_{\mathbf{m}} x_{i+1}$ for all $i = 1, \ldots, m - 1$, then all x_i's belong to different M_j's and hence $m \leq k$. At the same time, taking some $x_i \in M_i$, $i = 1, \ldots, k$, we have $x_i \tau_{\mathbf{m}} x_{i+1}$ for all $i = 1, \ldots, k - 1$. Hence the height of $\tau_{\mathbf{m}}$ equals k.

Next we claim that $\tau_{\mathbf{m}}$ is a maximal element of $\mathfrak{O}_n^{(k)}$. Indeed, let $\sigma \in \mathfrak{O}_n$ be such that $\tau_{\mathbf{m}} \subsetneq \sigma$. Then there exist $x, y \in \mathbf{N}$ such that $x \sigma y$, $x \in M_i$, $y \in M_j$ and $j \leq i$. Choose any $x_s \in M_s$, $s = 1, \ldots, i - 1$, and any $y_t \in M_t$, $t = j + 1, \ldots, k$. As $\tau_{\mathbf{m}} \subset \sigma$, for the sequence

$$z_1 = x_1, \ldots, z_{i-1} = x_{i-1}, z_i = x, z_{i+1} = y, z_{i+2} = y_{j+1}, \ldots, z_{i+k-j+1} = y_k$$

we have $z_s \sigma z_{s+1}$ for all $s = 1, \ldots, i+k-j$. As $j \leq i$, we have $i+k-j+1 \geq k+1$ and hence the height of σ is at least $k + 1$. Thus $\sigma \notin \mathfrak{O}_n^{(k)}$. This shows that $\tau_{\mathbf{m}}$ is a maximal element of $\mathfrak{O}_n^{(k)}$ and thus proves (i).

Now to prove (ii) we have to show that each $\sigma \in \mathfrak{O}_n^{(k)}$ is contained in some $\tau_{\mathbf{m}}$. Let $\sigma \in \mathfrak{O}_n^{(k)}$. Define M_1 as the set of all minimal elements of $\mathbf{N}$ with respect to σ, M_2 as the set of all minimal elements of $\mathbf{N} \backslash M_1$ with respect to σ, M_3 as the set of all minimal elements of $\mathbf{N} \backslash (M_1 \cup M_2)$ with respect to σ and so on.

We claim that $M_k \neq \varnothing$. Indeed, as $\sigma \in \mathfrak{O}_n^{(k)}$, there exist x_i, $i = 1, \ldots, k$, such that $x_i \sigma x_{i+1}$, $i = 1, \ldots, k - 1$. Assume $x_i \in M_{f(i)}$, $i = 1, \ldots, k$. Then $f(i) < f(j)$ if $i < j$ and hence $f(k) \geq k$. At the same time $M_{f(k)} \neq \varnothing$ and thus $M_k \neq \varnothing$.

Now we claim that $M_{k+1} = \varnothing$. Assume not and let $x_{k+1} \in M_{k+1}$. Then, by construction, there exist $x_i \in M_i$, $i = 1, \ldots, k$, such that $x_i \sigma x_{i+1}$ for all $i = 1, \ldots, k$. This contradicts the fact that the height of σ is k.

Finally, we claim that $\sigma \subset \tau_{\mathbf{m}}$, where $\mathbf{m} = (M_1, M_2, \ldots, M_k)$. Let $x \in M_i$, $y \in M_j$ and $i \leq j$. Then $y \sigma x$ would imply that x is not minimal in $\mathbf{N} \backslash (M_1 \cup \cdots \cup M_{i-1})$, which contradicts the definition of M_i. Hence for all x and y as above we have that $y \sigma x$ does not hold. This means that $\sigma \subset \tau_{\mathbf{m}}$, which proves the claim (ii).

The claim (iii) is obvious. $\square$

Now we are ready to formulate our main result of this chapter:

Theorem 8.2.3 *Let $k \in \mathbb{Z}$, $1 \leq k \leq n$.*

(i) *For each $\mathbf{m}$ as above the semigroups $N_{\tau_{\mathbf{m}}}$ and $N'_{\tau_{\mathbf{m}}}$ are maximal elements in $\mathrm{Nil}_k(\mathcal{PT}_n)$ and $\mathrm{Nil}_k(\mathcal{IS}_n)$, respectively.*

(ii) *Each maximal element of $\mathrm{Nil}_k(\mathcal{PT}_n)$ and $\mathrm{Nil}_k(\mathcal{IS}_n)$, has, respectively, the form $N_{\tau_{\mathbf{m}}}$ and $N'_{\tau_{\mathbf{m}}}$ for some $\mathbf{m}$ as above.*

(iii) *If $\mathbf{m} \neq \mathbf{n}$, then $N_{\tau_{\mathbf{m}}} \neq N_{\tau_{\mathbf{n}}}$ and $N'_{\tau_{\mathbf{m}}} \neq N'_{\tau_{\mathbf{n}}}$.*

Proof. Assume that $N_{\tau_{\mathbf{m}}} \subset T$ for some $T \in \mathrm{Nil}_k(\mathcal{PT}_n)$. Then $\tau_{N_{\tau_{\mathbf{m}}}} \subset \tau_T$ by Proposition 8.1.7(i). However, $\tau_{N_{\tau_{\mathbf{m}}}} = \tau_{\mathbf{m}}$ by Proposition 8.1.7(iii). As $\tau_{\mathbf{m}}$ is a maximal element in $\mathfrak{O}_n^{(k)}$ by Proposition 8.2.2(i) and $\tau_T \in \mathfrak{O}_n^{(k)}$ by Lemma 8.1.8(i), we get that $\tau_{\mathbf{m}} = \tau_T$. Hence $T \subset N_{\tau_{\mathbf{m}}}$ by the definition of $N_{\tau_{\mathbf{m}}}$. This implies $T = N_{\tau_{\mathbf{m}}}$ and proves (i) in the case of $\mathcal{PT}_n$. In the case of $\mathcal{IS}_n$ the proof is just the same.

Let now $T \in \mathrm{Nil}_k(\mathcal{PT}_n)$ be a maximal element. Then $\tau_T \in \mathfrak{O}_n^{(k)}$ and hence $\tau_T \subset \tau_{\mathbf{m}}$ for some $\mathbf{m}$ as above by Proposition 8.2.2(ii). Now Proposition 8.1.7(ii) implies that $T \subset N_{\tau_{\mathbf{m}}}$. Moreover, $N_{\tau_{\mathbf{m}}} \in \mathrm{Nil}_k(\mathcal{PT}_n)$ by Lemma 8.1.8(ii). Hence $T = N_{\tau_{\mathbf{m}}}$ by the maximality of T. This proves (ii) in the case of $\mathcal{PT}_n$. In the case of $\mathcal{IS}_n$ the proof is just the same.

Finally, let $\mathbf{m} \neq \mathbf{n}$. Without loss of generality we may assume that in this case there exist $x, y \in \mathbf{N}$ such that $x \tau_{\mathbf{m}} y$ holds while $x \tau_{\mathbf{n}} y$ does not hold. Consider the element $\alpha \in \mathcal{IS}_n$ of rank one, which is uniquely defined by the property $\alpha(y) = x$. Then $\alpha \in N_{\tau_{\mathbf{m}}}$ and $\alpha \in N'_{\tau_{\mathbf{m}}}$, while $\alpha \notin N_{\tau_{\mathbf{n}}}$ and $\alpha \notin N'_{\tau_{\mathbf{n}}}$. This completes the proof. $\square$

Example 8.2.4 If $M_1 = \{1\}$, $M_2 = \{2, 3\}$ and $\mathbf{m} = \{M_1, M_2\}$, then the semigroup $N_{\tau_{\mathbf{m}}}$ has the following elements:

$$\begin{pmatrix} 1 & 2 & 3 \\ \varnothing & \varnothing & \varnothing \end{pmatrix}, \quad \begin{pmatrix} 1 & 2 & 3 \\ \varnothing & \varnothing & 1 \end{pmatrix}, \quad \begin{pmatrix} 1 & 2 & 3 \\ \varnothing & 1 & \varnothing \end{pmatrix}, \quad \begin{pmatrix} 1 & 2 & 3 \\ \varnothing & 1 & 1 \end{pmatrix}.$$

The first three of these elements form the semigroup $N'_{\tau_{\mathbf{m}}}$.

Corollary 8.2.5 *If $k < n$, then each maximal element of $\mathrm{Nil}_k(\mathcal{PT}_n)$ and $\mathrm{Nil}_k(\mathcal{IS}_n)$ is contained in some maximal element of the sets $\mathrm{Nil}_{k+1}(\mathcal{PT}_n)$ and $\mathrm{Nil}_{k+1}(\mathcal{IS}_n)$, respectively.*

Proof. Let $\mathbf{m} = (M_1, M_2, \ldots, M_k)$. As $k < n$, by the Pigeonhole Principle there exists $i \in \{1, 2, \ldots, k\}$ such that $|M_i| > 1$. Let $x \in M_i$ and consider

$$\mathbf{n} = (M_1, \ldots, M_{i-1}, \{x\}, M_i \backslash \{x\}, M_{i+1}, \ldots, M_k).$$

Obviously $\tau_{\mathbf{m}} \subset \tau_{\mathbf{n}}$ and $\mathbf{n} \in \mathfrak{O}_n^{(k+1)}$. Hence, by Proposition 8.1.7(ii) we have $N_{\tau_{\mathbf{m}}} \subset N_{\tau_{\mathbf{n}}}$ and $N'_{\tau_{\mathbf{m}}} \subset N'_{\tau_{\mathbf{n}}}$. As by Theorem 8.2.3(ii) each maximal element of $\mathrm{Nil}_k(\mathcal{PT}_n)$ and $\mathrm{Nil}_k(\mathcal{IS}_n)$ has, respectively, the form $N_{\tau_{\mathbf{m}}}$ and $N'_{\tau_{\mathbf{m}}}$ for some $\mathbf{m}$ as above, the claim follows. $\qquad\square$

Corollary 8.2.6   (i) *The maximal elements of $\mathrm{Nil}(\mathcal{PT}_n)$ and $\mathrm{Nil}_n(\mathcal{PT}_n)$ (respectively of $\mathrm{Nil}(\mathcal{IS}_n)$ and $\mathrm{Nil}_n(\mathcal{IS}_n)$) coincide.*

(ii) *Both $\mathcal{PT}_n$ and $\mathcal{IS}_n$ contain $n!$ maximal nilpotent subsemigroups (that is, maximal elements in $\mathrm{Nil}(\mathcal{PT}_n)$ and $\mathrm{Nil}(\mathcal{IS}_n)$, respectively).*

(iii) *If T_1 and T_2 are maximal nilpotent subsemigroups of $\mathcal{PT}_n$ (or $\mathcal{IS}_n$), then there exists $\varphi \in \mathrm{Inn}(\mathcal{PT}_n)$ (resp. $\mathrm{Inn}(\mathcal{IS}_n)$) such that $\varphi(T_1) = T_2$. In particular, $T_1 \cong T_2$.*

Proof. The statement (i) follows immediately from Corollaries 8.2.5 and 8.1.9.

As $|\mathbf{N}| = n$, any element $\mathbf{m} = (M_1, \ldots, M_n) \in \mathfrak{O}_n^{(n)}$ is just a permutation of the elements of $\mathbf{N}$. Conversely, each permutation defines a unique element of the form $\mathbf{m} = (M_1, \ldots, M_n)$. Hence $\mathfrak{O}_n^{(n)}$ contains exactly $n!$ maximal elements. This proves (ii).

To prove (iii) consider $\mathbf{m} = (\{a_1\}, \ldots, \{a_n\})$ and $\mathbf{n} = (\{b_1\}, \ldots, \{b_n\})$. Let further $\alpha \in \mathcal{S}_n$ such that $\alpha(a_i) = b_i$, $i = 1, \ldots, n$. Then a direct calculation shows that $\alpha^{-1} N_{\tau_{\mathbf{n}}} \alpha = N_{\tau_{\mathbf{m}}}$, which proves (iii) and thus completes the proof. $\qquad\square$

Remark 8.2.7 Strictly speaking a "maximal nilpotent subsemigroup" is a maximal element of $\mathrm{Nil}(S)$. For example, in formulations of theorems we will use only this meaning. However, in the general discussion (for example, in the title of this section), it is often convenient to use this phrase meaning maximal elements of some $\mathrm{Nil}_k(S)$.

Exercise 8.2.8 Prove that $|\mathfrak{O}_n^{(n)}| = n!$.

Exercise 8.2.9 Let $\mathbf{m} = (M_1, \ldots, M_k)$ and $\mathbf{n} = (M_1', \ldots, M_k')$ be such that $|M_i| = |M_i'|$ for all $i = 1, \ldots, k$. Show that $N_{\tau_{\mathbf{m}}} \cong N_{\tau_{\mathbf{n}}}$ and $N'_{\tau_{\mathbf{m}}} \cong N'_{\tau_{\mathbf{n}}}$.

Corollary 8.2.10 *The number of maximal elements in both* $\mathrm{Nil}_k(\mathcal{PT}_n)$ *and* $\mathrm{Nil}_k(\mathcal{IS}_n)$ *equals*

$$\sum_{i=0}^{k}(-1)^i\binom{k}{i}(k-i)^n = k!\cdot \mathrm{S}(n,k).$$

Proof. By Theorem 8.2.3(ii), the number of maximal elements in both sets $\mathrm{Nil}_k(\mathcal{PT}_n)$ and $\mathrm{Nil}_k(\mathcal{IS}_n)$ equals the number X of ordered partitions $\mathbf{N} = M_1\cup\cdots\cup M_k$ into a disjoint union of nonempty subsets. From the definition of $\mathrm{S}(n,k)$ we immediately get $X = k!\mathrm{S}(n,k)$. In contrast, each ordered partition $M_1\cup\cdots\cup M_k$ of $\mathbf{N}$ as above defines a surjective function $f:\mathbf{N}\to\{1,2,\ldots,k\}$ via $f(i)=t$ such that $i\in M_t$, $i=1,\ldots,n$. This correspondence is obviously bijective and hence the statement follows by a standard application of the inclusion–exclusion formula. $\qquad\square$

8.3 Cardinalities of Maximal Nilpotent, Subsemigroups

Let $\mathbf{m}=(M_1,\ldots,M_k)$ be as in the previous section. From Exercise 8.2.9 it follows that $|N_{\tau\mathbf{m}}|$ and $|N'_{\tau\mathbf{m}}|$ depend only on the cardinalities $m_i=|M_i|$, $i=1,\ldots,k$. Set also $m_0=1$. We fix this notation for this section.

Proposition 8.3.1 *(i)*

$$|N_{\tau\mathbf{m}}| = \prod_{i=1}^{k}(m_0+m_1+m_2+\cdots+m_{i-1})^{m_i}.$$

(ii) Each maximal nilpotent subsemigroup of $\mathcal{PT}_n$ *has cardinality* $n!$.

Proof. To construct some $\alpha\in N_{\tau\mathbf{m}}$ we have to define $\alpha(x)$ for each $x\in\mathbf{N}$. If $x\in M_i$, then the definition of $N_{\tau\mathbf{m}}$ says that either $x\notin\mathrm{dom}(\alpha)$ or $\alpha(x)\in M_1\cup M_2\cup\cdots\cup M_{i-1}$. Hence we have $m_0+m_1+m_2+\cdots+m_{i-1}$ independent choices for $\alpha(x)$ for each such x. The statement (i) is now obtained applying the product rule.

The statement (ii) is a special case of (i) as for a maximal nilpotent subsemigroup we have $k=n$ and $m_i=1$ for all i. $\qquad\square$

In the case of $\mathcal{IS}_n$ the analogous question is much more interesting and nontrivial. To answer it we will need some preparation. We fix the following notation:

$$\mathrm{X}(m_1,\ldots,m_k)=|N'_{\tau\mathbf{m}}|.$$

If $f(x)=a_mx^m+a_{m-1}x^{m-1}+\cdots+a_1x+a_0$ is a polynomial with integer coefficients, we set

$$f(\mathrm{B})=a_m\mathrm{B}_m+a_{m-1}\mathrm{B}_{m-1}+\cdots+a_1\mathrm{B}_1+a_0,$$

where B_i denotes the i-th Bell number.

Exercise 8.3.2 Let $f(x)$ and $g(x)$ be two polynomials with integer coefficients and $r \in \mathbb{Z}$. Show that $(f + g)(\mathrm{B}) = f(\mathrm{B}) + g(\mathrm{B})$ and $(rg)(\mathrm{B}) = r(g(\mathrm{B}))$.

Finally, for $m \in \mathbb{N}$ we set $[x]_m = x(x - 1)(x - 2)\ldots(x - m + 1)$ and define the polynomial $F_{m_1,\ldots,m_k}(x)$ as follows:

$$F_{m_1,\ldots,m_k}(x) = [x]_{m_1}[x]_{m_2}\cdots\cdots[x]_{m_k}.$$

In the case of $\mathcal{IS}_n$ the cardinalities of $N'_{\tau_{\mathbf{m}}}$ are given by the following quite amazing result:

Theorem 8.3.3 *(i)* $|N'_{\tau_{\mathbf{m}}}| = F_{m_1,\ldots,m_k}(\mathrm{B})$.

(ii) Each maximal nilpotent subsemigroup of $\mathcal{IS}_n$ has cardinality B_n.

Proof. We start with the statement (ii). By Corollary 8.2.6(i), the maximal nilpotent subsemigroups of $\mathcal{IS}_n$ correspond to the case $m_i = 1$ for all i. So, we may assume $\mathbf{m} = (\{a_1\},\ldots,\{a_n\})$. Under such assumptions let $\alpha \in N'_{\tau_{\mathbf{m}}}$. Then the connected components of Γ_α form an unordered partition of $\mathcal{N}$ into a disjoint union of subsets. Conversely, let $X_1,\ldots,X_s$ be an unordered partition of $\mathcal{N}$ into a disjoint union of subsets. For each $j = 1,\ldots,s$ let $|X_j| = k_j$ and, further, let us write the elements of X_j in the same order as they appear in the sequence $a_1,\ldots,a_n$. Let $x_1^j, x_2^j,\ldots,x_{k_j}^j$ be the resulting sequence. Now define the element α as follows:

$$\mathrm{dom}(\alpha) = \bigcup_{j=1}^{s}\{x_2^j, x_3^j,\ldots,x_{k_j}^j\},$$

and $\alpha(x_t^j) = x_{t-1}^j$ for all $j = 1,\ldots,s$ and $t = 2,\ldots,k_j$. In other words, we define α as the element with the following graph Γ_α:

$$x_1^1 \xleftarrow{\ \alpha\ } x_2^1 \xleftarrow{\ \alpha\ } \cdots \xleftarrow{\ \alpha\ } x_{k_1}^1$$

$$x_1^2 \xleftarrow{\ \alpha\ } x_2^2 \xleftarrow{\ \alpha\ } \cdots \xleftarrow{\ \alpha\ } x_{k_2}^2$$

$$\vdots \qquad \vdots \qquad \vdots \qquad \vdots$$

$$x_1^s \xleftarrow{\ \alpha\ } x_2^s \xleftarrow{\ \alpha\ } \cdots \xleftarrow{\ \alpha\ } x_{k_s}^s$$

From the definition of $N'_{\tau_{\mathbf{m}}}$ we have $\alpha \in N'_{\tau_{\mathbf{m}}}$. Moreover, the connected components of Γ_α are exactly $X_1,\ldots,X_s$. It is easy to see that α is the only element of $N'_{\tau_{\mathbf{m}}}$ such that the connected components of Γ_α are exactly

$X_1, \ldots, X_s$. Hence we obtain a bijection between the elements of $N'_{\mathcal{T}\mathbf{m}}$ and partitions of $\mathbf{N}$ into unordered disjoint unions of nonempty subsets. This means that $|N'_{\mathcal{T}\mathbf{m}}| = \mathsf{B}_n$ in this case, proving (ii).

To prove the statement (i) we will need one auxiliary lemma.

Lemma 8.3.4

$$X(m_1, \ldots, m_{i-2}, m_{i-1}, 1, m_{i+1}, \ldots, m_k) =$$
$$= X(m_1, \ldots, m_{i-2}, m_{i-1}, m_{i+1} + 1, \ldots, m_k) +$$
$$+ m_{i+1} \cdot X(m_1, \ldots, m_{i-2}, m_{i-1}, m_{i+1}, \ldots, m_k).$$

Proof. Assume that $|M_i| = 1$, or in other words that

$$\mathbf{m} = (M_1, \ldots, M_{i-1}, \{a\}, M_{i+1}, \ldots, M_k).$$

We decompose $N'_{\mathcal{T}\mathbf{m}}$ into a disjoint union of two sets T_1 and T_2 as follows:

$$T_1 = \{\alpha \in N'_{\mathcal{T}\mathbf{m}} : a \notin \alpha(M_{i+1})\}, \quad T_2 = \{\alpha \in N'_{\mathcal{T}\mathbf{m}} : a \in \alpha(M_{i+1})\}.$$

By definition, the set T_1 coincides with $N'_{\mathcal{T}\mathbf{n}}$, where

$$\mathbf{n} = (M_1, \ldots, M_{i-1}, M_{i+1} \cup \{a\}, M_{i+2}, \ldots, M_k).$$

Let $M_{i+1} = \{a_1, \ldots, a_{m_{i+1}}\}$. For $j \in \{1, 2, \ldots, m_{i+1}\}$ define

$$T_2^j = \{\alpha \in T_2 : \alpha(a_j) = a\}.$$

Then $T_2 = T_2^1 \cup \cdots \cup T_2^{m_{i+1}}$ is a partition of T_2 into a disjoint union of nonempty subsets. Fix now $j \in \{1, 2, \ldots, m_{i+1}\}$. With every $\alpha \in T_2^j$, we associate the element $\alpha' \in N'_{\mathcal{T}\mathbf{k}}$, where $\mathbf{k} = (M_1, \ldots, M_{i-1}, M_{i+1}, \ldots, M_k)$, in the following way:

$$\alpha'(y) = \begin{cases} \alpha(y), & y \neq a_j; \\ \alpha(a), & y = a_j. \end{cases}$$

Obviously, the mapping $\alpha \mapsto \alpha'$ from T_2^j to $N'_{\mathcal{T}\mathbf{k}}$ is bijective. Hence $|T_2^j| = |N'_{\mathcal{T}\mathbf{k}}|$. This gives

$$X(m_1, \ldots, m_{i-1}, 1, m_{i+1}, m_{i+2}, \ldots, m_k) =$$
$$= |N'_{\mathcal{T}\mathbf{m}}| = |T_1| + |T_2| = |N'_{\mathcal{T}\mathbf{n}}| + m_{i+1}|N'_{\mathcal{T}\mathbf{k}}| =$$
$$= X(m_1, \ldots, m_{i-1}, m_{i+1} + 1, m_{i+2}, \ldots, m_k) +$$
$$+ m_{i+1} \cdot X(m_1, \ldots, m_{i-1}, m_{i+1}, m_{i+2}, \ldots, m_k)$$

and the claim follows.   □

We prove the statement (i) by induction on the sum m' of all m_i's such that $m_i > 1$. If $m' = 0$, then $m_i = 1$ for all i. In particular, $F_{1,\dots,1}(x) = x^n$ and $F_{1,\dots,1}(\mathsf{B}) = \mathsf{B}_n = |N'_{\tau \mathbf{m}}|$ according to the definitions and the statement (ii), proved above.

Assume now that $m' > 0$. Choose some i such that $m_i \geq 2$. Applying Lemma 8.3.4 and the inductive assumption we get

$$
\begin{aligned}
\mathsf{X}(m_1, \dots, m_{i-1}, m_i, \dots, m_k) = \\
= \mathsf{X}(m_1, \dots, m_{i-1}, 1, m_i - 1, m_{i+1}, \dots, m_k) - \\
- (m_i - 1) \cdot \mathsf{X}(m_1, \dots, m_{i-1}, m_i - 1, m_{i+1}, \dots, m_k) = \\
= F_{m_1, \dots, m_{i-1}, 1, m_i - 1, m_{i+1}, \dots, m_k}(\mathsf{B}) - \\
- (m_i - 1) \cdot F_{m_1, \dots, m_{i-1}, m_i - 1, m_{i+1}, \dots, m_k}(\mathsf{B}). \quad (8.2)
\end{aligned}
$$

Now we observe that

$$
[x]_m = (x - (m-1))[x]_{m-1} = [x]_{m-1}[x]_1 - (m-1)[x]_{m-1}.
$$

From this and the definition of the polynomials $F_{m_1, \dots, m_k}(x)$ we get

$$
\begin{aligned}
F_{m_1, \dots, m_{i-1}, m_i, \dots, m_k}(x) = F_{m_1, \dots, m_{i-1}, 1, m_i - 1, m_{i+1}, \dots, m_k}(x) - \\
- (m_i - 1) \cdot F_{m_1, \dots, m_{i-1}, m_i - 1, m_{i+1}, \dots, m_k}(x). \quad (8.3)
\end{aligned}
$$

Applying (8.3) to (8.2) and using Exercise 8.3.2 we get

$$
\mathsf{X}(m_1, \dots, m_{i-1}, m_i, \dots, m_k) = F_{m_1, \dots, m_{i-1}, m_i, \dots, m_k}(\mathsf{B}),
$$

completing the proof.   □

8.4 Combinatorics of Nilpotent Elements in $\mathcal{IS}_n$

In this section, we discover some combinatorial relations between different numbers, associated with the semigroup $\mathcal{IS}_n$, especially with nilpotent elements of $\mathcal{IS}_n$. Define the following numbers:

I_n = the cardinality of $\mathcal{IS}_n$
N_n = the total number of nilpotent elements in $\mathcal{IS}_n$
L_n = the total number of chains in the chain decompositions of all elements of $\mathcal{IS}_n$
M_n = the total number of chains in the chain decompositions of all nilpotent elements of $\mathcal{IS}_n$,
P_n = the total number of fixed points of all elements of $\mathcal{IS}_n$

Example 8.4.1 The semigroup $\mathcal{IS}_1$ has 2 elements:

$$\begin{pmatrix} 1 \\ 1 \end{pmatrix}, \quad \begin{pmatrix} 1 \\ \varnothing \end{pmatrix}.$$

For this semigroup, we have

$$I_1 = 2, \quad N_1 = 1, \quad L_1 = 1, \quad M_1 = 1, \quad P_1 = 1.$$

Example 8.4.2 The semigroup $\mathcal{IS}_2$ has 7 elements:

$$\begin{pmatrix} 1 & 2 \\ 1 & 2 \end{pmatrix}, \quad \begin{pmatrix} 1 & 2 \\ 2 & 1 \end{pmatrix}, \quad \begin{pmatrix} 1 & 2 \\ 1 & \varnothing \end{pmatrix}, \quad \begin{pmatrix} 1 & 2 \\ 2 & \varnothing \end{pmatrix},$$

$$\begin{pmatrix} 1 & 2 \\ \varnothing & 1 \end{pmatrix}, \quad \begin{pmatrix} 1 & 2 \\ \varnothing & 2 \end{pmatrix}, \quad \begin{pmatrix} 1 & 2 \\ \varnothing & \varnothing \end{pmatrix}.$$

For this semigroup, we have

$$I_2 = 7, \quad N_2 = 3, \quad L_2 = 6, \quad M_2 = 4, \quad P_2 = 4.$$

Now we will present some combinatorial relations involving I_n, N_n, L_n, M_n, and P_n.

Proposition 8.4.3
$$L_n = \sum_{\alpha \in \mathcal{IS}_n} \mathrm{strank}(\alpha).$$

Proof. Consider the sets

$$A = \{(\alpha, c, x) : \alpha \in \mathcal{IS}_n, \ c \text{ is a cycle of } \alpha, \ x \text{ is a point of } c\},$$
$$B = \{(\beta, l) : \beta \in \mathcal{IS}_n, \ l \text{ is a chain from the chain decomposition of } \beta\}.$$

Obviously, our statement is equivalent to the equality $|A| = |B|$.
Consider the mapping $f : A \to B$, defined as follows:

$$f((\alpha, (x, a, \ldots, b), x)) = (\beta, [x, a, \ldots, b]),$$

where β is obtained from α substituting the cycle $(x, a, \ldots, b)$ with the chain $[x, a, \ldots, b]$. Consider also the mapping $g : B \to A$, which is defined as follows: $g((\beta, [x, a, \ldots, b])) = (\alpha, (x, a, \ldots, b), x)$, where α is obtained from β substituting the chain $[x, a, \ldots, b]$ with the cycle $(x, a, \ldots, b)$. Obviously f and g are inverse to each other and thus $|A| = |B|$. $\square$

Theorem 8.4.4
$$P_n + \frac{1}{n}L_n = I_n. \tag{8.4}$$

Proof. Rewrite the equality (8.4) in the following form:

$$n\mathsf{P}_n + \mathsf{L}_n = n\mathsf{I}_n. \tag{8.5}$$

Consider the following sets:

$$A = \{(\alpha, x) \ : \ \alpha \in \mathcal{IS}_n, x \in \mathbf{N}\},$$
$$B = \{(\beta, l) \ : \ \beta \in \mathcal{IS}_n, l \text{ is a chain for the chain decomposition of } \beta\},$$
$$C = \{(\gamma, y, z) \ : \ \gamma \in \mathcal{IS}_n, y \text{ is a fixed point of } \gamma, z \in \mathbf{N}\}.$$

The equality (8.5) is equivalent to the equality $|C| + |B| = |A|$. To prove the latter equality, we decompose A into a disjoint union $A = A_1 \cup A_2$, where

$$A_1 = \{(\alpha, x) \in A \ : \ x \text{ belongs to some chain of } \alpha\},$$
$$A_2 = \{(\alpha, x) \in A \ : \ x \text{ belongs to some cycle of } \alpha\}.$$

Consider the transformation which maps the cycle $(x, a, \ldots, b)$ to the chain $[x, a, \ldots, b]$. As in the proof of Proposition 8.4.3 this transformation induces a bijection $A_2 \to B$. Hence $|A_2| = |B|$.

To prove $|A_1| = |C|$, we construct mutually inverse bijections $f : A_1 \to C$ and $g : C \to A_1$. Consider any element $(\alpha, x) \in A_1$. If x is the starting point of some chain $[x, a, \ldots, b]$ of length at least 2 from the chain decomposition of α, we define $f((\alpha, x)) = (\gamma, x, a)$, where γ is obtained from α substituting the chain $[x, a, \ldots, b]$ with the cycle (x) and the cycle $(a, \ldots, b)$. If x is the only point of the chain $[x]$, we define $f((\alpha, x)) = (\gamma, x, x)$, where γ is obtained from α substituting the chain $[x]$ with the cycle (x). Finally, if x is contained in some chain $[a, \ldots, b, x, c, \ldots, d]$ and is different from the starting point a of this chain, we define $f((\alpha, x)) = (\gamma, x, b)$, where γ is obtained from α substituting the chain $[a, \ldots, b, x, c, \ldots, d]$ with the cycle (x) and the chain $[a, \ldots, b, c, \ldots, d]$.

Let now $(\gamma, y, z) \in C$. If $y = z$, we define $g((\gamma, y, z)) = (\alpha, y)$, where α is obtained from γ substituting the cycle (y) with the chain $[y]$. If z is a point of some chain $[a_1, \ldots, a_s, z, b_1, \ldots, b_t]$ in the chain decomposition of γ, we set $g((\gamma, y, z)) = (\alpha, y)$, where α is obtained from γ substituting the cycle (y) and the chain $[a_1, \ldots, a_s, z, b_1, \ldots, b_t]$ with the chain $[a_1, \ldots, a_s, z, y, b_1, \ldots, b_t]$. Finally, if z is a point of some cycle $(z, a_1, \ldots, a_s)$ of γ, we set $g((\gamma, y, z)) = (\alpha, y)$, where α is obtained from γ substituting the cycles (y) and $(z, a_1, \ldots, a_s)$ with the chain $[y, z, a_1, \ldots, a_s]$.

Obviously, f and g are inverse to each other implying $|A_1| = |C|$, and the statement follows. $\square$

Let $\mathcal{N}_n$ denote the set of all nilpotent elements of $\mathcal{IS}_n$. We have $|\mathcal{N}_n| = \mathsf{N}_n$.

Proposition 8.4.5

$$\mathsf{I}_n = \mathsf{N}_n + \mathsf{P}_n.$$

Proof. The element $\alpha \in \mathcal{IS}_n$ can have some fixed points only in the case when α is not nilpotent. For every $\alpha \notin \mathcal{N}_n$ let $A_\alpha = \text{stim}(\alpha)$ and $\overline{A_\alpha} = N \backslash A_\alpha$. Consider the set

$$M_\alpha = \{\beta \in \mathcal{IS}_n : \text{stim}(\beta) = A_\alpha \text{ and } \alpha|_{\overline{A_\alpha}} = \beta|_{\overline{A_\alpha}}\}.$$

Obviously, $|M_\alpha| = \text{strank}(\alpha)!$ and either $M_\alpha = M_\beta$ or $M_\alpha \cap M_\beta = \varnothing$ for arbitrary $\alpha, \beta \in \mathcal{IS}_n$. Hence the sets M_α form a partition of $\mathcal{IS}_n \backslash \mathcal{N}_n$ into a disjoint union of subsets.

Lemma 8.4.6 *The total number of fixed points of all elements of $\mathcal{S}_n$ equals $n!$.*

Proof. Consider the following sets:

$$A = \{(\alpha, x, y) : \alpha \in \mathcal{S}_n, x, y \in \mathbf{N}, \alpha(x) = x\},$$
$$B = \{(\beta, z) : \beta \in \mathcal{S}_n, z \in \mathbf{N}\}.$$

The statement of lemma is obviously equivalent to the equality $|A| = |B|$.

Define the function $f : A \to B$ as follows: $f((\alpha, x, y)) = (\beta, y)$, where $\beta = \alpha$ if $x = y$, and β is obtained from α substituting the cycles (x) and $(y, a_1, \ldots, a_k)$ with the cycle $(x, y, a_1, \ldots, a_k)$ if $x \neq y$. Further define the function $g : B \to A$ as follows: $g((\beta, z)) = (\alpha, x, z)$, where $\alpha = \beta$ and $x = z$ if $\beta(z) = z$, and α is obtained from β substituting the cycle $(x, z, a_1, \ldots, a_k)$ with the cycles (x) and $(z, a_1, \ldots, a_k)$ in the case $\beta(z) \neq z$. Obviously f and g are mutually inverse bijections and hence $|A| = |B|$. $\square$

By Lemma 8.4.6 the total number of fixed points for all elements in M_α equals $\text{strank}(\alpha)!$ and hence equals the cardinality of M_α. Hence the total number P_n of all fixed points for all elements of $\mathcal{IS}_n$ equals the total number of those elements of $\mathcal{IS}_n$ which are not nilpotent. The claim follows. $\square$

Theorem 8.4.7 *(i)* $\mathsf{N}_n = \frac{1}{n}\mathsf{L}_n$.

(ii) $\mathsf{I}_n = \frac{1}{n+1}\mathsf{M}_{n+1}$.

Proof. Using Theorem 8.4.4 and Proposition 8.4.5 we have

$$\frac{1}{n}\mathsf{L}_n = \mathsf{I}_n - \mathsf{P}_n = \mathsf{I}_n - (\mathsf{I}_n - \mathsf{N}_n) = \mathsf{N}_n.$$

This proves (i).

Consider the following sets:

$$A = \{(x, \alpha) : x \in \{1, 2, \ldots, n+1\}, \alpha \in \mathcal{IS}(\{1, 2, \ldots, n+1\}\backslash\{x\})\}$$
$$B = \{(\beta, l) : \beta \in \mathcal{N}_{n+1}, l \text{ is a chain of } \beta\}$$

To prove (ii) it is enough to show that $|A| = |B|$.

Define the mapping $f : A \to B$ in the following way. Let $(x, \alpha) \in A$ and assume that

$$
\begin{pmatrix}
a_1 & \cdots & a_k \\
a_{i_1} & \cdots & a_{i_k}
\end{pmatrix}
$$

is the permutational part of α, where $a_1 < a_2 < \cdots < a_k$. Set $f((x, \alpha)) = (\beta, l) \in B$, where $l = [a_{i_1}, \ldots, a_{i_k}, x]$ and β is obtained from α substituting the permutational part with l.

Define the mapping $g : B \to A$, $g : (\beta, l) \mapsto (x, \alpha)$ in the following way: if $l = [a_1, \ldots, a_k, a_{k+1}]$, we set $x = a_{k+1}$ and α is obtained from β substituting l with the permutational part

$$
\begin{pmatrix}
a_{i_1} & \cdots & a_{i_k} \\
a_1 & \cdots & a_k
\end{pmatrix},
$$

where $a_{i_1} < a_{i_2} < \cdots < a_{i_k}$ are elements $a_1, \ldots, a_k$, written with respect to the natural increasing order. Obviously, f and g are mutually inverse bijections, which implies $|A| = |B|$ and completes the proof. $\qquad\square$

Theorem 8.4.8   *(i)* $N_n = I_{n-1} + L_{n-1}$.

(ii) $I_n = N_n + M_n$.

Proof. Consider the following sets:

$$
A = \{\alpha \in \mathcal{N}_n : n \notin \mathrm{im}(\alpha)\},
$$
$$
B = \{\alpha \in \mathcal{N}_n : n \in \mathrm{im}(\alpha)\},
$$
$$
C = \{(\beta, l) : \beta \in \mathcal{IS}_{n-1}, l \text{ is a chain of } \beta\}.
$$

We have $A \cup B = \mathcal{N}_n$, $A \cap B = \varnothing$, $|C| = L_{n-1}$. Hence the equality (i) would follow if we showed that $|A| = I_{n-1}$ and $|B| = |C|$.

Lemma 8.4.9 $|A| = I_{n-1}$.

Proof. This proof is rather similar to that of Theorem 8.4.7(ii). Define the function $f : A \to \mathcal{IS}_{n-1}$ in the following way: Let $\alpha \in A$ and $[n, a_1, \ldots, a_k]$ be the chain starting at n (it exists since $n \notin \mathrm{im}(\alpha)$). We set $f(\alpha) = \beta$, where $\beta(y) = \alpha(y)$ for all $y \notin \{a_1, \ldots, a_k\}$ and on $\{a_1, \ldots, a_k\}$ the element β is defined to be the following permutation:

$$
\begin{pmatrix}
a_{i_1} & \cdots & a_{i_k} \\
a_1 & \cdots & a_k
\end{pmatrix},
$$

where $a_{i_1} < a_{i_2} < \cdots < a_{i_k}$ are elements $a_1, \ldots, a_k$, written with respect to the natural increasing order.

Define further the function $g : \mathcal{IS}_{n-1} \to A$ as follows: For $\beta \in \mathcal{IS}_{n-1}$ let $\mathrm{stim}(\beta) = \{a_1, \ldots, a_k\}$, where $a_1 < a_2 < \cdots < a_k$. Assume further that on $\mathrm{stim}(\beta)$ the element β acts as the following permutation:

$$\begin{pmatrix} a_1 & \cdots & a_k \\ a_{i_1} & \cdots & a_{i_k} \end{pmatrix}.$$

We set $g(\beta) = \alpha$, where

$$\alpha(y) = \begin{cases} \beta(y), & y \notin \mathrm{stim}(\beta); \\ a_{i_1}, & y = n; \\ a_{i_{j+1}}, & y = a_{i_j} \text{ and } j < k; \\ \varnothing, & y = a_{i_k}. \end{cases}$$

Obviously, f and g are mutually inverse bijections, and hence $|A| = \mathbf{I}_{n-1}$. $\square$

Lemma 8.4.10 $|B| = |C|$.

Proof. Define the function $f : B \to C$ in the following way: Let $\alpha \in B$ and $[a_1, \ldots, a_k, n, b_1, \ldots, b_m]$ be the chain containing n (note that $k \geq 1$ since $n \in \mathrm{im}(\alpha)$). We set $f(\alpha) = (\beta, l)$, where β is obtained from α substituting the chain $[a_1, \ldots, a_k, n, b_1, \ldots, b_m]$ by the chain $[a_1, \ldots, a_k]$ and the permutation:

$$\begin{pmatrix} b_{i_1} & \cdots & b_{i_m} \\ b_1 & \cdots & b_m \end{pmatrix},$$

where $b_{i_1} < b_{i_2} < \cdots < b_{i_m}$ are elements $b_1, \ldots, b_m$, written with respect to the natural increasing order. Further, we set $l = [a_1, \ldots, a_k]$.

Define the function $g : C \to B$ as follows: For $(\beta, l) \in C$ let $l = [a_1, \ldots, a_k]$, $\mathrm{stim}(\beta) = \{b_1, \ldots, b_m\}$, where $b_1 < b_2 < \cdots < b_m$ and assume that on $\mathrm{stim}(\beta)$ the element β acts as the following permutation:

$$\begin{pmatrix} b_1 & \cdots & b_m \\ b_{i_1} & \cdots & b_{i_m} \end{pmatrix}.$$

We set $g(\beta, l) = \alpha$, where α is obtained from β substituting l and the above permutation by the chain $[a_1, \ldots, a_k, n, b_{i_1}, \ldots, b_{i_m}]$. Obviously, f and g are mutually inverse bijections, and hence $|B| = |C|$. $\square$

The statement (i) now follows from Lemmas 8.4.9 and 8.4.10.

Using Theorem 8.4.7(ii) and Proposition 8.4.5 we can rewrite the equality of (ii) as follows:

$$\mathbf{P}_n = n\mathbf{I}_{n-1}. \tag{8.6}$$

To prove (8.6) consider the following two sets:

$$D = \{(\alpha, x) \; : \; \alpha \in \mathcal{IS}_n, \; x \in \mathbf{N}, \; \alpha(x) = x\},$$
$$E = \{(\beta, y) \; : \; y \in \mathbf{N}, \; \beta \in \mathcal{IS}(\mathbf{N}\backslash\{y\})\}.$$

The equality (8.6) is equivalent to the equality $|D| = |E|$.

Define the mapping $f : D \to E$ in the following way: $f((\alpha, x)) = (\beta, x)$, where β is the restriction of α to $\mathbf{N}\backslash\{x\}$. Define the mapping $g : E \to D$ in the following way: $g((\beta, y)) = (\alpha, y)$, where

$$\alpha(z) = \begin{cases} \beta(z), & z \neq y; \\ y, & z = y. \end{cases}$$

Obviously, f and g are mutually inverse bijections and hence $|D| = |E|$. This proves (8.6) and completes the proof. $\qquad\square$

Corollary 8.4.11 $\mathsf{P}_n = \mathsf{M}_n$.

Proof. Follows from Proposition 8.4.5 and Theorem 8.4.8(ii). $\qquad\square$

Theorem 8.4.12 $\mathsf{N}_n = |\{\alpha \in \mathcal{IS}_n \; : \; 1 \notin \mathrm{dom}(\alpha)\}|$.

Proof. Set $A = \{\alpha \in \mathcal{IS}_n \; : \; 1 \notin \mathrm{dom}(\alpha)\}$ and consider the following decomposition of A into a disjoint union of subsets:

$$A = \bigcup_{X \subset \{2,3,\dots,n\}} A_X,$$

where $A_X = \{\alpha \in A \; : \; \mathrm{stim}(\alpha) = X\}$.

Consider also the following decomposition of $\mathcal{N}_n$ into a disjoint union of subsets:

$$\mathcal{N}_n = \bigcup_{X \subset \{2,3,\dots,n\}} B_X,$$

where

$$B_X = \{\beta \in \mathcal{N}_n \; : \; \beta \text{ contains a chain } [b_1, \dots, b_l, 1, a_1, \dots, a_k]$$
$$\text{and } \{a_1, \dots, a_k\} = X\}.$$

For a fixed $X = \{a_1, \dots, a_k\}$, $a_1 < a_2 < \cdots < a_k$, define the mapping $f : A_X \to B_X$ as follows: $f(\alpha) = \beta$, where β is obtained from α substituting the chain $[b_1, \dots, b_l, 1]$ of α (which exists as $1 \notin \mathrm{dom}(\alpha)$) and the permutational part

$$\begin{pmatrix} a_1 & \cdots & a_k \\ a_{i_1} & \cdots & a_{i_k} \end{pmatrix} \tag{8.7}$$

of α by the chain $[b_1, \dots, b_l, 1, a_{i_1}, \dots, a_{i_k}]$.

Define the mapping $g : B_X \to A_X$ as follows: $g(\beta) = \alpha$, where α is obtained from β substituting the chain $[b_1, \ldots, b_l, 1, a_{i_1}, \ldots, a_{i_k}]$ by the chain $[b_1, \ldots, b_l, 1]$ and the permutational part given by (8.7). The mappings f and g are obviously mutually inverse bijections and hence $|A_X| = |B_X|$ for all X. The statement of our theorem follows.　　　　　　　　　　　　　　　　$\square$

8.5　Addenda and Comments

8.5.1 In the context of ring theory, nilpotent semigroups were studied already in [Ko, Lev]. The oldest purely semigroup theoretical paper, dedicated to the study of nilpotent semigroups, which we managed to find, is [She].

8.5.2 An alternative notion of nilpotent semigroups was proposed in [Ma2]. This notion generalizes the notion of nilpotent groups. Finite nilpotent semigroups in our sense are sometimes referred to as *nil-semigroups*.

8.5.3 From the results of [KRS] it follows that almost all semigroups are nilpotent in the following sense: Let A_n denote the total number of semigroup structures on $\mathbf{N}$ and B_n denote the total number of nilpotent semigroup structures on $\mathbf{N}$, then $B_n/A_n \to 1$, $n \to \infty$. The proof of the main statement of [KRS] (which implies the above claim) is only outlined and no complete argument was ever published, so some specialists put this proof in question. Nevertheless, the experimental data obtained so far strongly suggest that the above statement is true. Thus in [SYT] it is shown that 99% of almost 2×10^9 semigroups consisting of 8 elements are nilpotent.

8.5.4 The idea to use partial orders for description of maximal nilpotent subsemigroups appears first in the papers [GK2, GK3, GK4]. In particular, the results of Sect. 8.1 and 8.2 are essentially proved in these papers. This idea has proved to be very useful and applicable to many different situations, see for example [GM1, GM2, KuMa3, Sh1, Sh2, St1, Ts4]. Furthermore, this idea led to the development of a very general and abstract approach to the study of nilpotent subsemigroups proposed in [GM6].

8.5.5 Theorem 8.3.3 is due to M. Pavlov. It appeared in [GP] for the first time, and can also be found in [GM4]. In [GM2] analogous ideas were successfully applied to determine the cardinalities of some maximal nilpotent subsemigroups of the semigroup $\mathcal{IO}_n$ of all partial order-preserving injections on the chain $1 < 2 < \cdots < n$.

8.5.6 The results of Sect. 8.4 are obtained in [GM5]. Some of these results were reproved in [JM] using generating functions.

8.5.7 Let $A \subset \mathbf{N}$ and $T \subset \mathcal{IS}_n$ be a nilpotent semigroup with the zero element ε_A.

Lemma 8.5.1 $\alpha(x) = x$ *for all* $\alpha \in T$ *and* $x \in A$.

Proof. Since ε_A is the zero element in T and $\varepsilon_A(x) = x$ for all $x \in A$, we have

$$\alpha(x) = \alpha(\varepsilon_A(x)) = (\alpha\varepsilon_A)(x) = \varepsilon_A(x) = x.$$

$\square$

Lemma 8.5.2 *Let* $\alpha \in T$ *and* $x \in \mathrm{dom}(\alpha) \cap (\mathbf{N} \backslash A)$. *Then* $\alpha(x) \in \mathbf{N} \backslash A$.

Proof. Assume that $\alpha(x) = y \in A$. Then $\alpha(x) = \alpha(y)$ by Lemma 8.5.1, which contradicts the fact that α is a partial injection. The claim follows. $\square$

By Lemma 8.5.2 we can consider the restriction map

$$T \quad \rightarrow \quad \mathcal{IS}(\mathbf{N} \backslash A)$$
$$\alpha \quad \mapsto \quad \alpha|_{\mathbf{N} \backslash A}.$$

By Lemma 8.5.1 this mapping is even injective. Since $\varepsilon_A|_{\mathbf{N} \backslash A}$ is the zero element of $\mathcal{IS}(\mathbf{N} \backslash A)$, the image of T under this restriction is a nilpotent subsemigroup of $\mathcal{IS}(\mathbf{N} \backslash A)$, now with the usual zero element. This reduces the study of all nilpotent subsemigroups of $\mathcal{IS}_n$ to the study of nilpotent subsemigroups with the usual zero element.

8.5.8 Define the mapping $\mathrm{i}_n : \mathcal{PT}_n \rightarrow \mathcal{T}_{n+1}$ in the following way: for $\alpha \in \mathcal{PT}_n$ and $x \in \{1, 2, \dots, n+1\}$ set

$$\mathrm{i}_n(\alpha)(x) = \begin{cases} \alpha(x), & x \in \mathrm{dom}(\alpha); \\ n+1, & \text{otherwise.} \end{cases}$$

A direct calculation shows that i_n is a homomorphism. Furthermore, i_n is obviously injective. The mapping i_n is called the *canonical inclusion of* $\mathcal{PT}_n$ to $\mathcal{T}_{n+1}$.

Proposition 8.5.3 *(i) Let* $T \subset \mathcal{PT}_n$ *be a nilpotent subsemigroup with the usual zero. Then* $\mathrm{i}_n(T)$ *is a nilpotent subsemigroup of* $\mathcal{T}_{n+1}$ *with the zero element* 0_{n+1}.

(ii) Let $T \subset \mathcal{T}_{n+1}$ *be a nilpotent subsemigroup with the zero element* 0_{n+1}. *Then* $T = \mathrm{i}_n(S)$ *for some nilpotent subsemigroup* $S \subset \mathcal{PT}_n$ *with the usual zero.*

Proof. The statement (i) follows from the facts that i_n is a homomorphism and $\mathrm{i}_n(\mathbf{0}) = 0_{n+1}$.

To prove (ii) consider any nilpotent subsemigroup $T \subset \mathcal{T}_{n+1}$ with the zero element 0_{n+1}. For each $\alpha \in T$ we have

$$\alpha(n+1) = \alpha(0_{n+1}(n+1)) = (\alpha 0_{n+1})(n+1) = 0_{n+1}(n+1) = n+1.$$

In particular, for $\alpha, \beta \in T$ we have $\alpha \neq \beta$ if and only if there exists $x \in \{1, 2, \ldots, n\}$ such that $\alpha(x) \neq \beta(x)$. Let S be the set of all $\gamma \in \mathcal{PT}_n$ such that $i_n(\gamma) \in T$. Then the previous argument implies that i_n induces a bijection between S and T. A direct calculation shows that S is in fact a subsemigroup of $\mathcal{PT}_n$. This completes the proof. $\qquad\square$

Proposition 8.5.3 reduces the study of nilpotent subsemigroups of $\mathcal{T}_{n+1}$ with the zero element 0_{n+1} to the study of nilpotent semigroups of $\mathcal{PT}_n$ with the usual zero. Since all left zeros of $\mathcal{T}_{n+1}$ can be mapped to 0_{n+1} using some inner automorphisms, we can reduce the study of all nilpotent subsemigroups of $\mathcal{T}_{n+1}$ with zero elements of the form 0_x, $x \in \{1, 2, \ldots, n+1\}$, to the study of nilpotent semigroups of $\mathcal{PT}_n$ with the usual zero.

8.5.9 Let now $\epsilon \in \mathcal{PT}_n$ be any idempotent and $T \subset \mathcal{PT}_n$ be any nilpotent subsemigroup of $\mathcal{PT}_n$ with the zero element ϵ. Note that if $\epsilon \in \mathcal{T}_n$, then from $\alpha^{(n+1)^n} = \epsilon$ for all $\alpha \in T$ it follows that $\mathrm{dom}(\alpha) = \mathbf{N}$ for all $\alpha \in T$, in particular, T is a nilpotent subsemigroup of $\mathcal{T}_n$. So the study of all nilpotent subsemigroups of $\mathcal{T}_n$ reduces to the study of all nilpotent subsemigroups of $\mathcal{PT}_n$.

Let $A = \mathrm{dom}(\epsilon)$, $\{a_1, \ldots, a_k\} = \mathrm{im}(\epsilon)$ and $A = A_1 \cup \cdots \cup A_k$ be the partition such that $\epsilon(x) = a_i$ for all $x \in A_i$.

Lemma 8.5.4 *Each $\alpha \in T$ has the following properties:*

(a) $\alpha(x) \in A_i$ for all i and all $x \in A_i$.

(b) $\alpha(a_i) = a_i$ for all i.

(c) $\alpha(x) \in \mathbf{N} \backslash A$ for all $x \in (\mathbf{N} \backslash A) \cap \mathrm{dom}(\alpha)$.

Proof. Let $x \in A_i$. Then for $\alpha(x)$ we have $\epsilon(\alpha(x)) = (\epsilon \alpha)(x) = \epsilon(x) = a_i$. Hence $\alpha(x) \in A_i$, which proves (a). The statement (c) is proved analogously.

For any $i = 1, \ldots, k$ we have $\alpha(a_i) = \alpha(\epsilon(a_i)) = (\alpha \epsilon)(a_i) = \epsilon(a_i) = a_i$. This proves (b) and completes the proof. $\qquad\square$

By Lemma 8.5.4 we can restrict each $\alpha \in T$ to each of the sets $A_1, \ldots, A_k$, $\mathbf{N} \backslash A$. These restrictions in fact define homomorphisms from T to $\mathcal{T}(A_i)$, $i = 1, \ldots, k$, and $\mathcal{PT}(\mathbf{N} \backslash A)$. If $\alpha, \beta \in T$ and $\alpha \neq \beta$, then the images of α and β with respect to at least one of these homomorphisms are different. The images of these homomorphisms are nilpotent subsemigroups of $\mathcal{T}(A_i)$, whose zero elements are some left zeros of $\mathcal{T}(A_i)$, and nilpotent subsemigroups of $\mathcal{PT}(\mathbf{N} \backslash A)$ with the usual zero. This reduces the study of nilpotent subsemigroups of $\mathcal{PT}_n$ with the zero element ϵ to the study of nilpotent subsemigroups of $\mathcal{PT}_n$ with the usual zero element and the study of nilpotent subsemigroups of $\mathcal{T}_n$, whose zero is a left zero of $\mathcal{T}_n$ (see 8.5.8). It might be a good exercise for the reader to write down all necessary details.

8.5.10 Let M be a finite set and G be a subgroup of $\mathcal{S}(M)$. Define the binary relation $\sim$ on M as follows: $x \sim y$ if and only if there exists $g \in G$ such that $x = g(y)$. It is easy to check that $\sim$ is an equivalence relation. The equivalence classes of this relation are called the *G-orbits* of M. Let $\mathfrak{O}_G$ denote the number of G-orbits. For $g \in G$ let $\mathrm{fi}(g)$ denote the number of fixed points of g. In the case $G = \mathcal{S}(M)$ the number of G-orbits on M is obviously 1. Hence Lemma 8.4.6 is a special case of the following famous statement, known as *Burnside's Lemma*, the *Cauchy-Frobenius Lemma* or the *Orbit-counting Theorem*:

Theorem 8.5.5
$$\mathfrak{O}_G = \frac{1}{|G|} \sum_{g \in G} \mathrm{fi}(g).$$

The proof of this statement can be found in many general books on combinatorics and group theory. We refer the reader to [Ne].

8.5.11 From Exercise 2.10.21 and Proposition 8.4.5 it follows that $\mathbf{P}_n / \mathbf{I}_n \to 1$, $n \to \infty$. This is an asymptotic analogue of Lemma 8.4.6 for $\mathcal{IS}_n$.

8.5.12 If $\mathbf{m} = (M_1, \ldots, M_k)$ and $\mathbf{n} = (M'_1, \ldots, M'_k)$, then the semigroups $N'_{\tau(\mathbf{m})}$ and $N'_{\tau(\mathbf{n})}$ are isomorphic if and only if $|M_i| = |M'_i|$ for all i. This is proved in [GK4].

8.6 Additional Exercises

8.6.1 For each $k \in \mathbb{N}$ construct a nilpotent semigroup S such that $|S| = \mathrm{nd}(S) = k$.

8.6.2 Show that every nilpotent semigroup contains exactly one idempotent.

8.6.3 Let S and T be two nilpotent semigroups. Consider the set $S \times T = \{(s,t) : s \in S, t \in T\}$ with the operation $(s,t) \cdot (s',t') = (ss', tt')$. Show that $S \times T$ is a nilpotent semigroup.

8.6.4 Let $\varphi : S \twoheadrightarrow T$ be an epimorphism of semigroups.

(a) Show that the semigroup T is nilpotent provided that S is nilpotent.

(b) Construct an example of S, T, and φ as above such that T is nilpotent and S is not nilpotent.

8.6.5 Let $\varphi : S \twoheadrightarrow T$ be an epimorphism of semigroups. Assume that T is nilpotent and $\mathrm{nd}(T) = k$. Let 0 be the zero element of T and assume that $X = \varphi^{-1}(0) = \{s \in S : \varphi(s) = 0\}$ is a nilpotent semigroup and $\mathrm{nd}(X) = l$. Show that S is a nilpotent semigroup and $\mathrm{nd}(S) \leq l \cdot k$.

8.6.6 ([Ko, Lev]) Let $\mathrm{Mat}_n(\mathbb{C})$ denote the semigroup of all $n \times n$ complex matrices with respect to the usual matrix multiplication. Let further $T \subset \mathrm{Mat}_n(\mathbb{C})$ be some nilpotent subsemigroup. Show that $\mathrm{nd}(T) \leq n$.

8.6.7 ([Ko, Lev, Fa, KuMa3]) Classify all maximal nilpotent subsemigroups of the semigroup $\mathrm{Mat}_n(\mathbb{C})$.

8.6.8 Let $T \subset \mathcal{PT}_n$ (or $T \subset \mathcal{IS}_n$) be a subsemigroup, containing $\mathcal{I}_1$. Show that for each $k > 0$ the mapping $X \mapsto T \cap X$ defines a bijection between the set of maximal elements in $\mathrm{Nil}_k(\mathcal{PT}_n)$ (resp. $\mathrm{Nil}_k(\mathcal{IS}_n)$) and the set $\mathrm{Nil}_k(T)$.

8.6.9 Construct a subsemigroup $T \subset \mathcal{IS}_n$, which contains $\mathcal{I}_1$ and for which there exist two nonisomorphic maximal nilpotent subsemigroups of nilpotency degree n.

8.6.10 For each $\mathbf{m}$ as in Sect. 8.2 determine the cardinality of the subsemigroup $N'_{\mathcal{T}\mathbf{m}} \cap \mathcal{I}_1$ of $\mathcal{IS}_n$.

8.6.11 Prove that N_n equals the number of all partial injections from $\mathbf{N}$ to $\{1, 2, \ldots, n-1\}$.

8.6.12 ([GM5]) Prove that

$$\sum_{k=0}^{n-1} (n-k) \binom{n}{k}^2 k! = \sum_{k=1}^{n} [n]_k \mathcal{I}_{n-k}.$$

8.6.13 ([GM5]) Prove that the following two sets have the same cardinality:

$$A = \{(\alpha, l) \ : \ \alpha \in \mathcal{IS}_n, \ l \text{ is a chain of } \alpha\},$$
$$B = \{(\beta, x) \ : \ \beta \in \mathcal{IS}_n, \ 1 \in \mathrm{dom}(\beta), \ x = \beta^k(1) \text{ for some } k \geq 0\}.$$

8.6.14 ([GM2, GM4]) Prove that the number of those $\alpha \in \mathcal{IS}_n$, which satisfy the following two conditions:

(a) $\alpha(x) < \alpha(y)$ for all $x, y \in \mathbf{N}$, $x < y$

(b) $\alpha(x) < x$ for all $x \in \mathbf{N}$

equals the n-th *Catalan number* $C_n = \frac{1}{n+1} \binom{2n}{n}$.

Chapter 9

Presentation

9.1 Defining Relations

Let A be a nonempty set, which we will call the *alphabet*. Elements of A will be called *letters*. Any finite nonempty sequence $a_1 a_2 \cdots a_k$ of elements from A will be called a *word* or a *word over A*. The set of all words over A is denoted A^+. If $u = a_1 a_2 \cdots a_k$ and $v = b_1 b_2 \cdots b_m$ are two words, we define their *product* or *concatenation* or *juxtaposition* as follows:

$$uv = a_1 a_2 \cdots a_k b_1 b_2 \cdots b_m.$$

It is obvious that this binary operation is associative, which turns A^+ into a semigroup. This semigroup is called the *free* semigroup with *base A*. Obviously, A is a generating system of A^+.

Let now S be a semigroup and A be a generating system for S. On the one hand, then every element $s \in S$ can be written as a product $s = a_1 a_2 \cdots a_k$ of some elements from A, however not uniquely in general. On the other hand, every product $a_1 a_2 \cdots a_k$ can be considered as an element of A^+. If there exist two words $u = a_1 a_2 \cdots a_k$ and $v = b_1 b_2 \cdots b_m$ in A^+, which determine the same element $s \in S$, we say that the *relation $u = v$* holds in S.

Proposition 9.1.1 *The binary relation*

$$\rho(S, A) = \{(u, v) \in A^+ \times A^+ : u = v \text{ is a relation in } S\}$$

is a congruence on A^+, moreover $A^+/\rho(S, A) \cong S$ canonically.

Proof. The mapping $\varphi : A^+ \to S$, which sends the word $a_1 a_2 \cdots a_k$ to the product $a_1 a_2 \cdots a_k$, is obviously a homomorphism of semigroups. It is surjective since A generates S. We further have $\rho(S, A) = \text{Ker}(\varphi)$ by definition. Hence $\rho(S, A)$ is a congruence on A^+. The claim $A^+/\rho(S, A) \cong S$ follows from Theorem 6.1.6. □

O. Ganyushkin, V. Mazorchuk, *Classical Finite Transformation Semigroups*, Algebra and Applications 9, DOI: 10.1007/978-1-84800-281-4_9, © Springer-Verlag London Limited 2009

Fixing some representatives in all classes of $\rho(S, A)$ determines a *canonical form* for each element of S with respect to the generating system A. Sometimes, it is convenient to identify the elements from S with their canonical forms. Such identification might be useful if the following two natural conditions are satisfied:

- There exists a constructable description for all canonical forms

- If the canonical forms of some elements $g, h \in S$ are known, then one can determine the canonical form of the element gh

Let $u = v$ be a relation in $S = \langle A \rangle$. We will say that some congruence ρ on A^+ *contains* the relation $u = v$ provided that $(u, v) \in \rho$. In particular, the uniform congruence $\omega_{A^+} = A^+ \times A^+$ contains all possible relations. As the intersection of an arbitrary family of congruences is a congruence, for each set Σ of relations, there exists an unique minimal congruence ρ_Σ on A^+, which contains all relations from Σ. If $\rho_\Sigma = \rho(S, A)$, the set Σ is called a *set* or a *system* of *defining relations* for S with respect to A. In this case, one says that S is a semigroup generated by A with the system Σ of defining relations. This is denoted $S = \langle A | \Sigma \rangle$. The pair $\langle A | \Sigma \rangle$ is called a *presentation* of S. Note that a system of defining relations for S with respect to A is not unique in general.

Let Σ be a subset of $A^+ \times A^+$. A pair $(u, v) \in A^+ \times A^+$ will be called a Σ-*pair* provided that $u = v$ or there exist decompositions $u = su_1t$ and $v = sv_1t$ (where $s, t, u_1, v_1 \in A^+$ or either s or t or both are empty), such that either (u_1, v_1) or (v_1, u_1) belongs to Σ.

Proposition 9.1.2 *Congruence ρ_Σ coincides with the set of all pairs (u, v) for which there exists a finite collection $w_1, w_2, \ldots, w_k \in A^+$ such that each of the pairs $(u, w_1), (w_1, w_2), \ldots, (w_{k-1}, w_k)$ and (w_k, v) is a Σ-pair.*

Proof. Let Ω denote the set of all pairs (u, v), which satisfy the condition of the proposition. Obviously, each Σ-pair is contained in ρ_Σ. As each congruence is an equivalence relation, in particular, is transitive, we get $\Omega \subset \rho_\Sigma$. To complete the proof it is enough to show that Ω is a congruence.

Let $(u, v) \in \Omega$ and $(u, w_1), (w_1, w_2), \ldots, (w_{k-1}, w_k), (w_k, v)$ be the corresponding collection of Σ-pairs. Then for every $w \in A^+$ we have that the pairs $(wu, ww_1), (ww_1, ww_2), \ldots, (ww_{k-1}, ww_k), (ww_k, wv)$ are Σ-pairs as well. Hence $(wu, wv) \in \Omega$ and Ω is left compatible. Analogously one shows that Ω is right compatible. This completes the proof. $\qquad \square$

A relation $u = v$ is said to *follow* from Σ provided that $(u, v) \in \rho_\Sigma$.

Remark 9.1.3 Assume that both $u = v$ and $v = w$ follow from Σ. Then, obviously, $u = w$ follows from Σ as well.

If S contains the unit element $\mathbf{1}$, we will always assume that every generating system A of S contains $\mathbf{1}$ and every system Σ of defining relations contains all relations of the form $a \cdot \mathbf{1} = a$ and $\mathbf{1} \cdot a = a$, $a \in A$. Such relations will be called *trivial*. As trivial relations for monoids should always be present, we always assume that they are present, and normally omit them.

Let now A and B be two generating systems for S. Then for every $a \in A$ there exists $v(a) \in B^+$ such that $a = v(a)$, and for every $b \in B$ there exists $w(b) \in A^+$ such that $b = w(b)$. We fix such $v(a)$, $a \in A$, and $w(b)$, $b \in B$. For arbitrary $x = a_1 a_2 \cdots a_k \in A^+$ and any $y = b_1 b_2 \cdots b_m \in B^+$ set

$$v(x) = v(a_1)v(a_2) \cdots v(a_k) \in B^+, \quad w(y) = w(b_1)w(b_2) \cdots w(b_m) \in A^+.$$

Let Σ be an arbitrary system of relations for the generating system A. Define a new system Σ_A^B of relations for the generating system B in the following way:

- If Σ contains the relation $a_1 \cdots a_k = a_1' \cdots a_m'$, then Σ_A^B contains the relation $v(a_1 \cdots a_k) = v(a_1' \cdots a_m')$

- If $b \in B$ and $w(b) = a_1 \cdots a_k$, then Σ_A^B contains the relation $b = v(a_1 \cdots a_k)$ (in the proof of Theorem 9.1.4 it will be convenient to denote the right-hand side of this relation by $f(b)$)

- Σ_A^B does not contain any other nontrivial relations

Theorem 9.1.4 *Let Σ be a system of defining relations of S with respect to the generating system A. Then Σ_A^B is a system of defining relations of S with respect to the generating system B.*

Proof. From the definitions we have $\Sigma_A^B \subset \rho(S, B)$. Hence, it is enough to show that every relation in $\rho(S, B)$ follows from Σ_A^B. Let

$$b_1 b_2 \cdots b_k = b_1' b_2' \cdots b_m' \tag{9.1}$$

be some relation in $\rho(S, B)$. For every i, $1 \le i \le k$, and every j, $1 \le j \le m$, the system Σ_A^B contains the relations $b_i = f(b_i)$ and $b_j' = f(b_j')$. This yields that the relations

$$b_1 b_2 \cdots b_k = f(b_1)f(b_2) \cdots f(b_k) \text{ and } b_1' b_2' \cdots b_m' = f(b_1')f(b_2') \cdots f(b_m')$$

follow from Σ_A^B (even belong to Σ_A^B).

From (9.1) it follows that $\rho(S, A)$ contains the relation $w(b_1 b_2 \cdots b_k) = w(b_1' b_2' \cdots b_m')$. As $\rho(S, A) = \rho_\Sigma$, by Proposition 9.1.2 there exists a finite collection $w_1, \ldots, w_r \in A^+$ such that all pairs

$$(w(b_1 b_2 \cdots b_k), w_1), (w_1, w_2), \ldots, (w_{r-1}, w_r), (w_r, w(b_1' b_2' \cdots b_m'))$$

are Σ-pairs. However, if (x, y) is a Σ-pair, then the pair $(v(x), v(y))$ is a Σ_A^B-pair by definition. This means that the relation

$$v(w(b_1 b_2 \cdots b_k)) = v(w(b_1' b_2' \cdots b_m'))$$

follows from Σ_A^B.

Observe now that the word $v(w(b_1 b_2 \cdots b_k))$ coincides with the word $f(b_1) f(b_2) \cdots f(b_k)$, and at the same time the word $v(w(b_1' b_2' \cdots b_m'))$ coincides with the word $f(b_1') f(b_2') \cdots f(b_m')$. Hence, the relation (9.1) follows from Σ_A^B by Remark 9.1.3. As every relation from $\rho(S, B)$ follows from Σ_A^B, we obtain that Σ_A^B is a system of defining relations of S with respect to the generating system B and the proof is complete. $\square$

9.2   A presentation for $\mathcal{IS}_n$

According to Lemma 3.1.1 and Theorem 3.1.4, a subset $A \subset \mathcal{IS}_n$ is a generating system for $\mathcal{IS}_n$ if and only if A contains some generating system A_1 of the symmetric group $\mathcal{S}_n$ and an element of rank $(n-1)$.

Lemma 9.2.1 *Transpositions* $(1, 2), (1, 3), \ldots, (1, n)$ *generate* $\mathcal{S}_n$.

Proof. This follows from the equality $(i, j) = (1, i)(1, j)(1, i)$ and the standard fact that the set of all transpositions generates $\mathcal{S}_n$ (see Exercise 3.3.2). $\square$

The above implies that we can choose for example the following generating system $\mathcal{A}$ for $\mathcal{IS}_n$: we take $(n-1)$ transpositions $\pi_k = (1, k)$, $k = 2, \ldots, n$, and the idempotent $\varepsilon_{(1)} = \varepsilon_{\{2,3,\ldots,n\}}$. Let Σ_1 denote an arbitrary system of defining relations for the group $\mathcal{S}_n$ with respect to the generating system $\mathcal{A}_1 = \{\pi_2, \ldots, \pi_n\}$. We also set $\varepsilon_{(k)} = \pi_k \varepsilon_{(1)} \pi_k$ for all $k = 2, \ldots, n$.

Theorem 9.2.2 *Let $n \geq 4$ and Σ denote the union of Σ_1 with the set of the following relations:*

(a) $\varepsilon_{(1)}^2 = \varepsilon_{(1)}$

(b) $\varepsilon_{(1)} \varepsilon_{(2)} = \varepsilon_{(2)} \varepsilon_{(1)}$

(c) $\varepsilon_{(2)} \pi_k = \pi_k \varepsilon_{(2)}$, $k = 3, 4, \ldots, n$

(d) $\varepsilon_{(k)} \pi_2 = \pi_2 \varepsilon_{(k)}$, $k = 3, 4, \ldots, n$

(e) $\pi_2 \varepsilon_{(1)} \varepsilon_{(2)} = \varepsilon_{(1)} \varepsilon_{(2)}$

Then Σ is a system of defining relations for the semigroup $\mathcal{IS}_n$ with respect to the generating system $\mathcal{A}$.

To prove this theorem we will need several lemmas. Note that, since Σ_1 is a system of defining relations for $\mathcal{S}_n$, we may use arbitrary relations from $\mathcal{S}_n$ in the proof.

Lemma 9.2.3 *The relation $\varepsilon_{(i)}^2 = \varepsilon_{(i)}$ follows from Σ for any i.*

Proof. As $\pi_i^2 = \varepsilon$ and $\varepsilon_{(1)}^2 = \varepsilon_{(1)}$ belongs to Σ, we have

$$
\begin{aligned}
\varepsilon_{(i)}^2 &= \pi_i \varepsilon_{(1)} \pi_i \cdot \pi_i \varepsilon_{(1)} \pi_i \\
&= \pi_i \varepsilon_{(1)} \cdot \varepsilon_{(1)} \pi_i \\
&= \pi_i \varepsilon_{(1)} \pi_i \\
&= \varepsilon_{(i)}.
\end{aligned}
$$

$\square$

Lemma 9.2.4 *For any $k \neq 1$ and $k \neq i$ the relation $\pi_i \varepsilon_{(k)} = \varepsilon_{(k)} \pi_i$ follows from Σ.*

Proof. If $k = 2$ or $i = 2$, the claim follows from the relations (c) and (d), respectively. If both $k \neq 2$ and $i \neq 2$, then we have

$$
\begin{aligned}
\pi_i \varepsilon_{(k)} &= \pi_i \pi_k \pi_2 \varepsilon_{(2)} \pi_2 \pi_k \\
&= \pi_k \pi_2 \pi_k \pi_i \pi_k \varepsilon_{(2)} \pi_2 \pi_k \\
\text{(by (c))} \quad &= \pi_k \pi_2 \varepsilon_{(2)} \pi_k \pi_i \pi_k \pi_2 \pi_k \\
&= \pi_k \pi_2 \varepsilon_{(2)} \pi_2 \pi_k \pi_i \\
&= \varepsilon_{(k)} \pi_i
\end{aligned}
$$

using the relations $\pi_i \pi_k \pi_2 = \pi_k \pi_2 \pi_k \pi_i \pi_k$ and $\pi_2 \pi_k \pi_i = \pi_k \pi_i \pi_k \pi_2 \pi_k$ in $\mathcal{S}_n$.
$\square$

Lemma 9.2.5 *Let $k \neq 1$. If $\pi \in \mathcal{S}_n$ is a cyclic permutation such that $\pi(i) = k$, then π can be written as $\pi = \tau(1, k)(1, i)$, where $\tau(k) = k$.*

Proof. For $i \neq 1$ we consider all possible cases:

(1) if $i = k$, then $\pi = \pi(1, k)(1, i)$

(2) if $\pi = (i, k)$, then $\pi = (1, i)(1, k)(1, i)$

(3) if $\pi = (i, k, a, \ldots, b)$, $a \neq 1$, $b \neq 1$, then $\pi = (1, a, \ldots, b, i)(1, k)(1, i)$

(4) if $\pi = (1, i, k, a, \ldots, b)$, then $\pi = (1, a, \ldots, b)(1, k)(1, i)$

(5) if $\pi = (i, k, 1, a, \ldots, b)$, then $\pi = (i, a, \ldots, b)(1, k)(1, i)$

If $i = 1$, then the only possible cases are (2) and (3). The case (2) is trivial. In the case (3) we have $\pi = (1, a, \ldots, b)(1, k)$. $\square$

Lemma 9.2.6 *Let $k \neq 1$ and $\pi \in S_n$ be such that $\pi(k) = k$. Then π may be written as $\pi = \pi_{j_1} \pi_{j_2} \cdots \pi_{j_m}$ such that $k \notin \{j_1, j_2, \ldots, j_m\}$.*

Proof. This statement is obvious.   □

Lemma 9.2.7 *Let $\pi \in S_n$ be such that $\pi(i) = k$. Then the relation $\pi \varepsilon_{(i)} = \varepsilon_{(k)} \pi$ follows from Σ.*

Proof. We first consider the case $k, i \neq 1$. From Lemmas 9.2.5 and 9.2.6 it follows that π can be written as a product $\pi = \pi_{j_1} \pi_{j_2} \cdots \pi_{j_m} \pi_k \pi_i$ such that $k \notin \{j_1, \ldots, j_m\}$. Then

$$
\begin{aligned}
\pi \varepsilon_{(i)} &= \pi_{j_1} \pi_{j_2} \cdots \pi_{j_m} \pi_k \pi_i \cdot \pi_i \varepsilon_{(1)} \pi_i \\
&= \pi_{j_1} \pi_{j_2} \cdots \pi_{j_m} \pi_k \varepsilon_{(1)} \pi_i \\
&= \pi_{j_1} \pi_{j_2} \cdots \pi_{j_m} \pi_k \varepsilon_{(1)} \pi_k \cdot \pi_k \pi_i \\
&= \pi_{j_1} \pi_{j_2} \cdots \pi_{j_m} \varepsilon_{(k)} \pi_k \pi_i \\
\text{(by Lemma 9.2.4)} \quad &= \varepsilon_{(k)} \pi_{j_1} \pi_{j_2} \cdots \pi_{j_m} \pi_k \pi_i \\
&= \varepsilon_{(k)} \pi.
\end{aligned}
$$

If $k = 1$ and $i \neq 1$, for the permutation $\tau = \pi_2 \pi$ we have $\tau(i) = 2$ and hence $\tau \varepsilon_{(i)} = \varepsilon_{(2)} \tau$ by the arguments above. From this we obtain

$$
\begin{aligned}
\pi \varepsilon_{(i)} &= \pi_2 \pi_2 \pi \varepsilon_{(i)} \\
&= \pi_2 \tau \varepsilon_{(i)} \\
&= \pi_2 \varepsilon_{(2)} \tau \\
&= \pi_2 \varepsilon_{(2)} \pi_2 \pi \\
&= \varepsilon_{(1)} \pi.
\end{aligned}
$$

The case $i = 1$ now follows by symmetry.   □

Lemma 9.2.8 *The relation $\varepsilon_{(i)} \varepsilon_{(k)} = \varepsilon_{(k)} \varepsilon_{(i)}$ follows from Σ.*

Proof. The statement is obvious in the case $i = k$. We split the rest into three different cases:

First we consider the case when $i = 2$. Because of the relation (b) we may assume $k > 2$. We have

$$
\begin{aligned}
\varepsilon_{(2)} \varepsilon_{(k)} &= \varepsilon_{(2)} \pi_k \varepsilon_{(1)} \pi_k \\
\text{(by Lemma 9.2.4)} \quad &= \pi_k \varepsilon_{(2)} \varepsilon_{(1)} \pi_k \\
\text{(by (b))} \quad &= \pi_k \varepsilon_{(1)} \varepsilon_{(2)} \pi_k \\
&= \pi_k \varepsilon_{(1)} \pi_k \cdot \pi_k \varepsilon_{(2)} \pi_k \\
&= \varepsilon_{(k)} \pi_k \varepsilon_{(2)} \pi_k \\
\text{(by (c))} \quad &= \varepsilon_{(k)} \varepsilon_{(2)} \pi_k \pi_k \\
&= \varepsilon_{(k)} \varepsilon_{(2)}.
\end{aligned}
$$

Now we consider the case $i = 1$. Because of the relation (b) we may assume $k > 2$. We have

$$
\begin{aligned}
\varepsilon_{(1)}\varepsilon_{(k)} &= \pi_2\varepsilon_{(2)}\pi_2\varepsilon_{(k)} \\
\text{(by Lemma 9.2.4)} &= \pi_2\varepsilon_{(2)}\varepsilon_{(k)}\pi_2 \\
\text{(using the case } i = 2) &= \pi_2\varepsilon_{(k)}\varepsilon_{(2)}\pi_2 \\
\text{(by Lemma 9.2.4)} &= \varepsilon_{(k)}\pi_2\varepsilon_{(2)}\pi_2 \\
&= \varepsilon_{(k)}\varepsilon_{(1)}.
\end{aligned}
$$

Finally we consider the case $i, k > 2$. We have

$$
\begin{aligned}
\varepsilon_{(i)}\varepsilon_{(k)} &= \varepsilon_{(i)}\pi_k\varepsilon_{(1)}\pi_k \\
\text{(by Lemma 9.2.4)} &= \pi_k\varepsilon_{(i)}\varepsilon_{(1)}\pi_k \\
\text{(using the case } i = 1) &= \pi_k\varepsilon_{(1)}\varepsilon_{(i)}\pi_k \\
\text{(by Lemma 9.2.4)} &= \pi_k\varepsilon_{(1)}\pi_k\varepsilon_{(i)} \\
&= \varepsilon_{(k)}\varepsilon_{(i)}.
\end{aligned}
$$

$\square$

Lemma 9.2.9 *The relation* $\pi_i\varepsilon_{(1)}\varepsilon_{(i)} = \varepsilon_{(1)}\varepsilon_{(i)}$ *follows from* Σ.

Proof. For $i = 2$ the statement coincides with the relation (e). For $i > 2$ we first observe that

$$
\begin{aligned}
\varepsilon_{(1)}\varepsilon_{(2)}\pi_2 &= \varepsilon_{(2)}\varepsilon_{(1)}\pi_2 \\
&= \pi_2\varepsilon_{(1)}\pi_2 \cdot \pi_2\varepsilon_{(2)}\pi_2 \cdot \pi_2 \\
&= \pi_2\varepsilon_{(1)}\varepsilon_{(2)} \\
\text{(by (e))} &= \varepsilon_{(1)}\varepsilon_{(2)}.
\end{aligned}
$$

We further have:

$$
\begin{aligned}
\pi_i\varepsilon_{(1)}\varepsilon_{(i)} &= \pi_i\varepsilon_{(1)}\pi_i\varepsilon_{(1)}\pi_i \\
&= \pi_i\pi_2\varepsilon_{(2)}\pi_2\pi_i\varepsilon_{(1)}\pi_2\pi_2\pi_i \\
\text{(using } \pi_2\pi_i = \pi_i\pi_2\pi_i\pi_2) &= \pi_i\pi_2\varepsilon_{(2)}\pi_i\pi_2\pi_i\pi_2\varepsilon_{(1)}\pi_2\pi_2\pi_i \\
\text{(by Lemma 9.2.4)} &= \pi_i\pi_2\pi_i\varepsilon_{(2)}\pi_2\pi_i\varepsilon_{(2)}\pi_2\pi_i \\
\text{(by Lemma 9.2.4)} &= \pi_i\pi_2\pi_i\pi_2\pi_2\varepsilon_{(2)}\pi_2\varepsilon_{(2)}\pi_i\pi_2\pi_i \\
\text{(using } \pi_2\pi_i = \pi_i\pi_2\pi_i\pi_2) &= \pi_2\pi_i\varepsilon_{(1)}\varepsilon_{(2)}\pi_2\pi_i\pi_2 \\
\text{(using the observation above)} &= \pi_2\pi_i\varepsilon_{(1)}\varepsilon_{(2)}\pi_i\pi_2 \\
\text{(by Lemma 9.2.4)} &= \pi_2\pi_i\varepsilon_{(1)}\pi_i\varepsilon_{(2)}\pi_2 \\
&= \pi_2\varepsilon_{(i)}\varepsilon_{(2)}\pi_2 \\
\text{(by Lemma 9.2.4)} &= \varepsilon_{(i)}\pi_2\varepsilon_{(2)}\pi_2 \\
&= \varepsilon_{(i)}\varepsilon_{(1)}.
\end{aligned}
$$

$\square$

Lemma 9.2.10 *For any transposition* $(i,j) = \pi_j \pi_i \pi_j$ *we have that the relation* $(i,j)\varepsilon_{(i)}\varepsilon_{(j)} = \varepsilon_{(i)}\varepsilon_{(j)}$ *follows from* Σ.

Proof. Using Lemma 9.2.4 we have

$$
\begin{aligned}
\pi_j \pi_i \pi_j \varepsilon_{(i)} \varepsilon_{(j)} &= \pi_j \pi_i \varepsilon_{(i)} \pi_j \varepsilon_{(j)} \pi_j \pi_j \\
&= \pi_j \pi_i \varepsilon_{(i)} \varepsilon_{(1)} \pi_j \\
\text{(by Lemmas 9.2.8 and 9.2.9)} \quad &= \pi_j \varepsilon_{(i)} \varepsilon_{(1)} \pi_j \\
\text{(by Lemma 9.2.4)} \quad &= \varepsilon_{(i)} \pi_j \varepsilon_{(1)} \pi_j \\
&= \varepsilon_{(i)} \varepsilon_{(j)}.
\end{aligned}
$$

$\square$

Lemma 9.2.11 *Let* $\{1, 2, \ldots, n\} = \{i_1, \ldots, i_m\} \cup \{j_1, \ldots, j_{n-m}\}$ *and assume* $m \geq 2$. *Let further* $\pi, \tau \in \mathcal{S}_n$ *be such that* $\pi(j_k) = \tau(j_k)$ *for all* $k = 1, \ldots, n - m$. *Then the relation*

$$
\pi \varepsilon_{(i_1)} \cdots \varepsilon_{(i_m)} = \tau \varepsilon_{(i_1)} \cdots \varepsilon_{(i_m)} \tag{9.2}
$$

follows from Σ.

Proof. For π and τ as above we have $\pi = \tau \mu$, where μ acts on the set $\{j_1, \ldots, j_{n-m}\}$ as the identity transformation. Hence we can write $\mu = \mu_r \cdots \mu_1$, where each factor is a transposition on $\{i_1, \ldots, i_m\}$. Let $\mu_1 = (a, b)$. As $\varepsilon_{(i_1)}, \ldots, \varepsilon_{(i_m)}$ commute by Lemma 9.2.8 and are idempotents by Lemma 9.2.3, we have

$$
\varepsilon_{(i_1)} \cdots \varepsilon_{(i_m)} = \varepsilon_{(a)} \varepsilon_{(b)} \varepsilon_{(i_1)} \cdots \varepsilon_{(i_m)}. \tag{9.3}
$$

Now we compute:

$$
\begin{aligned}
\pi \varepsilon_{(i_1)} \cdots \varepsilon_{(i_m)} &= \tau \mu_r \cdots \mu_1 \varepsilon_{(i_1)} \cdots \varepsilon_{(i_m)} \\
\text{(by (9.3))} \quad &= \tau \mu_r \cdots \mu_1 \varepsilon_{(a)} \varepsilon_{(b)} \varepsilon_{(i_1)} \cdots \varepsilon_{(i_m)} \\
\text{(by Lemma 9.2.10)} \quad &= \tau \mu_r \cdots \mu_2 \varepsilon_{(a)} \varepsilon_{(b)} \varepsilon_{(i_1)} \cdots \varepsilon_{(i_m)} \\
\text{(by (9.3))} \quad &= \tau \mu_r \cdots \mu_2 \varepsilon_{(i_1)} \cdots \varepsilon_{(i_m)}.
\end{aligned}
$$

Now the statement follows by induction. $\square$

Proof of Theorem 9.2.2. A straightforward calculation shows that the relations (a)–(e) hold in $\mathcal{IS}_n$.

From Lemmas 9.2.4 and 9.2.7, it follows that every word over the alphabet $\mathcal{A}$ can be written in the form

$$
\pi \varepsilon_{(j_1)} \varepsilon_{(j_2)} \cdots \varepsilon_{(j_m)}, \tag{9.4}
$$

where $m \geq 0$ and $\pi \in \mathcal{S}_n$. Using Lemmas 9.2.3 and 9.2.8, the word (9.4) can be written in the form

$$\pi \varepsilon_{(i_1)} \varepsilon_{(i_2)} \cdots \varepsilon_{(i_k)},$$

where $k \geq 0$ and $1 \leq i_1 < i_2 < \cdots < i_k \leq n$.

It is left to show that every relation in $\mathcal{IS}_n$ of the form

$$\pi \varepsilon_{(i_1)} \varepsilon_{(i_2)} \cdots \varepsilon_{(i_k)} = \tau \varepsilon_{(j_1)} \varepsilon_{(j_2)} \cdots \varepsilon_{(j_m)}, \qquad (9.5)$$

where $\pi, \tau \in \mathcal{S}_n$, $k \geq 0$, $m \geq 0$, $1 \leq i_1 < i_2 < \cdots < i_k \leq n$ and $1 \leq j_1 < j_2 < \cdots < j_m \leq n$, follows from Σ.

For $\alpha = \pi \varepsilon_{(i_1)} \varepsilon_{(i_2)} \cdots \varepsilon_{(i_k)}$ we have $\overline{\mathrm{dom}}(\alpha) = \{i_1, \ldots, i_k\}$. Analogously for $\beta = \tau \varepsilon_{(j_1)} \varepsilon_{(j_2)} \cdots \varepsilon_{(j_m)}$ we have $\overline{\mathrm{dom}}(\beta) = \{j_1, \ldots, j_m\}$. Hence (9.5) implies $k = m$, $i_1 = j_1, \ldots, i_k = j_k$ and also $\pi(l) = \tau(l)$ for any element $l \in \mathbf{N} \setminus \{i_1, \ldots, i_k\}$. Hence the relation (9.5) has the form

$$\pi \varepsilon_{(i_1)} \varepsilon_{(i_2)} \cdots \varepsilon_{(i_k)} = \tau \varepsilon_{(i_1)} \varepsilon_{(i_2)} \cdots \varepsilon_{(i_k)}. \qquad (9.6)$$

If $k = 0$ or $k = 1$, the permutations π and τ coincide and thus $\pi = \tau$ follows from Σ_1, implying that (9.6) follows from Σ. If $k \geq 2$, the relation (9.6) follows from Σ by Lemma 9.2.11. This completes the proof. $\square$

9.3 A Presentation for $\mathcal{T}_n$

Recall that in Sect. 5.3 we denoted by $\varepsilon_{m,k}$ the idempotent of rank $(n-1)$ in $\mathcal{T}_n$ such that $\varepsilon_{m,k}(m) = \varepsilon_{m,k}(k) = m$. From Lemma 3.1.1 and Theorem 3.1.3 it follows that the set $\mathcal{A}$, which consists of ε, all transpositions $(i,j) \in \mathcal{S}_n$ and all idempotents $\varepsilon_{m,k}$ of rank $(n-1)$, is a generating system of $\mathcal{T}_n$.

Let Σ_1 denote an arbitrary system of defining relations for the group $\mathcal{S}_n$ with respect to the generating system $\mathcal{A}_1$, consisting of the identity element and all transpositions. Denote by Σ the system of relations obtained by adding to Σ_1 the following relations where the letters i, j, k, l denote different elements of $\mathbf{N}$:

(a) $(k,l)\varepsilon_{i,j}(k,l) = \varepsilon_{i,j}$

(b) $(j,k)\varepsilon_{i,j}(j,k) = \varepsilon_{i,k}$

(c) $(k,i)\varepsilon_{i,j}(k,i) = \varepsilon_{k,j}$

(d) $(i,j)\varepsilon_{i,j}(i,j) = \varepsilon_{j,i}$

(e) $\varepsilon_{i,j}\varepsilon_{l,k} = \varepsilon_{l,k}\varepsilon_{i,j}$

(f) $\varepsilon_{i,j}\varepsilon_{i,k} = \varepsilon_{i,k}\varepsilon_{i,j} = \varepsilon_{i,j}\varepsilon_{j,k}$

(g) $\varepsilon_{i,j}\varepsilon_{k,i} = \varepsilon_{k,j}(i,j)$

(h) $\varepsilon_{i,j}\varepsilon_{k,j} = \varepsilon_{k,j}$

(i) $\varepsilon_{i,j}\varepsilon_{i,j} = \varepsilon_{i,j}$

(j) $\varepsilon_{i,j}\varepsilon_{j,i} = \varepsilon_{i,j}$

(k) $\varepsilon_{i,j}(i,j) = \varepsilon_{i,j}$

Theorem 9.3.1 *The system Σ is a system of defining relations for $\mathcal{T}_n$ with respect to the generating system $\mathcal{A}$.*

As in the previous section, to prove this theorem we will need several technical lemmas. By the same arguments as in the previous section we will freely use all relations in the symmetric group.

Lemma 9.3.2 *The relation $\pi\varepsilon_{i,j}\pi^{-1} = \varepsilon_{\pi(i),\pi(j)}$ follows from Σ for every $\pi \in \mathcal{S}_n$.*

Proof. If π is the identity element, the claim is obvious. If π is a transposition, the claim follows from the relations (a)–(d). In the general case the claim follows by inductions since every permutation can be written as a product of transpositions. $\square$

Lemma 9.3.3 *Let $k > 1$ and $f = \varepsilon_{i_k,j_k}\varepsilon_{i_{k-1},j_{k-1}} \cdots \varepsilon_{i_1,j_1}$ be such that*

$$\{i_k,j_k\} \cap \{j_1,\ldots,j_{k-1}\} \neq \varnothing \quad \text{and} \quad \{i_m,j_m\} \cap \{j_1,\ldots,j_{m-1}\} = \varnothing$$

for all $m < k$. Then the relation

$$f = \begin{cases} \varepsilon_{i_{k-1},j_{k-1}}\varepsilon_{i_{k-2},j_{k-2}} \cdots \varepsilon_{i_1,j_1}, & j_k \in \{j_1,\ldots,j_{k-1}\}, \\ (i_k,j_k)\varepsilon_{i_{k-1},j_{k-1}} \cdots \varepsilon_{i_1,j_1}, & otherwise \end{cases}$$

follows from Σ.

Proof. We prove the statement by induction on k. For $k = 2$ the claim follows from the relations (h) and (i) if $j_2 = j_1$ or the relations (g), (b) and Lemma 9.3.2 if $j_2 \neq j_1$.

Assume now that $k \geq 3$ and that the statement is proved if the number of factors is less than k. Consider first the case $j_k \in \{j_1,\ldots,j_{k-1}\}$. Let $j_k = j_m$ for some $m < k$. If $m > 1$, then from the inductive assumption we have

$$\varepsilon_{i_k,j_k}\varepsilon_{i_{k-1},j_{k-1}} \cdots \varepsilon_{i_m,j_m} = \varepsilon_{i_{k-1},j_{k-1}} \cdots \varepsilon_{i_m,j_m},$$

which implies $f = \varepsilon_{i_{k-1},j_{k-1}}\varepsilon_{i_{k-2},j_{k-2}} \cdots \varepsilon_{i_1,j_1}$ as required.

Let now $m = 1$. If ε_{i_2,j_2} and ε_{i_1,j_1} commute, we have

$$\varepsilon_{i_k,j_k}\varepsilon_{i_{k-1},j_{k-1}} \cdots \varepsilon_{i_2,j_2}\varepsilon_{i_1,j_1} = \varepsilon_{i_k,j_k}\varepsilon_{i_{k-1},j_{k-1}} \cdots \varepsilon_{i_3,j_3}\varepsilon_{i_1,j_1}\varepsilon_{i_2,j_2}$$
$$\text{(by the inductive assumption)} = \varepsilon_{i_{k-1},j_{k-1}} \cdots \varepsilon_{i_3,j_3}\varepsilon_{i_1,j_1}\varepsilon_{i_2,j_2}$$
$$= \varepsilon_{i_{k-1},j_{k-1}} \cdots \varepsilon_{i_3,j_3}\varepsilon_{i_2,j_2}\varepsilon_{i_1,j_1}.$$

Finally, assume that ε_{i_2,j_2} and ε_{i_1,j_1} do not commute. Because of (e), this is possible only in the case $\{i_1,j_1\} \cap \{i_2,j_2\} \neq \varnothing$. We have $j_1 \notin \{i_2,j_2\}$ by assumption. The case $i_1 = i_2$ is not possible because of the relation (f). Hence $i_1 = j_2$. In this case the relation (f) yields

$$\varepsilon_{i_2,j_2}\varepsilon_{i_1,j_1} = \varepsilon_{i_2,j_1}\varepsilon_{i_2,j_2}. \tag{9.7}$$

Using (9.7) we compute:

$$
\begin{aligned}
\varepsilon_{i_k,j_k}\varepsilon_{i_{k-1},j_{k-1}} \cdots \varepsilon_{i_3,j_3}\varepsilon_{i_2,j_2}\varepsilon_{i_1,j_1} &= \varepsilon_{i_k,j_k}\varepsilon_{i_{k-1},j_{k-1}} \cdots \varepsilon_{i_3,j_3}\varepsilon_{i_2,j_1}\varepsilon_{i_2,j_2} \\
\text{(by the inductive assumption)} &= \varepsilon_{i_{k-1},j_{k-1}} \cdots \varepsilon_{i_3,j_3}\varepsilon_{i_2,j_1}\varepsilon_{i_2,j_2} \\
\text{(by (9.7))} &= \varepsilon_{i_{k-1},j_{k-1}} \cdots \varepsilon_{i_3,j_3}\varepsilon_{i_2,j_2}\varepsilon_{i_1,j_1}.
\end{aligned}
$$

If $j_k \notin \{j_1,\ldots,j_{k-1}\}$, then $i_k \in \{j_1,\ldots,j_{k-1}\}$. This case is dealt with by analogous arguments and is left to the reader. □

Let $J = \{j_1,\ldots,j_k\} \subset \mathbf{N}$ be nonempty and $i \in \mathbf{N}\backslash J$. The relation (f) implies that the elements $\varepsilon_{i,j_1},\ldots,\varepsilon_{i,j_k}$ pairwise commute. Hence the product $\varepsilon_{i,j_1}\varepsilon_{i,j_2}\cdots\varepsilon_{i,j_k}$ does not depend on the order of the factors. Let us denote this product by $\varepsilon_{i,J}$. Note that from (e) we have $\varepsilon_{i_1,J_1}\varepsilon_{i_2,J_2} = \varepsilon_{i_2,J_2}\varepsilon_{i_1,J_1}$ provided that $(J_1 \cup \{i_1\}) \cap (J_2 \cup \{i_2\}) = \varnothing$.

Lemma 9.3.4 *Let $k \geq 1$ and $f = \varepsilon_{i_k,j_k}\varepsilon_{i_{k-1},j_{k-1}} \cdots \varepsilon_{i_1,j_1}$ be such that*

$$\{i_m,j_m\} \cap \{j_1,\ldots,j_{m-1}\} = \varnothing$$

for all $m \leq k$. Then there exists a partition $\{j_1,\ldots,j_k\} = K_1 \cup \cdots \cup K_r$ into disjoint nonempty blocks and elements $\{k_1,\ldots,k_r\} \in \{i_1,\ldots,i_k\}$ such that

$$(K_p \cup \{k_p\}) \cap (K_q \cup \{k_q\}) = \varnothing \text{ if } p \neq q$$

and the relation $f = \varepsilon_{k_r,K_r}\varepsilon_{k_{r-1},K_{r-1}} \cdots \varepsilon_{k_1,K_1}$ follows from Σ.

Proof. For $k = 1$ the claim is obvious. The rest is proved by induction on k. From the inductive assumption we have

$$\varepsilon_{i_{k-1},j_{k-1}} \cdots \varepsilon_{i_1,j_1} = \varepsilon_{l_s,L_s} \cdots \varepsilon_{l_1,L_1}, \tag{9.8}$$

where $L_1 \cup \cdots \cup L_s = \{j_1,\ldots,j_{k-1}\}$, $l_1,\ldots,l_s \in \{i_1,\ldots,i_{k-1}\}$ and also $(L_p \cup \{l_p\}) \cap (L_q \cup \{l_q\}) = \varnothing$ for $p \neq q$.

We first assume that $i_k \in \{l_1,\ldots,l_s\}$. As all factors on the right-hand side of (9.8) commute, we may assume $i_k = l_s$. In this case we have

$$
\begin{aligned}
\varepsilon_{i_k,j_k}\varepsilon_{l_s,L_s} &= \varepsilon_{l_s,j_k}\varepsilon_{l_s,L_s} \\
&= \varepsilon_{l_s,L'_s},
\end{aligned}
$$

where $L'_s = L_s \cup \{j_k\}$. This gives the necessary decomposition for the product f: $f = \varepsilon_{l_s,L'_s}\varepsilon_{l_{s-1},L_{s-1}} \cdots \varepsilon_{l_1,L_1}$.

Assume $i_k \notin \{l_1, \ldots, l_s\}$. If $j_k \notin \{l_1, \ldots, l_s\}$, then we set $l_{s+1} = i_k$, $L_{s+1} = \{j_k\}$, and have the necessary decomposition for the product f: $f = \varepsilon_{l_{s+1},L_{s+1}} \varepsilon_{l_s,L_s} \cdots \varepsilon_{l_1,L_1}$. If $j_k \in \{l_1, \ldots, l_s\}$, then as before we may assume $j_k = l_s$. Let $L_s = \{x_1, \ldots, x_t\}$. Then we compute:

$$
\begin{aligned}
\varepsilon_{i_k,j_k} \varepsilon_{l_s,L_s} &= \varepsilon_{i_k,l_s} \varepsilon_{l_s,x_1} \cdots \varepsilon_{l_s,x_t} \\
\text{(by the relation (f))} &= \varepsilon_{i_k,x_1} \varepsilon_{i_k,l_s} \varepsilon_{l_s,x_2} \cdots \varepsilon_{l_s,x_t} \\
&= \cdots \\
&= \varepsilon_{i_k,x_1} \cdots \varepsilon_{i_k,x_t} \varepsilon_{i_k,l_s}.
\end{aligned}
$$

If we now set $l'_s = i_k$ and $L'_s = L_s \cup \{l_s\}$, we again obtain the necessary decomposition for the product f: $f = \varepsilon_{l'_s,L'_s} \varepsilon_{l_{s-1},L_{s-1}} \cdots \varepsilon_{l_1,L_1}$. $\square$

As in Exercise 2.10.24 for $\alpha \in \mathcal{T}_n$ we let $\mathsf{t}(\alpha) = (\mathsf{t}_0(\alpha), \ldots, \mathsf{t}_n(\alpha))$ denote the type of α.

Lemma 9.3.5 Let $\Delta = (t_0, t_1, \ldots, t_n)$ be a vector with nonnegative integer coefficients. Then the following conditions are equivalent:

(i) There exists $\alpha \in \mathcal{T}_n$ such that $\Delta = \mathsf{t}(\alpha)$.

(ii) $\displaystyle\sum_{k=0}^{n} t_k = n$ and $\displaystyle\sum_{k=0}^{n} k t_k = n$.

Proof. The implication (i)$\Rightarrow$(ii) is Exercise 1.5.12. To prove the implication (ii)$\Rightarrow$(i) let Δ be such that the condition (ii) is satisfied. Define the idempotent transformation π_Δ in the following way: Consider the usual order on $\mathbf{N}$. Partition $\mathbf{N}$ into an ordered collection of disjoint nonempty subsets as follows: first take t_1 subsets $K_1 = \{1\}$, $K_2 = \{2\}, \ldots, K_{t_1} = \{t_1\}$; then take t_2 subsets $L_1 = \{t_1 + 1, t_1 + 2\}, \ldots, L_{t_2} = \{t_1 + 2t_2 - 1, t_1 + 2t_2\}$; then take t_3 subsets $M_1 = \{t_1 + 2t_2 + 1, t_1 + 2t_2 + 2, t_1 + 2t_2 + 3\}, \ldots, M_{t_3} = \{t_1 + 2t_2 + 3t_3 - 2, t_1 + 2t_2 + 3t_3 - 1, t_1 + 2t_2 + 3t_3\}$; and so on. Let $k_1, \ldots, k_{t_1}$, $l_1, \ldots, l_{t_2}, m_1, \ldots, m_{t_3}$ and so on denote the corresponding minimal elements in these subsets. Set

$$
\pi_\Delta = \begin{pmatrix} K_1 & \cdots & K_{t_1} & L_1 & \cdots & L_{t_2} & M_1 & \cdots & M_{t_3} & \cdots \\ k_1 & \cdots & k_{t_1} & l_1 & \cdots & l_{t_2} & m_1 & \cdots & m_{t_3} & \cdots \end{pmatrix}. \tag{9.9}
$$

It follows from the construction that $\mathsf{t}(\pi_\Delta) = \Delta$. $\square$

Remark 9.3.6 The transformation π_Δ from Lemma 9.3.5 can be constructed in a different way: For every $x \in \mathbf{N}$ there exists a unique number $k(x)$ such that $\displaystyle\sum_{i<k(x)} i t_i < x$ and $\displaystyle\sum_{i \le k(x)} i t_i \ge x$. Then a straightforward calculation shows that the assignment

$$
\pi_\Delta(x) = \sum_{i<k(x)} i t_i + k(x) \left\lfloor \frac{x - 1 - \sum_{i<k(x)} i t_i}{k(x)} \right\rfloor + 1
$$

defines π_Δ.

Using (9.9), to each vector Δ, which satisfies Lemma 9.3.5(ii), we associate the word

$$\varepsilon_\Delta = \varepsilon_{l_1, L_1 \setminus \{l_1\}} \cdots \varepsilon_{l_{t_2}, L_{t_2} \setminus \{l_{t_2}\}} \varepsilon_{m_1, M_1 \setminus \{m_1\}} \cdots \varepsilon_{m_{t_3}, M_{t_3} \setminus \{m_{t_3}\}} \cdots . \qquad (9.10)$$

Lemma 9.3.7 *Let* $\mu, \eta \in \mathcal{E}(\mathcal{T}_n)$ *be such that* $\mathsf{t}(\mu) = \mathsf{t}(\eta)$. *Then there exists* $\sigma \in \mathcal{S}_n$ *such that* $\eta = \sigma \mu \sigma^{-1}$.

Proof. Set

$$M_k^\mu = \{x \in \mathbf{N} : |\{y \in \mathbf{N} : \mu(y) = x\}| = k\}.$$

Similarly define M_k^η. As $|M_k^\mu| = |M_k^\eta|$ for each k, we can define bijections $\tau_k : M_k^\mu \to M_k^\eta$ for all $k > 0$. For every $x \in M_k^\mu$ each of the sets $P_k^x = \{y \in \mathbf{N} : \mu(y) = x\}$ and $Q_k^x = \{y \in \mathbf{N} : \eta(y) = \tau_k(x)\}$ contains exactly k elements. Moreover, $x \in P_k^x$ and $\tau_k(x) \in Q_k^x$. Hence we can define bijections $\Theta_k^x : P_k^x \to Q_k^x$ such that $\theta_k^x(x) = \tau_k(x)$.

As $\bigcup_{k>0} \bigcup_x P_k^x = \mathbf{N}$ and $P_{k_1}^{x_1} \cap P_{k_2}^{x_2} = \varnothing$ if $(k_1, x_1) \neq (k_2, x_2)$, the bijections $\{\theta_k^x\}$ define a permutation $\sigma \in \mathcal{S}_n$. The equality $\eta = \sigma \mu \sigma^{-1}$ is checked by a direct calculation. □

Lemma 9.3.8 *For each word* u *over* $\mathcal{A}$ *there exist* $\alpha, \beta \in \mathcal{S}_n$ *and a uniquely defined vector* Δ *satisfying Lemma 9.3.5(ii) such that the relation* $u = \alpha \varepsilon_\Delta \beta$ *follows from* Σ.

Proof. The word u has the form $\gamma_1 \gamma_2 \cdots \gamma_m$, where each γ is either an element from $\mathcal{S}_n$ or has the form $\varepsilon_{i,j}$. By Lemma 9.3.2, the word u can be reduced to the form

$$u = \varepsilon_{i_k, j_k} \cdots \varepsilon_{i_1, j_1} \mu, \qquad (9.11)$$

where $\mu \in \mathcal{S}_n$. Among all possible relations of this form we choose the one with minimal possible k. Then we claim that $\{i_m, j_m\} \cap \{j_1, \ldots, j_{m-1}\} = \varnothing$ for all $m \leq k$. Indeed, if not, then there exists $t \leq k$ such that $\{i_t, j_t\} \cap \{j_1, \ldots, j_{t-1}\} \neq \varnothing$ and we may assume that t is minimal having this property. Then from Lemma 9.3.3 we obtain that

$$u = \varepsilon_{i_k, j_k} \cdots \varepsilon_{i_{t+1}, j_{t+1}} \nu \varepsilon_{i_{t-1}, j_{t-1}} \cdots \varepsilon_{i_1, j_1} \mu, \qquad (9.12)$$

where $\nu \in \mathcal{S}_n$. Applying Lemma 9.3.2 again we reduce the right-hand side of (9.12) to the form (9.11) with $k - 1$ factors of the form $\varepsilon_{i,j}$. This contradicts the minimality of k.

Thus $\{i_m, j_m\} \cap \{j_1, \ldots, j_{m-1}\} = \varnothing$ for all $m \leq k$. By Lemma 9.3.4, we can reduce u to the form

$$u = \varepsilon_{k_r, K_r} \cdots \varepsilon_{k_1, K_1} \mu,$$

where $(K_p \cup \{k_p\}) \cap (K_q \cup \{k_q\}) = \varnothing$ for $p \neq q$.

The product $\gamma = \varepsilon_{k_r, K_r} \cdots \varepsilon_{k_1, K_1}$ is an idempotent of $\mathcal{T}_n$. Let $\Delta = \mathsf{t}(\gamma)$. By Lemma 9.3.7 there exists $\alpha \in \mathcal{S}_n$ such that $\gamma = \alpha \varepsilon_\Delta \alpha^{-1}$. If we now set $\beta = \alpha^{-1}\mu$, the relation $u = \alpha \varepsilon_\Delta \beta$ follows from Σ by Lemma 9.3.2.

From Exercise 2.10.24 we have that $u = \alpha_1 \varepsilon_{\Delta_1} \beta_1 = \alpha_2 \varepsilon_{\Delta_2} \beta_2$ implies $\mathsf{t}(\varepsilon_{\Delta_1}) = \mathsf{t}(\varepsilon_{\Delta_2})$. The proof is now completed by the observation that ε_Δ is uniquely determined by its type. $\qquad\square$

Lemma 9.3.9 *Let $\alpha \in \mathcal{T}_n$ and $\mathsf{t}(\alpha) = (t_0, t_1, \ldots, t_n)$. Then*

$$|\mathcal{S}_n \alpha \mathcal{S}_n| = \frac{(n!)^2}{t_0! t_1! \cdots t_n! (1!)^{t_1} \cdots (n!)^{t_n}}.$$

Proof. From Exercise 2.10.24 we have that $\beta \in \mathcal{S}_n \alpha \mathcal{S}_n$ if and only if $\mathsf{t}(\beta) = \mathsf{t}(\alpha)$. In particular, $\operatorname{rank}(\beta) = t_1 + t_2 + \cdots + t_n$. Each such element β can be obtained in the following way: Take any partition of $\mathbf{N}$ into t_1 blocks with one element, t_2 blocks with two elements and so on. After this we choose $\operatorname{im}(\beta)$ (which we can do in $\binom{n}{\operatorname{rank}(\beta)}$ different ways) and define an arbitrary bijection between our blocks and $\operatorname{im}(\beta)$ (which we can do in $(\operatorname{rank}(\beta))!$ different ways).

To get all necessary partitions of $\mathbf{N}$ consider some permutation i_1, $i_2, \ldots, i_n$ of the elements from $\mathbf{N}$ and say that the blocks with one element are $\{i_1\}, \{i_2\}, \ldots, \{i_{t_1}\}$; the blocks with two elements are $\{i_{t_1+1}, i_{t_1+2}\}, \ldots, \{i_{t_1+2t_2-1}, i_{t_1+2t_2}\}$; and so on. As the order of blocks of the same cardinality and the order of elements in each block are not important, each partition will be obtained exactly $t_1! \cdots t_n! (1!)^{t_1} \cdots (n!)^{t_n}$ times. Hence

$$|\mathcal{S}_n \alpha \mathcal{S}_n| = \frac{n!}{t_1! \cdots t_n! (1!)^{t_1} \cdots (n!)^{t_n}} \binom{n}{t_1 + \cdots + t_n} (t_1 + \cdots + t_n)!$$

$$= \frac{(n!)^2}{t_0! t_1! \cdots t_n! (1!)^{t_1} \cdots (n!)^{t_n}}.$$

$\qquad\square$

Lemma 9.3.10 *Let $\alpha \in \mathcal{T}_n$ and $\Delta = \mathsf{t}(\alpha) = (t_0, t_1, \ldots, t_n)$. Then the number of congruence classes in ρ_Σ, which contain words of the form $\alpha \varepsilon_\Delta \beta$, $\alpha, \beta \in \mathcal{S}_n$, does not exceed*

$$\frac{(n!)^2}{t_0! t_1! \cdots t_n! (1!)^{t_1} \cdots (n!)^{t_n}}.$$

Proof. Let π_Δ be as in (9.9). Denote by P_Δ the set of all elements from $\mathcal{S}_n$, for which the sets $B_1 = \{k_1, k_2, \ldots, k_{t_1}\}$, $B_2 = \{l_1, \ldots, l_{t_2}\}$ and so on are invariant. Set $B_0 = \mathbf{N} \backslash (B_1 \cup \cdots \cup B_n)$. Obviously P_Δ is a subgroup of $\mathcal{S}_n$ of order $p_\Delta = t_0! t_1! \cdots t_n!$.

Let us show that for any $\nu \in P_\Delta$, there exists $\mu \in \mathcal{S}_n$ such that the relation $\nu \varepsilon_\Delta \mu = \varepsilon_\Delta$ follows from Σ. Note that P_Δ is generated by those transpositions (i, j) for which both i and j belong to the same B_k, $k = 0, \ldots, n$. Hence, it is enough to consider the case when $\nu = (i, j)$ such that $i, j \in B_k$.

Consider the set $\{x \in \mathbf{N} : \pi_\Delta(x) = i\} =: \{i_1, \ldots, i_k\}$. As $i \in \{i_1, \ldots, i_k\}$, we may assume $i = i_1$. Analogously, we define the set $\{j_1 = j, j_2, \ldots, j_k\}$. Set $\mu = (i_1, j_1) \cdots (i_k, j_k)$ and let us show that $\nu \varepsilon_\Delta \mu = \varepsilon_\Delta$ follows from Σ.

If i, j, k, l are pairwise different, we have

$$\begin{aligned}
(k, l)\varepsilon_{i,j} &= (k, l)\varepsilon_{i,j}(k, l)^2 \\
\text{(by (a))} &= \varepsilon_{i,j}(k, l).
\end{aligned}$$

From this and the relation (9.10) it follows that it is enough to prove the relation

$$(i_1, j_1)\varepsilon_{i_1,i_2} \cdots \varepsilon_{i_1,i_k}\varepsilon_{j_1,j_2} \cdots \varepsilon_{j_1,j_k}(i_1, j_1) \cdots (i_k, j_k) =$$
$$= \varepsilon_{i_1,i_2} \cdots \varepsilon_{i_1,i_k}\varepsilon_{j_1,j_2} \cdots \varepsilon_{j_1,j_k} \quad (9.13)$$

Using Lemma 9.3.2, the left-hand side of (9.13) can be reduced to the following:

$$\varepsilon_{j_1,i_2} \cdots \varepsilon_{j_1,i_k}\varepsilon_{i_1,j_2} \cdots \varepsilon_{i_1,j_k}(i_2, j_2) \cdots (i_k, j_k).$$

Using the relation (e), the latter can be reduced to

$$\varepsilon_{j_1,i_2}\varepsilon_{i_1,j_2}\varepsilon_{j_1,i_3}\varepsilon_{i_1,j_3} \cdots \varepsilon_{j_1,i_k}\varepsilon_{i_1,j_k}(i_2, j_2) \cdots (i_k, j_k).$$

Finally, using (a) the last expression reduces to

$$\varepsilon_{j_1,i_2}\varepsilon_{i_1,j_2}(i_2, j_2)\varepsilon_{j_1,i_3}\varepsilon_{i_1,j_3}(i_3, j_3) \cdots \varepsilon_{j_1,i_k}\varepsilon_{i_1,j_k}(i_k, j_k). \quad (9.14)$$

As soon as, using (g), we have

$$\begin{aligned}
\varepsilon_{k,j}\varepsilon_{i,l}(l, j) &= \varepsilon_{k,j}\varepsilon_{j,l}\varepsilon_{i,j} \\
\text{(by (f))} &= \varepsilon_{k,j}\varepsilon_{k,l}\varepsilon_{i,j} \\
\text{(by (f))} &= \varepsilon_{k,l}\varepsilon_{k,j}\varepsilon_{i,j} \\
\text{(by (h))} &= \varepsilon_{k,l}\varepsilon_{i,j},
\end{aligned}$$

the word (9.14) reduces to the form

$$\varepsilon_{j_1,j_2}\varepsilon_{i_1,i_2}\varepsilon_{j_1,j_3}\varepsilon_{i_1,i_3} \cdots \varepsilon_{j_1,j_k}\varepsilon_{i_1,i_k}.$$

The latter reduces to the right-hand side of (9.13) using (e).

Now let Q_Δ denote the set of all permutations from $\mathcal{S}_n$, with respect to which all sets in the upper row of (9.9) are invariant. Then Q_Δ is a subgroup of $\mathcal{S}_n$ of order $q_\Delta = (1!)^{t_1} \cdots (n!)^{t_n}$. Let us show that for any $\theta \in Q_\Delta$ the relation $\varepsilon_\Delta \theta = \varepsilon_\Delta$ follows from Σ. As above, it is enough to consider the case $\theta = (i,j)$, where both i and j belong to the same set R from the upper row of (9.9). Let $R = \{i_1, \ldots, i_k\}$. From (a) and (9.10) it follows that it is enough to prove the relation

$$\varepsilon_{i_1,i_2} \cdots \varepsilon_{i_1,i_k}(i,j) = \varepsilon_{i_1,i_2} \cdots \varepsilon_{i_1,i_k}. \tag{9.15}$$

From (f) we have that the ε_{i_1,i_m}'s commute. If $i = i_1$ or $j = i_1$, we can permute the factors such that the factor ε_{i_1,i_m} is such that $\{i_1, i_m\} = \{i,j\}$. In this case (9.15) follows from (k). If $i,j \neq i_1$, we have

$$
\begin{aligned}
\varepsilon_{i_1,i_2} \cdots \varepsilon_{i_1,i_k}(i,j) &= \cdots \varepsilon_{i_1,i}\varepsilon_{i_1,j}(i,j) \\
(\text{by (f)}) &= \cdots \varepsilon_{i_1,i}\varepsilon_{i,j}(i,j) \\
(\text{by (k)}) &= \cdots \varepsilon_{i_1,i}\varepsilon_{i,j} \\
(\text{by (f)}) &= \cdots \varepsilon_{i_1,i}\varepsilon_{i_1,j}.
\end{aligned}
$$

For every $\nu \in P_\Delta$ let μ_ν be such that the relation $\nu\varepsilon_\Delta\mu_\nu = \varepsilon_\Delta$ follows from Σ. Then for arbitrary $\alpha, \beta \in S_n$ the congruence class of ρ_Σ, containing $\alpha\varepsilon_\Delta\beta$, will contain all $\alpha\nu\varepsilon_\Delta\theta\mu_\nu\beta$, where $\nu \in P_\Delta$ and $\theta \in Q_\Delta$. Hence this class contains at least $|P_\Delta| \cdot |Q_\Delta| = p_\Delta q_\Delta$ words from $S_n\varepsilon_\Delta S_n$. This implies that the number of congruence classes of ρ_Σ which contain words from $S_n\varepsilon_\Delta S_n$ does not exceed

$$\frac{|S_n|^2}{p_\Delta q_\Delta} = \frac{(n!)^2}{t_0!t_1! \cdots t_n!(1!)^{t_1} \cdots (n!)^{t_n}}.$$

$\square$

Proof of Theorem 9.3.1. The mapping $\varphi : \mathcal{A}^+ \to \mathcal{T}_n$ for which the letters are mapped to the corresponding elements of $\mathcal{T}_n$ is, obviously, an epimorphism. The kernel of this epimorphism coincides with $\rho(\mathcal{T}_n, \mathcal{A})$. By Theorem 6.1.6, we have $\mathcal{T}_n \cong \mathcal{A}^+/\text{Ker}(\varphi)$. It is easy to check that all relations (a)–(k) hold in $\mathcal{T}_n$. Hence $\text{Ker}(\varphi) \supset \rho_\Sigma$. If $\text{Ker}(\varphi) \neq \rho_\Sigma$, we would have

$$|\mathcal{T}_n| = |\mathcal{A}^+/\text{Ker}(\varphi)| < |\mathcal{A}^+/\rho_\Sigma|.$$

However, from Lemmas 9.3.8–9.3.10, it follows that $|\mathcal{A}^+/\rho_\Sigma| \leq |\mathcal{T}_n|$. Hence $\text{Ker}(\varphi) = \rho_\Sigma$ and the proof is complete. $\square$

Remark 9.3.11 From Theorem 3.1.3 it follows that $\mathcal{T}_n$ is already generated by the subset $\mathcal{A}'$ of $\mathcal{A}$, consisting of transpositions $(1,2), \ldots, (1,n)$, and the idempotent $\varepsilon_{1,2}$. Using Theorem 9.1.4 the system Σ of defining relations with respect to $\mathcal{A}$ can be transformed into a system Σ' of defining relations with respect to $\mathcal{A}'$.

9.4 A presentation for $\mathcal{PT}_n$

For the semigroup $\mathcal{PT}_n$ we consider the set

$$\mathcal{A} = \{(1,2), (1,3), \ldots, (1,n), \varepsilon_{1,2}, \varepsilon_{(1)}\},$$

which is a generating system by Theorem 3.1.5. Let $\tilde{\Sigma}$ denote the system of relations, which consists of the system Σ from Theorem 9.2.2, the system Σ' from Remark 9.3.11 and the following relations:

(a) $\varepsilon_{(2)}\varepsilon_{1,2} = \varepsilon_{1,2}$, $\varepsilon_{1,2}\varepsilon_{(2)} = \varepsilon_{(2)}$

(b) $\varepsilon_{(1)}\varepsilon_{1,2} = \varepsilon_{1,2}\varepsilon_{(1)}\varepsilon_{(2)}$

(c) $\varepsilon_{(3)}\varepsilon_{1,2} = \varepsilon_{1,2}\varepsilon_{(3)}$

Theorem 9.4.1 *For $n \geq 4$ the system $\tilde{\Sigma}$ is a system of defining relations for $\mathcal{PT}_n$ with respect to the generating system $\mathcal{A}$.*

We again will need several lemmas. Taking into account that $\tilde{\Sigma}$ contains systems of defining relations for both $\mathcal{IS}_n$ and $\mathcal{T}_n$, we will freely use relations between elements of each of these semigroups.

Lemma 9.4.2 *The relation $\varepsilon_{i,j}\varepsilon_{(i)}\varepsilon_{(j)} = \varepsilon_{(i)}\varepsilon_{i,j}$ follows from $\tilde{\Sigma}$.*

Proof. Let us first show that $\varepsilon_{1,j}\varepsilon_{(1)}\varepsilon_{(j)} = \varepsilon_{(1)}\varepsilon_{1,j}$ follows from $\tilde{\Sigma}$. We have

$$
\begin{aligned}
\varepsilon_{1,j}\varepsilon_{(1)}\varepsilon_{(j)} &= \\
(\text{from } \mathcal{T}_n) &= (2,j)\varepsilon_{1,2}(2,j)\varepsilon_{(1)}\varepsilon_{(j)} \\
(\text{from } \mathcal{IS}_n) &= (2,j)\varepsilon_{1,2}\varepsilon_{(1)}\varepsilon_{(2)}(2,j) \\
(\text{by (b)}) &= (2,j)\varepsilon_{(1)}\varepsilon_{1,2}(2,j) \\
(\text{from } \mathcal{IS}_n) &= \varepsilon_{(1)}(2,j)\varepsilon_{1,2}(2,j) \\
(\text{from } \mathcal{T}_n) &= \varepsilon_{(1)}\varepsilon_{1,j}.
\end{aligned}
$$

Using this we have

$$
\begin{aligned}
\varepsilon_{i,j}\varepsilon_{(i)}\varepsilon_{(j)} &= \\
(\text{from } \mathcal{T}_n) &= (1,i)\varepsilon_{1,j}(1,i)\varepsilon_{(i)}\varepsilon_{(j)} \\
(\text{from } \mathcal{IS}_n) &= (1,i)\varepsilon_{1,j}\varepsilon_{(1)}\varepsilon_{(j)}(1,i) \\
(\text{previous relation}) &= (1,i)\varepsilon_{(1)}\varepsilon_{1,j}(1,i) \\
(\text{from } \mathcal{IS}_n) &= \varepsilon_{(i)}(1,i)\varepsilon_{1,j}(1,i) \\
(\text{from } \mathcal{T}_n) &= \varepsilon_{(i)}\varepsilon_{i,j}.
\end{aligned}
$$

$\square$

Lemma 9.4.3 *If $k \neq i$ and $k \neq j$, then the relation $\varepsilon_{i,j}\varepsilon_{(k)} = \varepsilon_{(k)}\varepsilon_{i,j}$ follows from $\tilde{\Sigma}$.*

Proof. Let us first show that for $k > 2$ the relation $\varepsilon_{1,2}\varepsilon_{(k)} = \varepsilon_{(k)}\varepsilon_{1,2}$ follows from $\tilde{\Sigma}$. If $k = 3$, the relation coincides with (c). For $k > 3$ we have

$$
\begin{aligned}
\varepsilon_{1,2}\varepsilon_{(k)} &= \\
(\text{from } \mathcal{IS}_n) &= \varepsilon_{1,2}(1,k)(1,3)(1,k)\cdot(1,k)\varepsilon_{(3)}\cdot(1,3)(1,k) \\
(\text{from } \mathcal{IS}_n \text{ and } \mathcal{T}_n) &= (1,k)(1,3)(1,k)\varepsilon_{1,2}\cdot\varepsilon_{(3)}(1,k)\cdot(1,3)(1,k) \\
(\text{by (c)}) &= (1,k)(1,3)(1,k)\cdot\varepsilon_{(3)}\varepsilon_{1,2}\cdot(1,k)(1,3)(1,k) \\
(\text{from } \mathcal{IS}_n \text{ and } \mathcal{T}_n) &= (1,k)(1,3)\cdot\varepsilon_{(3)}(1,k)\cdot(1,k)(1,3)(1,k)\varepsilon_{1,2} \\
(\text{from } \mathcal{IS}_n) &= \varepsilon_{(k)}\varepsilon_{1,2}.
\end{aligned}
$$

To prove the general case choose $t \neq 1,2$ and $\mu \in \mathcal{S}_n$ such that $\mu(1) = i$, $\mu(2) = j$ and $\mu(t) = k$. Using the previous relation we have

$$
\begin{aligned}
\varepsilon_{i,j}\varepsilon_{(k)} &= \mu\varepsilon_{1,2}\mu^{-1}\varepsilon_{(k)} \\
(\text{from } \mathcal{IS}_n) &= \mu\varepsilon_{1,2}\varepsilon_{(t)}\mu^{-1} \\
(\text{previous relation}) &= \mu\varepsilon_{(t)}\varepsilon_{1,2}\mu^{-1} \\
(\text{from } \mathcal{IS}_n) &= \varepsilon_{(k)}\mu\varepsilon_{1,2}\mu^{-1} \\
&= \varepsilon_{(k)}\varepsilon_{i,j}.
\end{aligned}
$$

$\square$

Lemma 9.4.4 *The relation $\varepsilon_{i,j}\varepsilon_{(j)} = \varepsilon_{(j)}$ follows from $\tilde{\Sigma}$.*

Proof. Let $\mu \in \mathcal{S}_n$ be such that $\mu(1) = i$, $\mu(2) = j$. We have

$$
\begin{aligned}
\varepsilon_{i,j}\varepsilon_{(j)} &= \mu\varepsilon_{1,2}\mu^{-1}\varepsilon_{(j)} \\
(\text{from } \mathcal{IS}_n) &= \mu\varepsilon_{1,2}\varepsilon_{(2)}\mu^{-1} \\
(\text{by (a)}) &= \mu\varepsilon_{(2)}\mu^{-1} \\
(\text{from } \mathcal{IS}_n) &= \varepsilon_{(j)}.
\end{aligned}
$$

$\square$

Lemma 9.4.5 *Let $\mathbf{N} = \{i_1, \ldots, i_{n-m}\} \cup \{j_1, \ldots, j_m\}$, where $m \geq 1$. Then for the element*

$$
\gamma = \begin{pmatrix} i_1 & \cdots & i_{n-m} & \{j_1, \ldots, j_m\} \\ i_1 & \cdots & i_{n-m} & i_1 \end{pmatrix} \in \mathcal{T}_n
$$

the relation $\gamma\varepsilon_{(j_1)}\cdots\varepsilon_{(j_m)} = \varepsilon_{(j_1)}\cdots\varepsilon_{(j_m)}$ follows from $\tilde{\Sigma}$.

Proof. Observe that $\gamma = \varepsilon_{i_1,j_1}\varepsilon_{j_1,j_2}\varepsilon_{j_2,j_3}\cdots\varepsilon_{j_{m-1},j_m}$. As the $\varepsilon_{(t)}$'s commute, the statement of the lemma follows from Lemma 9.4.4. $\square$

Lemma 9.4.6 Let $\mathbf{N} = \{i_1, \ldots, i_{n-m}\} \cup \{j_1, \ldots, j_m\}$, where $m \geq 1$. If $\alpha, \beta \in \mathcal{T}_n$ are such that $\alpha(i_k) = \beta(i_k)$ for all k, $1 \leq k \leq n - m$, then the relation

$$\alpha\varepsilon_{(j_1)} \cdots \varepsilon_{(jm)} = \beta\varepsilon_{(j_1)} \cdots \varepsilon_{(jm)} \tag{9.16}$$

follows from $\tilde{\Sigma}$.

Proof. Let γ be as in Lemma 9.4.5. Then the relation (9.16) can be rewritten as follows:

$$\alpha\gamma\varepsilon_{(j_1)} \cdots \varepsilon_{(jm)} = \beta\gamma\varepsilon_{(j_1)} \cdots \varepsilon_{(jm)}.$$

For α and β, satisfying our assumptions, we have $\alpha\gamma = \beta\gamma$ in $\mathcal{T}_n$. Hence $\alpha\gamma = \beta\gamma$ follows from $\tilde{\Sigma}$, which yields the statement of the lemma. $\square$

Proof of Theorem 9.4.1. One easily checks the relations (a)–(c) for the corresponding elements in $\mathcal{PT}_n$.

From the $\mathcal{IS}_n$ case (Lemmas 9.2.4 and 9.2.7) we know how to commute the $\varepsilon_{(i)}$'s with transpositions. From Lemmas 9.4.2 and 9.4.3 we know how to commute the $\varepsilon_{(i)}$'s with the $\varepsilon_{k,j}$'s. Using these rules, every word in the alphabet $\mathcal{A}$ can be written in the form

$$\alpha\varepsilon_{(j_1)} \cdots \varepsilon_{(jm)}, \ m \geq 0, \ \alpha \in \mathcal{T}_n.$$

Since the $\varepsilon_{(i)}$'s commute, we may assume $1 \leq j_1 < \cdots < j_m \leq n$.

It is left to show that in the semigroup $\mathcal{PT}_n$ every relation of the form

$$\alpha\varepsilon_{(j_1)} \cdots \varepsilon_{(jm)} = \beta\varepsilon_{(i_1)} \cdots \varepsilon_{(i_k)}, \tag{9.17}$$

where $\alpha, \beta \in \mathcal{T}_n$, $m, k \geq 0$, $1 \leq j_1 < \cdots < j_m \leq n$ and $1 \leq i_1 < \cdots < i_k \leq n$, follows from $\tilde{\Sigma}$. From the equalities

$$\overline{\mathrm{dom}}(\alpha\varepsilon_{(j_1)} \cdots \varepsilon_{(jm)}) = \{j_1, \ldots, j_m\},$$
$$\overline{\mathrm{dom}}(\beta\varepsilon_{(i_1)} \cdots \varepsilon_{(i_k)}) = \{i_1, \ldots, i_k\}$$

it follows that $k = m$ and $i_s = j_s$ for all $s = 1, \ldots, k$. Further from (9.17) it follows that $\alpha(l) = \beta(l)$ for any $l \in \mathbf{N} \backslash \{i_1, \ldots, i_k\}$. Hence the relation (9.17) has the form

$$\alpha\varepsilon_{(j_1)} \cdots \varepsilon_{(jm)} = \beta\varepsilon_{(j_1)} \cdots \varepsilon_{(jm)}. \tag{9.18}$$

If $m = 0$, we have an equality of two elements in $\mathcal{T}_n$, which follows from $\tilde{\Sigma}$ as the latter contains a system of defining relations for $\mathcal{T}_n$. If $m > 0$, the relation (9.18) follows from $\tilde{\Sigma}$ by Lemma 9.4.6. This completes the proof. $\square$

9.5   Addenda and Comments

9.5.1 Theorem 9.1.4 is proved by Aizenshtat in [Ai1]. Theorems 9.2.2 and 9.4.1 are taken from Sutov's announcement [Sut1], where proofs are only outlined. Theorem 9.3.1 is proved in [II].

9.5.2 The group $\mathcal{S}_n$ contains lots of generating systems, even irreducible ones (see for example Exercises 3.3.2 and 3.3.3). For many of them some systems of defining relations are known. In particular, for the generating system $\{\pi_i : i = 2, \ldots, n\}$ from Sect. 9.2 one of the systems of defining relations is proposed by Carmichael, see [Ca, p. 169]. This system looks as follows:

$$\pi_2^2 = \cdots = \pi_n^2 = (\pi_2\pi_3)^3 = \cdots = (\pi_{n-1}\pi_n)^3 = (\pi_n\pi_2)^3 = \varepsilon;$$
$$(\pi_i\pi_{i+1}\pi_i\pi_j)^2 = \varepsilon, \ 2 \le i,j \le n, j \notin \{i, i+1\}, \pi_{n+1} = \pi_2.$$

9.5.3 In representation theory one mostly uses the following system of *Coxeter* generators of $\mathcal{S}_n$:

$$\tau_1 = (1,2), \tau_2 = (2,3), \ldots, \tau_{n-1} = (n-1,n).$$

The name refers to the fact that $\mathcal{S}_n$ is an example of a *Coxeter group*, the latter being defined axiomatically via a fixed presentation. For $\mathcal{S}_n$ the corresponding system of defining relations looks as follows:

$$\tau_1^2 = \cdots = \tau_{n-1}^2 = \varepsilon;$$
$$(\tau_i\tau_j)^2 = \varepsilon, \ i \le j - 2;$$
$$(\tau_i\tau_{i+1})^3 = \varepsilon, \ 1 \le i \le n - 2.$$

This system was proposed by Moore in [Moo]. Using the first relation, the two last relations can be rewritten as follows:

$$\tau_i\tau_j = \tau_j\tau_i, \ i \le j - 2; \quad \tau_i\tau_{i+1}\tau_i = \tau_{i+1}\tau_i\tau_{i+1}.$$

The latter relations are known as *braid relations*.

9.5.4 Independently of Sutov, Popova in 1961 proposed a system of defining relations for $\mathcal{IS}_n$ with respect to the generating system $\{\tau_1, \ldots, \tau_{n-1}, \varepsilon_{(1)}\}$ (see [Pp]). This system consists of the relations from 9.5.3 and the following relations:

$$\varepsilon_{(1)} = \varepsilon_{(1)}^2 = \tau_2\varepsilon_{(1)}\tau_2 = \tau_{n-1}\tau_{n-2}\cdots\tau_2\varepsilon_{(1)}\tau_2\cdots\tau_{n-2}\tau_{n-1};$$
$$(\varepsilon_{(1)}\tau_1)^2 = \varepsilon_{(1)}\tau_1\varepsilon_{(1)} = (\tau_1\varepsilon_{(1)})^2. \tag{9.19}$$

Another system of defining relations for the same generating system can be found in [Li, Chap. 9]. It again consists of the relations from 9.5.3 and the following relations:

$$\varepsilon_{(1)}^2 = \varepsilon_{(1)}; \ (\varepsilon_{(1)}\tau_1)^2 = (\varepsilon_{(1)}\tau_1)^3; \ \varepsilon_{(1)}\tau_k = \tau_k\varepsilon_{(1)}, 2 \le k < n. \tag{9.20}$$

Solomon in [So] proposes a presentation for $\mathcal{IS}_n$ with respect to the generating system $\{\tau_1, \ldots, \tau_{n-1}, \nu = [1, 2, \ldots, n]\}$. It is obtained by adding to the relations from 9.5.3 the following relations (here $1 \leq i \leq n - 1$):

$$\tau_i \nu^{i+1} = \nu^{i+1}; \ \nu^{n-i+1}\tau_i = \nu^{n-i+1}; \ \nu\tau_i = \tau_{i+1}\nu;$$

$$\nu\tau_{n-1}\tau_{n-2}\cdots\tau_1\nu = \nu.$$

Delgado and Fernandes show in [DF] that the following collection of relations is a system of defining relations for $\mathcal{IS}_n$ with respect to the generating system $\{\pi = (1, 2), \tau = (1, 2, \ldots, n), \varepsilon_{(1)}\}$:

$$\pi^2 = \tau^n = (\tau\pi)^{n-1} = (\pi\tau^{n-1}\pi\tau)^3 = \varepsilon;$$

$$(\pi\tau^{n-k}\pi\tau^k)^2 = \varepsilon, \ 2 \leq k \leq n - 2;$$

$$\tau^{n-1}\pi\tau\varepsilon_{(1)}\tau^{n-1}\pi\tau = \tau\pi\varepsilon_{(1)}\pi\tau^{n-1} = \varepsilon_{(1)} = \varepsilon_{(1)}^2;$$

$$(\varepsilon_{(1)}\pi)^2 = \varepsilon_{(1)}\pi\varepsilon_{(1)} = (\pi\varepsilon_{(1)})^2.$$

9.5.5 Let S be a semigroup. A presentation $S = \langle A|\Sigma\rangle$ of S is called *irreducible* provided that A is an irreducible generating system and Σ is a minimal system of relations in the sense that any proper subset of Σ is not a system of defining relations for S with respect to A. Obviously, if A is irreducible and Σ is finite, then we can always find a subset Σ' of Σ such that the presentation $S = \langle A|\Sigma'\rangle$ is irreducible.

For the semigroups $\mathcal{IS}_n$, $\mathcal{PT}_n$, and $\mathcal{T}_n$, it is clear that every irreducible presentation of each of these semigroups contains an irreducible presentation of the symmetric group $\mathcal{S}_n$.

Irreducible presentations are of course "most effective". However, rather often some other properties, for example simplicity, existence of canonical forms, existence of effective algorithms for reduction to canonical forms, etc., are more important than irreducibility. With respect to irreducible presentations of $\mathcal{T}_n$ the following result of Aizenshtat from [Ai1] looks rather remarkable:

Theorem 9.5.1 *For every irreducible generating system of $\mathcal{T}_n$ there is an irreducible presentation of $\mathcal{T}_n$ with respect to this system, which is obtained adding not more than seven relations to the corresponding system of defining relations for $\mathcal{S}_n$.*

9.6 Additional Exercises

9.6.1 Prove that A is the only irreducible generating system for A^+.

9.6.2 Describe the congruence $\rho(S, A)$ for the cyclic semigroup $S = \langle a \rangle$, generated by the element a of type (k, m).

9.6.3 Let S be a semigroup, $A = S$, and Σ consist of all relations $s \cdot t = st$, $s, t \in S$. Prove that $S = \langle A | \Sigma \rangle$ is a presentation of S.

9.6.4 Let Σ denote some system of defining relations for the semigroup S with respect to the generating system A. Let Σ' be another system of relations with respect to A. Prove that Σ' is a system of defining relations if and only if every relation from Σ follows from Σ'.

9.6.5 ([KuMa2]) Consider for $\mathcal{IS}_n$ the generating system

$$A = \{\tau_1, \ldots, \tau_{n-1}, \varepsilon_{(1)}, \ldots, \varepsilon_{(n)}\}.$$

Show that a system of defining relations for A can be obtained by adding to the relations from 9.5.3 the following relations:

$$\varepsilon_{(i)}^2 = \varepsilon_{(i)}; \quad \varepsilon_{(i)}\varepsilon_{(j)} = \varepsilon_{(j)}\varepsilon_{(i)}; \quad \tau_i \varepsilon_{(i)}\varepsilon_{(i+1)} = \varepsilon_{(i)}\varepsilon_{(i+1)};$$
$$\tau_i \varepsilon_{(i)} = \varepsilon_{(i+1)}\tau_i; \quad \tau_i \varepsilon_{(j)} = \varepsilon_{(j)}\tau_i, \ j \neq i, i+1.$$

9.6.6 Prove that

$$\langle \varepsilon, \varepsilon_{(1)} | \varepsilon^2 = \varepsilon, \varepsilon_{(1)}^2 = \varepsilon_{(1)}, \varepsilon\varepsilon_{(1)} = \varepsilon_{(1)}\varepsilon = \varepsilon_{(1)} \rangle$$

is a presentation of $\mathcal{IS}_1$.

9.6.7 Show that the relation $(1,2)^2 = \varepsilon$ together with the relations

$$\varepsilon_{(1)}^2 = \varepsilon_{(1)}; \ \varepsilon_{(1)}\varepsilon_{(2)} = \varepsilon_{(2)}\varepsilon_{(1)}; \ (1,2)\varepsilon_{(1)}\varepsilon_{(2)} = \varepsilon_{(1)}\varepsilon_{(2)}; \ (1,2)\varepsilon_{(1)} = \varepsilon_{(2)}(1,2)$$

are defining relations for $\mathcal{IS}_2$ with respect to the generating system $\varepsilon_{(1)}$, $(1,2)$.

9.6.8 Write down a system of defining relations of $\mathcal{IS}_3$ with respect to the generating system $\{\varepsilon, \pi_2, \pi_3, [1,2]\}$.

9.6.9 ([Sut1]) Let

$$\alpha = \begin{pmatrix} \{i_1, \ldots, i_m\} & B_1 & \cdots & B_k \\ i & c_1 & \cdots & c_k \end{pmatrix} \in \mathcal{T}_n,$$

where $i \notin \{c_1, \ldots, c_k\}$. Using Lemmas 9.4.2 and 9.4.3 show that the relation $\alpha \varepsilon_{(i_1)} \cdots \varepsilon_{(i_m)} = \varepsilon_{(i)}\alpha$ follows from $\tilde{\Sigma}$.

9.6.10 Prove that every relation from (9.19) follows from relations (9.20).

Chapter 10

Transitive Actions

10.1 Action of a Semigroup on a Set

Let S be a semigroup and M be a nonempty set. An *action of S on M* is, roughly speaking, a realization of each element from S as a (partial) transformation of M such that the product in S corresponds to the composition of (partial) transformations. More formally, an *action of S on M* is a homomorphism from S to one of the semigroups $\mathcal{T}(M)$, $\mathcal{PT}(M)$, or $\mathcal{IS}(M)$. Depending on the choice of the latter semigroup, one speaks of the action of S on M by transformations, by partial transformations, or by partial permutations, respectively.

If S has a unit element, then it is usually required that it is represented by the identity transformation of M. In particular, if S is a group, then each element of S is represented by an invertible transformation of M and thus an *action of a group* on M is just a homomorphism from this group to the group of all invertible transformations of M (the symmetric group on M).

Let φ and ψ be actions of S on some sets M and N, respectively (we allow $N = M$). These actions are called *equivalent* or *similar* if they can be reduced to each other by some bijection between elements of M and N, that is, if there exists a bijection $\pi : M \to N$ such that the following diagram commutes for each $s \in S$:

$$
\begin{array}{ccc}
M & \xrightarrow{\ \varphi(s)\ } & M \\
\pi \downarrow & & \downarrow \pi \\
N & \xrightarrow{\ \psi(s)\ } & N
\end{array}
$$

(that is, $\pi\varphi(s) = \psi(s)\pi$ for all $s \in S$). If $M = N$, from Theorem 7.1.3 we get that an equivalent requirement is $\varphi = \Lambda_\pi \psi$, where Λ_π is the inner automorphism of $\mathcal{T}_n$, $\mathcal{PT}_n$, or $\mathcal{IS}_n$, respectively, which is induced by π.

To describe all actions of a given semigroup on a given set, even up to similarity, is usually a difficult problem. It is thus natural to consider

O. Ganyushkin, V. Mazorchuk, *Classical Finite Transformation Semigroups*, Algebra and Applications 9, DOI: 10.1007/978-1-84800-281-4_10,

some classes of actions, defined by some natural properties. One of the most natural properties of an action is transitivity. An action φ of S on M is called *transitive* provided that for any $x, y \in M$ (note that $x = y$ is allowed) there exists $s \in S$ such that $\varphi(s)$ maps x to y.

Let φ be an action of S on M and $N \subset M$. The subset N is said to be *invariant* with respect to φ provided that for every $m \in N$ and for every $s \in S$ such that $\varphi(s)(m)$ is defined we have $\varphi(s)(m) \in N$. The whole M and the empty subset are invariant with respect to any action. The action φ is called *quasi-transitive* provided that M cannot be written as a disjoint union of two nonempty subsets, invariant with respect to φ. Each transitive action is obviously quasi-transitive. The converse is true for groups; however, for semigroups it need not be true in the general case.

Exercise 10.1.1 Let S be a semigroup acting on M. Show that this action is transitive if and only if the only invariant subsets with respect to this action are M and the empty subset.

Exercise 10.1.2 Construct an example of a quasi-transitive but not transitive action of a finite semigroup S on a finite set M.

An action φ is called *faithful* provided that φ is injective. By Cayley's Theorem (Theorem 2.4.3), each finite semigroup admits a faithful action on a finite set by transformations. Analogously, by the Preston-Wagner Theorem (Theorem 2.9.5), each inverse semigroup admits a faithful action on a finite set by partial permutations.

Example 10.1.3 The identity maps define the actions of $\mathcal{S}_n$, $\mathcal{T}_n$, $\mathcal{PT}_n$, and $\mathcal{IS}_n$ on $\mathbf{N}$, which are called *natural*. These actions are obviously transitive and faithful. More general, any automorphism of $\mathcal{S}_n$, $\mathcal{T}_n$, $\mathcal{PT}_n$, and $\mathcal{IS}_n$ defines a faithful action of the respective semigroup on $\mathbf{N}$. For $\mathcal{T}_n$, $\mathcal{PT}_n$, and $\mathcal{IS}_n$ such actions are all similar to the natural action by Theorem 7.1.3. For $\mathcal{S}_n$ such actions are similar to the natural action unless $n = 6$, see 7.6.2.

Example 10.1.4 Mapping all elements of a semigroup S to the identity transformation of $\{1\}$ defines an action triv_S of S, which is called *trivial*. This action is obviously transitive; however, it is faithful if and only if $|S| = 1$.

Exercise 10.1.5 Let φ be an action of S on M. Show that φ is transitive if and only if for every $x, y \in M$ there exists a sequence $x = x_0, x_1, x_2, \ldots, x_k = y$ (where k may depend on both x and y) of elements from M and a sequence $s_1, s_2, \ldots, s_k$ of elements from S such that $\varphi(s_i)(x_{i-1}) = x_i$ for all i.

In the present chapter, we will describe all transitive actions of $\mathcal{T}_n$, $\mathcal{PT}_n$, and $\mathcal{IS}_n$ on finite sets by appropriate transformations up to similarity. Without loss of generality, we may assume that our semigroups act on the set $\mathbf{N}_k$ for some k. To start with we recall the description of transitive actions for groups.

10.2 Transitive Actions of Groups

Let G be a group acting on a set M via φ. To simplify the notation, if the action φ is fixed, for $g \in G$ and $m \in M$ we will write $g \cdot m$ instead of $\varphi(g)(m)$ (and we will use this notation for semigroups as well). The kernel $\mathrm{Ker}(\varphi)$ has the form ρ_H with respect to some normal subgroup $H \triangleleft G$ by Theorem 6.2.5. Abusing both the language and notation we call H the *kernel* of φ and denote it by $\mathrm{Ker}(\varphi)$.

Let $m \in M$. Then the set

$$\mathrm{St}_G(m) = \{g \in G \ : \ g \cdot m = m\}$$

is called the *stabilizer* of m with respect to the action φ. Note that $\mathrm{St}_G(m)$ is nonempty as it contains the identity element of G. From the definition we also have that

$$\mathrm{Ker}(\varphi) = \bigcap_{m \in M} \mathrm{St}_G(m). \tag{10.1}$$

Example 10.2.1 Consider the natural action of the symmetric group $\mathcal{S}_n$ on $\mathbf{N}$. Then the stabilizer of the element $n \in \mathbf{N}$ consists by definition of all $\pi \in \mathcal{S}_n$ such that $\pi(n) = n$. This stabilizer can be canonically identified with $\mathcal{S}_{n-1}$ via the restriction to the subset $\{1, 2, \ldots, n-1\}$, which is invariant with respect to all $\pi \in \mathrm{St}_{\mathcal{S}_n}(n)$.

Lemma 10.2.2 *Let G be a group acting on a set M.*

(i) For every $m \in M$ the set $\mathrm{St}_G(m)$ is a subgroup of G.

(ii) For every $m \in M$ and $g \in G$ we have $\mathrm{St}_G(m) = g^{-1}\mathrm{St}_G(g \cdot m)g$.

Proof. As we have already noted, the set $\mathrm{St}_G(m)$ contains the identity element $\mathbf{1}$ of G. If $g, h \in \mathrm{St}_G(m)$, then $g \cdot (h \cdot m) = g \cdot m = m$ and hence $gh \in \mathrm{St}_G(m)$. Finally, if $g \in \mathrm{St}_G(m)$, then applying g^{-1} to $m = g \cdot m$ we get

$$g^{-1} \cdot m = g^{-1} \cdot (g \cdot m) = g^{-1}g \cdot m = \mathbf{1} \cdot m = m \tag{10.2}$$

and hence $g^{-1} \in \mathrm{St}_G(m)$. The statement (i) follows.

Let $h \in \mathrm{St}_G(g \cdot m)$. Then, using (10.2) we get

$$g^{-1}hg \cdot m = g^{-1} \cdot (h \cdot (g \cdot m)) = g^{-1} \cdot (g \cdot m) = m,$$

hence $g^{-1}hg \in \mathrm{St}_G(m)$ and thus $g^{-1}\mathrm{St}_G(g \cdot m)g \subset \mathrm{St}_G(m)$. Analogously, we have $ghg^{-1} \in \mathrm{St}_G(g \cdot m)$ provided that $h \in \mathrm{St}_G(m)$, which implies $g\mathrm{St}_G(m)g^{-1} \subset \mathrm{St}_G(g \cdot m)$. Multiplying the latter inclusion with g^{-1} from the left and with g from the right, we get $\mathrm{St}_G(m) \subset g^{-1}\mathrm{St}_G(g \cdot m)g$ and the statement (ii) follows. $\square$

Let G be a group and H be a subgroup of G. Consider the set $G/H = \{gH : g \in G\}$ of all left cosets of G modulo H (see Sect. 6.2).

Proposition 10.2.3   *(i)* For $x, g \in G$ the assignment $x \cdot_H gH := xgH$ defines a transitive action of G on G/H.

(ii) The kernel of this action is the normal subgroup $\bigcap\limits_{g \in G} g^{-1} H g$ of G.

Proof. Let $\mathbf{1}$ denote the identity element of G. Then $\mathbf{1} \cdot_H gH := \mathbf{1}gH = gH$, and hence $\mathbf{1}$ acts as the identity transformation on G/H. Further, if $g, x, y \in G$, we have

$$x \cdot_H (y \cdot_H gH) = x \cdot_H ygH = xygH = xy \cdot_H gH,$$

which proves that $\cdot_H$ is indeed an action. As for any $x, y \in G$ we have $(xy^{-1}) \cdot_H yH = xy^{-1}yH = xH$, it follows that $\cdot_H$ is transitive. This proves the statement (i).

As $\mathbf{1} \cdot_H \mathbf{1}H = \mathbf{1}H$ and $\mathrm{St}_G(\mathbf{1}H) = H$, the statement (ii) follows from (10.1) and Lemma 10.2.2(ii). $\square$

Let G act transitively on M and let $m \in M$ be fixed. Set $H = \mathrm{St}_G(m)$ and for each $k \in M$ choose some $g_k \in G$ such that $g_k(m) = k$. Using these conventions, we have the following complete description of transitive group actions up to similarity:

Theorem 10.2.4   *(i)* The action of G on M is similar to the action $\cdot_H$ of G on G/H, defined in Proposition 10.2.3, via the mapping $f : M \to G/H$ defined as follows: $f(k) = g_k H$.

(ii) If H and F are subgroups of G, then the actions $\cdot_H$ and $\cdot_F$ are similar if and only if there exists $g \in G$ such that $H = g^{-1}Fg$ (in other words if and only if H and F are conjugate).

Proof. Let $g \in G$ and $k \in M$. On the one hand, we have $g \cdot_H f(k) = g \cdot_H g_k H = gg_k H$. On the other hand, let $g \cdot k = l$ for some $l \in M$. Then $f(g \cdot k) = f(l) = g_l H$. At the same time,

$$g_l^{-1} gg_k \cdot m = g_l^{-1} \cdot (g \cdot (g_k \cdot m)) = g_l^{-1} \cdot (g \cdot k) = g_l^{-1} \cdot l = m$$

by (10.2). This implies $g_l^{-1} gg_k \in H$ and thus $gg_k H = g_l H$. Therefore $g \cdot_H f(k) = f(g \cdot k)$, which proves the statement (i).

We first prove the sufficiency part of the statement (ii). Assume that $H = g^{-1}Fg$ for some $g \in G$. With respect to the action $\cdot_F$ we have $F = \mathrm{St}_G(F)$ by definition. Applying Lemma 10.2.2(ii) for the same action we obtain $\mathrm{St}_G(g^{-1}F) = g^{-1}\mathrm{St}_G(F)g = H$. Now the fact that $\cdot_H$ and $\cdot_F$ are similar follows from (i).

To prove the necessity part of (ii) assume that $\cdot_H$ and $\cdot_F$ are similar via some mapping $\alpha : G/H \rightarrow G/F$. Let $\alpha(H) = gF$ for some $g \in G$. Then for any $x \in H$ we have

$$x \cdot_F (gF) = x \cdot_F (\alpha(H)) = \alpha(x \cdot_H H) = \alpha(xH) = \alpha(H) = gF.$$

In particular, $H \subset \mathrm{St}_G(gF)$. If $x \notin H$, then $xH \neq H$ and thus $\alpha(xH) \neq gF$ since α is bijective, which implies $x \notin \mathrm{St}_G(gF)$. This means that $H = \mathrm{St}_G(gF)$. From Lemma 10.2.2(ii) it follows that H and F are conjugate. This completes the proof. $\square$

From Theorem 10.2.4, we see that all groups, in particular, the symmetric group $\mathcal{S}_n$, admit a lot of different transitive actions on finite sets. Moreover, from Proposition 10.2.3(ii) and 6.5.5 it follows that, apart from few exceptions, all such actions of $\mathcal{S}_n$ are faithful. It will be interesting to compare this result with the corresponding results for the semigroups $\mathcal{T}_n$, $\mathcal{PT}_n$, and $\mathcal{IS}_n$, which we will obtain in the next sections.

10.3   Transitive Actions of $\mathcal{T}_n$

In this section, we determine, up to similarity, all transitive actions of $\mathcal{T}_n$. As $\mathcal{T}_n$ is finite, it is enough to consider finite sets. The answer is given by the following:

Theorem 10.3.1 *Let* φ *be a transitive action of* $\mathcal{T}_n$ *on* $\mathbf{N}_m$ *for some* m. *Then either* $m = 1$ *and* φ *is the trivial action or* $m = n$ *and* φ *is similar to the natural action.*

To prove this theorem we will need the following lemma:

Lemma 10.3.2 *Let* ψ *be an action of* $\mathcal{T}_n$ *on* $\mathbf{N}_m$.

(i) *For every* $a \in \mathbf{N}$ *each block of the partition* $\rho_{\psi(0_a)}$ *of* $\mathbf{N}_m$ *is invariant with respect to* ψ.

(ii) *The set*

$$\bigcup_{a \in \mathbf{N}} \mathrm{im}(\psi(0_a))$$

is invariant with respect to ψ.

Proof. Let $x \in \mathbf{N}_m$ and $\alpha \in \mathcal{T}_n$. As 0_a is a left zero, we have

$$0_a \cdot (\alpha \cdot x) = (0_a \alpha) \cdot x = 0_a \cdot x$$

and the statement (i) follows.

Let $\alpha \in \mathcal{T}_n$, $a \in \mathbf{N}$ and $x = 0_a \cdot y$ for some $y \in \mathbf{N}_m$. Then we have

$$\alpha \cdot x = \alpha \cdot (0_a \cdot y) = (\alpha 0_a) \cdot y = 0_{\alpha(a)} \cdot y$$

and the statement (ii) follows. $\square$

Proof of Theorem 10.3.1. Assume first that φ is faithful. To distinguish the elements of $\mathcal{T}_n$ and $\mathcal{T}_m$ in this proof, we will use the notation $\alpha^{(m)}$ to emphasize that α is an element of $\mathcal{T}_m$. Because of Lemma 10.3.2(i), the transitivity of φ implies that for every $a \in \mathbf{N}$ the element $\varphi(0_a)$ must have rank one, that is, there exists $f(a) \in \mathbf{N}_m$ such that $\varphi(0_a) = 0_{f(a)}^{(m)}$. Because of Lemma 10.3.2(ii), the transitivity of φ implies that

$$\mathbf{N}_m = \bigcup_{a \in \mathbf{N}} \{f(a)\}.$$

As φ is faithful, all $f(a)$'s are different and we get $m = n$. Hence φ determines an automorphism of $\mathcal{T}_n$. By Theorem 7.1.3 every automorphism of $\mathcal{T}_n$ is a conjugation by some element from $\mathcal{S}_n$. By definition, the means that φ is similar to the natural action.

Now assume that φ is not faithful. Then, by Theorem 6.3.10 we have $\varphi(0_a) = \varphi(0_b)$ for all $a, b \in \mathbf{N}$. The same arguments as in the previous paragraph imply first that for all $a \in \mathbf{N}$ we have $\varphi(0_a) = 0_z^{(m)}$ for some fixed $z \in \mathbf{N}_m$; and then even that $\mathbf{N}_m = \{z\}$. Hence φ is the trivial action in this case. This completes the proof.   □

The following corollary now shows quite a striking difference with the symmetric group $\mathcal{S}_n$:

Corollary 10.3.3 *Every transitive and faithful action of $\mathcal{T}_n$ is similar to the natural action.*

10.4   Actions Associated with $\mathcal{L}$-Classes

In this section, we present one general construction of transitive actions by partial transformation for arbitrary semigroups.

Let S be a semigroup and $e \in \mathcal{E}(S)$. From Theorem 4.4.11 we know that the $\mathcal{H}$-class $\mathcal{H}(e)$ is a group. Let H be a subgroup of $\mathcal{H}(e)$. Consider the $\mathcal{L}$-class $\mathcal{L}(e)$. It is a disjoint union of $\mathcal{H}$-classes, so we can pick a representative in each such $\mathcal{H}$-class. Let $\mathfrak{a} = \{a_i : i \in I\}$ denote this collection of representatives. Note that $a_i \in S^1 e$ by definition and hence $a_i e = a_i$ for all $i \in I$. As we also have $e \in S^1 a_i$ by definition, there exists $a_i' \in S^1$ such that $a_i' a_i = e$. In this situation from Green's lemma we have that the assignment $x \mapsto a_i x$ defines a bijective mapping from $\mathcal{H}(e)$ to $\mathcal{H}(a_i)$, whose inverse is given by $y \mapsto a_i' y$. In particular, every element from $\mathcal{L}(e)$ admits a unique presentation as a product $a_i g$, where $i \in I$ and $g \in \mathcal{H}(e)$. Note further that $\{gH : g \in \mathcal{H}(e)\} = z\{gH : g \in \mathcal{H}(e)\}$ for any $z \in \mathcal{H}(e)$. This, in particular, means that the following set:

$$\mathcal{M}(e, H) = \{a_i g H : i \in I, g \in \mathcal{H}(e)\}$$

does not depend on the choice of $\mathfrak{a}$.

For every $s \in S$ we define the partial transformation $\xi_{e,H}(s)$ of the set $\mathcal{M}(e, H)$ by prescribing its value on $a_i g H$, $g \in \mathcal{H}(e)$, in the following way:

$$s \cdot a_i g H = \begin{cases} a_j g' g H, & s a_i = a_j g' \text{for some} j \in I, g' \in \mathcal{H}(e); \\ \varnothing, & s a_i \notin \mathcal{L}(e). \end{cases}$$

Theorem 10.4.1 *(i) $\xi_{e,H}$ is a transitive action of the semigroup S by partial transformations on the set $\mathcal{M}(e, H)$.*

(ii) If $\hat{e} \in \mathcal{E}(S)$ is such that $\hat{e} D e$, then there exists an appropriate subgroup $\hat{H}$ of $\mathcal{H}(\hat{e})$ such that the actions $\xi_{e,H}$ and $\xi_{\hat{e},\hat{H}}$ are similar.

Proof. Let $s, t \in S$ and $i \in I$. Then $st \cdot a_i g H$ is defined if and only if $S^1 s t a_i = S^1 a_i$. As $S^1 s t a_i \subset S^1 t a_i \subset S^1 a_i$, we have $S^1 s t a_i = S^1 a_i$ if and only if $S^1 s t a_i = S^1 t a_i$ and $S^1 t a_i = S^1 a_i$, in other words if and only if both $t \cdot a_i g H$ and $s \cdot t a_i g H$ are defined.

Now assume that $s t a_i, t a_i \in \mathcal{L}(e)$ and further that $t a_i = a_j g$, $s a_j = a_k h$ and $(st) a_i = a_l z$ for appropriate $j, k, l \in I$ and $g, h, z \in H$. We have

$$a_l z = (st) a_i = s(t a_i) = s(a_j g) = (s a_j) g = (a_k h) g = a_k(hg), \qquad (10.3)$$

which implies $k = l$ and $z = hg$ because of the uniqueness of the presentation of $s t a_i$. This proves that $\xi_{e,H}$ is indeed an action by partial transformations. The transitivity of this action follows immediately from the definitions and Green's lemma. The statement (i) follows.

To prove (ii) let us first assume that $\hat{e} \mathcal{L} e$. For such $\hat{e}$ from $\hat{e} \in S^1 e$ it follows that $\hat{e} e = \hat{e}$ and similarly $e \hat{e} = e$. In this case for $x, y \in \mathcal{H}(e)$ we have

$$\hat{e}(xy) = \hat{e} x e y = \hat{e} x e \hat{e} y = \hat{e} x \hat{e} y = (\hat{e} x)(\hat{e} y),$$

which means that the mapping $x \mapsto \hat{e} x$ from $\mathcal{H}(e)$ to $\mathcal{H}(\hat{e})$ is a group isomorphism. As $x \hat{e} = x$ for all $x \in \mathcal{L}(\hat{e})$, we have

$$\{a_i g H \ : \ g \in \mathcal{H}(e)\} = \{a_i \hat{e} g \hat{e} H \ : \ g \in \mathcal{H}(e)\} = \{a_i(\hat{e} g)(\hat{e} H) \ : \ \hat{e} g \in \mathcal{H}(\hat{e})\}.$$

This means that $\mathcal{M}(e, H) = \mathcal{M}(\hat{e}, \hat{e} H)$, which proves the statement (ii) in the case $\hat{e} \mathcal{L} e$.

If $\hat{e} \notin \mathcal{L}(e)$, we can pick some element $a \in \mathcal{D}(e)$ such that $a \mathcal{R} e$ and $a \mathcal{L} \hat{e}$. Let $u, v \in S$ be such that $au = e$ and $ev = a$. Then Green's lemma says that the mapping $\lambda_v : \mathcal{L}(e) \to \mathcal{L}(\hat{e})$, $x \mapsto xv$ is a bijection. Let $\mathcal{H}(\hat{e}) = \lambda_v(\mathcal{H}(a_i))$. From Theorem 4.7.5 it follows that the composition of λ_v with $x \mapsto a_i x$ induces a group isomorphism from $\mathcal{H}(e)$ to $\mathcal{H}(\hat{e})$. Let $\hat{H}$ denote the image of H under this isomorphism. Note that λ_v maps the set $a_k g H$ to the set

$$a_k g H v = a_k e g H v = a_k a_i' a_i e g H v = a_k a_i' a_i g H v = a_k a_i' a_i g v a_i H v.$$

As the element $\hat{g} = a_i g v$ belongs to $\mathcal{H}(e)$, we have $\hat{g} = \hat{e} \hat{g}$. Hence $a_k g H v = a_k a_i' \hat{e}(\hat{g} \hat{H})$, moreover, the factor $a_k a_i' \hat{e} = a_k a_i' a_i e v = a_k e v = a_k v$ belongs to $\mathcal{L}(\hat{e})$. Then λ_v induces a bijection from $\mathcal{M}(e, H)$ to $\mathcal{M}(\hat{e}, \hat{e}H)$, which by definition commutes with the left multiplication with elements from S. The statement (ii) follows and the proof is complete. $\qquad\square$

From Theorem 10.4.1 it follows that, up to similarity, the action $\xi_{e,H}$ defined above depends only on the choice of a regular $\mathcal{D}$-class and a subgroup H in the abstract group, which is isomorphic to every maximal subgroup of S, contained in this $\mathcal{D}$-class.

Remark 10.4.2 If $s \in S$ is such that the two-sided ideal $S^1 s S^1$ does not intersect the $\mathcal{L}$-class $\mathcal{L}_e$, then the domain of the transformation $\xi_{e,H}(s)$ is empty. In particular, if the set

$$\{s \in S \,:\, S^1 s S^1 \cap \mathcal{L}_e = \varnothing\}$$

contains more than one element, then the action $\xi_{e,H}$ is not faithful.

10.5 Transitive Actions of $\mathcal{PT}_n$ and $\mathcal{IS}_n$

In the previous section, we constructed some natural transitive actions of semigroups. In this section we will show that for the semigroups $\mathcal{PT}_n$ and $\mathcal{IS}_n$ these are all transitive actions.

Theorem 10.5.1   (i) Let S be one of the semigroups $\mathcal{PT}_n$ or $\mathcal{IS}_n$. Then each transitive action of S is either trivial or similar to $\xi_{e,H}$ for some $e \in \mathcal{E}(S)$ and a subgroup H of $\mathcal{H}(e)$.

(ii) For the semigroup $\mathcal{IS}_n$ each action $\xi_{e,H}$ is an action by partial permutations.

Proof. Let φ be a transitive action of S on some set M. As S is finite and φ is transitive, we have that M is finite as well. We split the proof of the statement (i) into a sequence of observation. We start with the following easy one:

Lemma 10.5.2 *Every $x \in \mathrm{im}(\varphi(\mathbf{0}))$ is invariant with respect to φ.*

Proof. Let $\alpha \in S$ and $x = \mathbf{0} \cdot y$ for some $x, y \in M$. Then

$$\alpha \cdot x = \alpha \cdot (\mathbf{0} \cdot y) = (\alpha \mathbf{0}) \cdot y = \mathbf{0} \cdot y = x$$

and the claim follows. $\qquad\square$

Assume now that $\mathrm{im}(\varphi(\mathbf{0})) \neq \varnothing$. In this case from Lemma 10.5.2 and the transitivity of φ we derive $M = \mathrm{im}(\varphi(\mathbf{0}))$, moreover, M consists of exactly one element. Hence φ is the trivial action.

From now on we assume that $\mathrm{im}(\varphi(\mathbf{0})) = \varnothing$. Let $k > 0$ be minimal for which there exists an idempotent e of rank k such that $\mathrm{im}(\varphi(e)) \neq \varnothing$. Our principal observation is the following:

Lemma 10.5.3 M *is a disjoint union of* $\mathrm{im}(\varphi(\varepsilon_A))$, $A \subset \mathbf{N}$, $|A| = k$.
Moreover, $\mathrm{im}(\varphi(\varepsilon_A)) \neq \varnothing$ *for all such* A.

Proof. First we claim that $\mathrm{im}(\varphi(\varepsilon_A)) \neq \varnothing$ for each $A \subset \mathbf{N}$, $|A| = k$. Indeed, if we assume that $\mathrm{im}(\varphi(\varepsilon_A)) = \varnothing$, then $\mathrm{im}(\varphi(\alpha)) = \varnothing$ for all $\alpha \in S\varepsilon_A S$. However, $e \in S\varepsilon_A S$ by Theorem 4.2.8, which contradicts our assumption $\mathrm{im}(\varphi(e)) \neq \varnothing$.

Now we claim that $\mathrm{im}(\varphi(\varepsilon_A)) \cap \mathrm{im}(\varphi(\varepsilon_B)) = \varnothing$ for any $A \neq B \subset \mathbf{N}$, $|A| = |B| = k$. Indeed, as $A \neq B$ and $|A| = |B| = k$ we have $|A \cap B| < k$ and thus for any $x \in \mathrm{im}(\varphi(\varepsilon_B))$ we obtain

$$\varepsilon_A \cdot (\varepsilon_B \cdot x) = (\varepsilon_A \varepsilon_B) \cdot x = \varepsilon_{A \cap B} \cdot x = \varnothing$$

by our assumptions. This implies $\mathrm{im}(\varphi(\varepsilon_A)) \cap \mathrm{im}(\varphi(\varepsilon_B)) = \varnothing$.

Finally we claim that the set

$$N = \bigcup_{A \subset \mathbf{N}, |A|=k} \mathrm{im}(\varphi(\varepsilon_A))$$

is invariant with respect to φ. Let $\alpha \in S$ and $x \in \mathrm{im}(\varphi(\varepsilon_A))$ for some $A \subset \mathbf{N}$ such that $|A| = k$. Then $\alpha \cdot x = \alpha \cdot (\varepsilon_A \cdot x) = \alpha\varepsilon_A \cdot x$. If $\mathrm{rank}(\alpha\varepsilon_A) < k$, then from $\alpha\varepsilon_A = \varepsilon_{\mathrm{im}(\alpha\varepsilon_A)}\alpha\varepsilon_A$ it follows that $\mathrm{im}(\varphi(\alpha\varepsilon_A)) \subset \mathrm{im}(\varphi(\varepsilon_{\mathrm{im}(\alpha\varepsilon_A)})) = \varnothing$, which means that $\alpha \cdot x = \varnothing$. Hence we may assume $\mathrm{rank}(\alpha\varepsilon_A) = k$. Then for $B = \mathrm{im}(\alpha\varepsilon_A)$ we have $|B| = k$, implying $\alpha\varepsilon_A = \varepsilon_B \alpha\varepsilon_A$. From this we obtain

$$\varepsilon_B \cdot (\alpha \cdot x) = \varepsilon_B \cdot (\alpha\varepsilon_A \cdot x) = (\varepsilon_B \alpha\varepsilon_A) \cdot x = \alpha\varepsilon_A \cdot x = \alpha \cdot (\varepsilon_A \cdot x) = \alpha \cdot x,$$

implying $\alpha \cdot x \in \mathrm{im}(\varphi(\varepsilon_B))$.

The statement of the lemma now follows from the transitivity of φ. This completes the proof. $\square$

Our next step is the following:

Lemma 10.5.4 *Let* $A \subset \mathbf{N}$, $|A| = k$. *Then* φ *induces a transitive action of the group* $\mathcal{H}(\varepsilon_A)$ *on* $\mathrm{im}(\varphi(\varepsilon_A))$.

Proof. By Exercise 10.1.1 we have to prove that the only subsets of the set $\mathrm{im}(\varphi(\varepsilon_A))$ invariant with respect to the restriction of φ to $\mathcal{H}(\varepsilon_A)$ are the whole $\mathrm{im}(\varphi(\varepsilon_A))$ and the empty set. Assume that this is not the case and let $X \subset \mathrm{im}(\varphi(\varepsilon_A))$ be a proper invariant subset.

Let $\alpha \in S$ be such that $\mathrm{rank}(\alpha\varepsilon_A) = k$. Then either $\mathrm{im}(\alpha\varepsilon_A) = A$, in which case $\alpha\varepsilon_A \in \mathcal{H}(\varepsilon_A)$, or $\mathrm{im}(\alpha\varepsilon_A) = B \neq A$ for some $B \subset \mathbf{N}$, $|B| = k$. In the second case, from $\alpha\varepsilon_A = \varepsilon_B\alpha\varepsilon_A$ it follows that $\alpha \cdot \mathrm{im}(\varepsilon_A) \subset \mathrm{im}(\varepsilon_B)$. As X is invariant with respect to the action of $\mathcal{H}(\varepsilon_A)$, for any $x \in X$ we have

$$\{\alpha \cdot x : \alpha \in S\} \subset X \cup \bigcup_{B \subset \mathbf{N}, |B|=k, B \neq A} \mathrm{im}(\varphi(\varepsilon_B)) \neq M$$

by our assumptions and Lemma 10.5.3. This contradicts the transitivity of
the action φ. $\qquad\square$

Let $A = \{1, 2, \ldots, k\}$ and $e = \varepsilon_A$. For $B \subset \mathbf{N}$, $|B| = k$, let α_B denote
the unique increasing bijection from A to B. Note that all α_B are elements
of $\mathcal{IS}_n$, in particular, we have their inverse elements α_B^{-1} in $\mathcal{IS}_n$. Then
$\mathfrak{a} = \{\alpha_B : B \subset \mathbf{N}, |B| = k\}$ contains exactly one representative of each
$\mathcal{H}$-class in $\mathcal{L}(e)$. Finally, fix some $x \in \mathrm{im}(\varphi(\varepsilon_A))$ and set $H = \mathrm{St}_{\mathcal{H}(\varepsilon_A)}(x)$.
Our final step in the proof of the statement (i) is the following:

Lemma 10.5.5 *The mapping* $f : \mathcal{M}(e, H) \to M$, $\alpha_B \beta H \mapsto \alpha_B \beta \cdot x$ *(here*
$\alpha_B \in \mathfrak{a}$ *and* $\beta \in \mathcal{H}(e)$*) produces a similarity between the actions* $\xi_{e,H}$ *and* φ.

Proof. For $\alpha \in S$ we have $\alpha \cdot x = \alpha \cdot (e \cdot x) = \alpha e \cdot x$. If $\alpha \cdot x \neq \varnothing$, then
$\mathrm{rank}(\alpha e) = k$, in particular, $\alpha e \in \mathcal{L}(e)$. This implies $M = S \cdot x = \mathcal{L}(e) \cdot x$.
Since each element of $\mathcal{L}(e)$ belongs to some $\alpha_i \beta H$, from the transitivity of
φ we get that f is surjective.

Assume $\alpha_B \beta \cdot x = \alpha_C \gamma \cdot x$. Then $B = C$ follows from Lemma 10.5.3.
We further have $\alpha_B^{-1} \alpha_B \beta \cdot x = \alpha_B^{-1} \alpha_B \gamma \cdot x$. Note that $\alpha_B^{-1} \alpha_B = \varepsilon_A$ by
construction, which yields $\beta \cdot x = \gamma \cdot x$. Hence $\gamma^{-1} \beta \in H$ and $\beta H = \gamma H$.
This implies that f is injective, and hence bijective.

Let now $\alpha \in S$. If $\mathrm{rank}(\alpha \alpha_B) < k$, then both $\alpha \cdot \alpha_B \beta H = \varnothing$ and
$\alpha \cdot (\alpha_B \beta \cdot x) = (\alpha \alpha_B \beta) \cdot x = \varnothing$. If $\mathrm{rank}(\alpha \alpha_B) = k$, then $\alpha \alpha_B \in \mathcal{L}(e)$ can
be uniquely written as $\alpha \alpha_B = \alpha_C \gamma$ for some $\gamma \in \mathcal{H}(e)$. Then from the
definitions we have

$$f(\alpha \cdot \alpha_B \beta H) = f(\alpha_C \gamma \beta H) = \alpha_C \gamma \beta \cdot x = \alpha \alpha_B \beta \cdot x =$$
$$= \alpha \cdot (\alpha_B \beta \cdot x) = \alpha \cdot f(\alpha_B \beta H),$$

which completes the proof of the lemma. $\qquad\square$

The statement (i) is proved.

Now we prove the statement (ii). We retain the notation from the paragraph before Lemma 10.5.5. By Theorem 10.5.1 we may assume $e = \varepsilon_A$. Let
$\beta \in \mathcal{IS}_n$ be such that $\beta \alpha_B \pi H = \beta \alpha_C \tau H \in \mathcal{M}(e, H)$ for some $\alpha_B, \alpha_C \in \mathfrak{a}$
and $\pi, \tau \in \mathcal{H}(\varepsilon_A)$. Then

$$\beta^{-1} \beta \alpha_B \pi H = \beta^{-1} \beta \alpha_C \tau H \in \mathcal{M}(e, H) \qquad (10.4)$$

as well since $\beta^{-1} \beta \alpha_B \pi H \notin \mathcal{M}(e, H)$ would imply for any $h \in H$ the inequality $\mathrm{rank}(\beta^{-1} \beta \alpha_B \pi h) < \mathrm{rank}(\beta \alpha_B \pi h)$ and thus

$$\mathrm{rank}(\beta \alpha_B \pi h) = \mathrm{rank}(\beta \beta^{-1} \beta \alpha_B \pi h) \leq \mathrm{rank}(\beta^{-1} \beta \alpha_B \pi h) < \mathrm{rank}(\beta \alpha_B \pi h),$$

a contradiction.

We have $\beta^{-1} \beta = \varepsilon_D$ for some $D \subset \mathbf{N}$. As both α_B and α_C are elements
of rank k, from (10.4) it follows that both $\varepsilon_D \alpha_B$ and $\varepsilon_D \alpha_C$ have rank k

as well. This yields $\varepsilon_D \alpha_B = \alpha_B$ and $\varepsilon_D \alpha_C = \alpha_C$. Now the equality (10.4) implies $\alpha_B \pi H = \alpha_C \tau H$, which shows that $\xi_{e,H}(\beta)$ is a partial permutation. This completes the proof. $\qquad\square$

Corollary 10.5.6 *Let S be one of the semigroups $\mathcal{PT}_n$ or $\mathcal{IS}_n$. Then every transitive and faithful action of S is similar to the natural action.*

Proof. By Theorem 10.5.1, any action of S is equivalent to some $\xi_{e,H}$. From Theorem 4.2.8 and the construction of the action $\xi_{e,H}$ it follows that for any $\alpha \in S$ such that $\operatorname{rank}(\alpha) < \operatorname{rank}(e)$ we have $S^1 \alpha S^1 \cap \mathcal{L}(e) = \varnothing$. Hence from Remark 10.4.2 we derive that $\xi_{e,H}$ is not faithful unless $\operatorname{rank}(e) = 1$.

If $\operatorname{rank}(e) = 1$, then $\mathcal{H}(e) \cong \mathcal{S}_1$ by Theorem 5.1.4(ii). As $\mathcal{S}_1$ contains exactly one subgroup, from Theorem 10.5.1 we obtain that S has, up to similarity, at most one transitive and faithful action. Our statement now follows from the obvious observation that the natural action is both transitive and faithful (see Example 10.1.3). $\qquad\square$

10.6 Addenda and Comments

10.6.1 The notion of similarity admits a categorical definition. Consider the category $\mathfrak{A}$ defined as follows: The objects of $\mathfrak{A}$ are triples (S, M, φ), where S is a semigroup, M is a nonempty set, and φ is an action of S on M by some fixed type of transformations. If (S, M, φ) and (S', M', φ') are two objects, then $\mathfrak{A}\big((S, M, \varphi), (S', M', \varphi')\big)$ is the set of all pairs (α, f), where $\alpha : S \to S'$ is a homomorphism and $f : M \to M'$ is a mapping, such that for all $s \in S$ the following diagram commutes:

$$
\begin{array}{ccc}
M & \xrightarrow{\varphi(s)} & M \\
{\scriptstyle f}\downarrow & & \downarrow{\scriptstyle f} \\
M' & \xrightarrow{\varphi'(\alpha(s))} & M'.
\end{array}
$$

The composition of two morphisms $(\alpha, f) \in \mathfrak{A}\big((S, M, \varphi), (S', M', \varphi')\big)$ and $(\alpha', f') \in \mathfrak{A}\big((S', M', \varphi'), (S'', M'', \varphi'')\big)$ is defined to be $(\alpha'\alpha, f'f)$. For a fixed S let $\mathfrak{A}^S$ denote the subcategory whose objects are (S, M, φ) for all possible M and φ, and morphisms are (id, f) for all possible f. Using the categorical language, similar actions of S are exactly the actions which can be obtained from each other by composing with an isomorphism in $\mathfrak{A}^S$. Both categories $\mathfrak{A}$ and $\mathfrak{A}^S$ seem to be quite complicated and not much is known about them.

10.6.2 The results presented in Sects. 10.3–10.5 are special cases of a more general result obtained by Ponizovskiy in [Po1, Po2]. A special case of Corollary 10.5.6 for $\mathcal{IS}_n$ appears also in [Vl1]. In fact, Ponizovskiy classified all transitive actions by partial transformations of all finite semigroups. His

answer of course includes the actions $\xi_{e,H}$ constructed in Sect. 10.4; however, in the general case one has to consider technically more complicated
actions on $\mathcal{L}$-classes as well. These actions generalize the action $\xi_{e,H}$ and
have the form $\mathcal{L}(e)/\rho$, where ρ is a left compatible equivalence relation on
$\mathcal{L}(e)$. Combinatorial description of such relations even for the semigroup $\mathcal{T}_n$
is complicated.

10.6.3 The actions considered in this chapter are *left* actions. One defines
right actions in the obvious dual way. Left actions of S are naturally the same
as right actions of $\overleftarrow{S}$. As $\mathcal{IS}_n$ is inverse, description of transitive right $\mathcal{IS}_n$-
actions is simply the dual to the description of the transitive left actions,
presented in Theorem 10.5.1. For the semigroups $\mathcal{T}_n$ and $\mathcal{PT}_n$ description
of right actions by partial transformations is much more complicated and
requires the general approach of [Po1]. At the same time an analogue of
Theorem 10.3.1 for the semigroup $\overleftarrow{\mathcal{T}_n}$ trivializes, see Exercise 10.7.7. This
shows that right actions for transformation semigroups are less natural than
left actions.

10.6.4 There is a rather advanced and well developed abstract theory of
semigroups acting on sets, see for example [CP2, Chap. 11]. There is also a
well-developed abstract theory of semigroup actions, in particular, transitive
actions, by binary relations. This one was developed by Schein, see [Sc2, Sc3].

10.6.5 Schein also developed in [Sc1] a general theory of actions of inverse
semigroups in terms of the so-called *closed subsemigroups*. Let S be an inverse semigroup. A subsemigroup H of S is said to be *closed* provided that
$\alpha \in H$ implies $\beta \in H$ for every $\beta \in S$ such that $\alpha^{-1}\alpha = \beta^{-1}\alpha$ (for elements of $\mathcal{IS}_n$ the latter condition is equivalent to $\mathrm{dom}(\alpha) \subset \mathrm{dom}(\beta)$ and
$\alpha(x) = \beta(x)$ for any $x \in \mathrm{dom}(\alpha)$). For a closed inverse subsemigroup H of a
semigroup S Schein naturally defines cosets of S modulo H and constructs
a transitive action of S by partial permutations on the set of all cosets.
Applications of this approach to $\mathcal{IS}_n$ are discussed in [Vl1].

The above notion of closed subsemigroups admits a natural generaliza-
tion to $\mathcal{PT}_n$: a subsemigroup H of $\mathcal{PT}_n$ is said to be *closed* provided that
$\alpha \in H$ implies $\beta \in H$ for every $\beta \in S$ such that $\mathrm{dom}(\alpha) \subset \mathrm{dom}(\beta)$ and
$\alpha(x) = \beta(x)$ for any $x \in \mathrm{dom}(\alpha)$. However, as far as we know, there is
no classification of closed subsemigroups of $\mathcal{PT}_n$ similar to that of closed
inverse subsemigroups of $\mathcal{IS}_n$, given in [Vl1, Theorem 1] (see also Exercise 10.7.10 below). We also do not know of any description of arbitrary
closed subsemigroups of $\mathcal{PT}_n$.

10.6.6 Let φ be an action of some semigroup S on some set M. The action
φ is said to be *semitransitive* provided that for any $x, y \in M$ (we allow
$x = y$) there exists some $s \in S$ such that either $s \cdot x = y$ or $s \cdot y = x$.
Each transitive action is obviously semitransitive. The converse is not true

in general. The easiest general example is the following: Let S be a finite monoid with unit $\mathbf{1}$ and $G = G_\mathbf{1} \neq S$. Let further φ be an action of G on some set M. Consider the set $\overline{M} = M \cup \{*\}$ (we assume $* \notin M$) and define the action $\overline{\varphi}$ of S on $\overline{M}$ as follows:

$$\overline{\varphi}(s)(m) = \begin{cases} \varphi(s)(m), & s \in G,\ m \in M; \\ *, & \text{otherwise.} \end{cases}$$

It is straightforward to verify that $\overline{\varphi}$ is an action. It is further easy to see that $\overline{\varphi}$ is semitransitive if and only if φ is transitive.

The notion of a semitransitive action is a natural generalization of that of a transitive action. However, it seems that for the moment not much is known about such actions, even for classical semigroups. Some first results in the study of semitransitive actions of inverse semigroups are obtained in [C-O], where the reader will also find some references to the literature, where such actions naturally appear.

10.7 Additional Exercises

10.7.1 Let S denote one of the semigroups $\mathcal{S}_n$, $\mathcal{T}_n$, $\mathcal{PT}_n$, or $\mathcal{IS}_n$. For $\alpha \in S$ and $X \in \mathcal{B}(\mathbf{N})$ define

$$\alpha \cdot X = \{\alpha(x) : x \in X \cap \text{dom}(\alpha)\}.$$

Show that this defines an action of S on $\mathcal{B}(\mathbf{N})$. Will this action be faithful? transitive? quasi-transitive?

10.7.2 Is it true that a quasi-transitive action of an inverse semigroup by partial permutations is transitive?

10.7.3 Let S denote one of the semigroups $\mathcal{S}_n$, $\mathcal{T}_n$, $\mathcal{PT}_n$, or $\mathcal{IS}_n$. The group $\text{Aut}(S)$ acts on S in the natural way. Is this action faithful? transitive? quasi-transitive?

10.7.4 Let S denote one of the semigroups $\mathcal{S}_n$, $\mathcal{T}_n$, $\mathcal{PT}_n$, or $\mathcal{IS}_n$. The semigroup $\text{End}(S)$ acts on S in the natural way. Is this action faithful? transitive? quasi-transitive?

10.7.5 Show that the assignment $\alpha \cdot \pi = \alpha\pi\alpha^{-1}$ defines an action of $\mathcal{IS}_n$ on itself by transformations. Is this action faithful? transitive? quasi-transitive?

10.7.6 Let S be a semigroup. Show that the assignment $s \cdot x = sx$ defines an action of S on itself by transformations. Prove that this action is transitive if and only if S has exactly one $\mathcal{L}$-class.

10.7.7 Describe all transitive actions of $\overleftarrow{\mathcal{T}_n}$ by transformations.

10.7.8 Show that any semitransitive action of a group is transitive.

10.7.9 ([KuMa4]) Let S be a finite inverse monoid. For $x \in S$ let e_x denote the idempotent, which has the form x^k for some k.

(a) Prove that the assignment

$$a \cdot x = \begin{cases} axa^{-1}, & a^{-1}ae_x = e_x; \\ \varnothing, & \text{otherwise,} \end{cases}$$

where $a, x \in S$, defines an action of S by partial transformations on itself.

(b) Prove that for $x, y \in S$ the elements x and y are conjugate (that is, $x \sim_S y$) if and only if there exist $a, b, z \in S$ such that $a \cdot x = b \cdot y = z$.

10.7.10 ([Vl1]) Let $A \subset \mathbf{N}$ and H be a subgroup of $\mathcal{H}(\varepsilon_A)$. Denote by $C(A, H)$ the subsemigroup of $\mathcal{IS}_n$, which consists of all α for which there exists $\beta \in H$ such that $\alpha(x) = \beta(x)$ for all $x \in A$.

(a) Show that $C(A, H)$ is a closed inverse subsemigroup of $\mathcal{IS}_n$.

(b) Show that every closed inverse subsemigroup of the semigroup $\mathcal{IS}_n$ is of the form $C(A, H)$ for appropriate A and H.

10.7.11 ([Vl2, Vl3]) Describe all closed inverse subsemigroups and all transitive actions (up to similarity) for every finite inverse semigroup S satisfying the following condition: there exists $a \in S$ such that the minimal inverse subsemigroup of S containing a coincides with S.

10.7.12 Classify up to similarity all transitive actions of $\mathcal{T}_n$ by partial permutations on finite sets.

10.7.13 Classify up to similarity all transitive actions of $\mathcal{PT}_n$ by partial permutations on finite sets.

10.7.14 Prove that every transitive action of $\mathcal{IS}_n$ by partial permutations on a finite set remains transitive when restricted to $\mathcal{S}_n$.

10.7.15 ([Po2]) (a) Let φ be an action of a finite inverse semigroup S by partial permutations on a finite set M. Prove that there exists a decomposition $M = M_1 \cup M_2 \cup \cdots \cup M_k$ into invariant subsets such that for any $i \in \{1, 2, \ldots, k\}$ either $|M_i| = 1$ or for any $x, y \in M_i$ there exists $s \in S$ such that $\varphi(s)(x) = y$.

(b) Prove that the statement (a) is no longer true in general if instead of an action by partial permutations one considers an action by partial transformations.

Chapter 11

Linear Representations

11.1 Representations and Modules

In this chapter, we work over the field $\mathbb{C}$ of complex numbers, in particular, all vector spaces are vector spaces over $\mathbb{C}$.

Let S be a semigroup and V a vector space (which we always assume to be finite-dimensional). A *representation* of S in V is a homomorphism φ from S to $\mathrm{End}_{\mathbb{C}}(V)$, the semigroup of all linear transformations of V. If S is a monoid, one additionally requires that φ maps the identity element from S to the identity transformation on V.

An alternative but equivalent notion is that of a *module*. If V is a vector space, then one says that V has the structure of an *S-module* (alternatively is a *module over S*) if we are given a mapping $S \times V \to V$, $(s, v) \mapsto s \cdot v$ such that the following conditions are satisfied:

- $s \cdot (v + w) = s \cdot v + s \cdot w$ for all $s \in S$, $v, w \in V$

- $s \cdot (\lambda v) = \lambda(s \cdot v)$ for all $s \in S$, $v \in V$ and $\lambda \in \mathbb{C}$

- $s \cdot (t \cdot v) = st \cdot v$ for all $s, t \in S$ and $v \in V$

- $\mathbf{1} \cdot v = v$ for all $v \in V$ if $\mathbf{1}$ is the identity element of S

Exercise 11.1.1 Show that the notion of a representation is equivalent to that of a module.

The language of modules usually requires simpler notation and hence we will mostly use it. Let V be an S-module. A subspace $W \subset V$ is called a *submodule* of V provided that $s \cdot w \in W$ for all $w \in W$ and $s \in S$. Obviously, the whole V and the zero vector space 0 are submodules of any module. An S-module V is called *simple* provided that it is nonzero and the only submodules of V are V and 0. The corresponding representation is called *irreducible*.

O. Ganyushkin, V. Mazorchuk, *Classical Finite Transformation Semigroups*, Algebra and Applications 9, DOI: 10.1007/978-1-84800-281-4_11,
© Springer-Verlag London Limited 2009

If V is an S-module and W is a submodule of V, then the quotient space V/W carries the natural structure of an S-module given by

$$s \cdot (v + W) = sv + W. \tag{11.1}$$

This module is called the *quotient* of V modulo W.

Exercise 11.1.2 Verify that (11.1) indeed defines an S-module structure on V/W.

Example 11.1.3 Let S be any semigroup. Then the assignment $s \cdot c = c$, $c \in \mathbb{C}$, defines on $\mathbb{C}$ the structure of an S-module. This module is called the *trivial S-module* and is usually denoted by $\mathbb{C}_{\mathrm{triv}}$. The trivial module is simple because $\mathbb{C}$, being one-dimensional, has only two subspaces, namely, $\mathbb{C}$ and 0. The same argument implies that every one-dimensional S-module is simple.

Example 11.1.4 Let S denote one of the semigroups $\mathcal{S}_n$, $\mathcal{IS}_n$, $\mathcal{T}_n$, or $\mathcal{PT}_n$. For $\alpha \in S$ let $\varphi(\alpha) = (m_{i,j})_{i,j=1}^n$ denote the $n \times n$-matrix such that

$$m_{i,j} = \begin{cases} 1, & \alpha(j) = i; \\ 0, & \text{otherwise.} \end{cases}$$

For example, we have

$$\varphi\left(\left(\begin{array}{cccccc} 1 & 2 & 3 & 4 & 5 & 6 \\ 2 & 1 & 1 & \varnothing & 5 & 3 \end{array}\right)\right) = \begin{pmatrix} 0 & 1 & 1 & 0 & 0 & 0 \\ 1 & 0 & 0 & 0 & 0 & 0 \\ 0 & 0 & 0 & 0 & 0 & 1 \\ 0 & 0 & 0 & 0 & 0 & 0 \\ 0 & 0 & 0 & 0 & 1 & 0 \\ 0 & 0 & 0 & 0 & 0 & 0 \end{pmatrix}.$$

As α is a (partial) transformation, we obtain that each column of the matrix $\varphi(\alpha)$ contains at most one nonzero element. It is easy to check that for $v \in \mathbb{C}^n$ the assignment $\alpha \cdot v = \varphi(\alpha)v$ defines on $\mathbb{C}^n$ the structure of an S-module. This module is called the *natural S-module*.

If $\alpha \in \mathcal{S}_n$, then every row and every column of $\varphi(\alpha)$ contains exactly one nonzero element. If $\alpha \in \mathcal{IS}_n$, then every row and every column of $\varphi(\alpha)$ contains at most one nonzero element. The latter corresponds to placements of nonattacking rooks on an $n \times n$ chessboard. Because of this interpretation, the semigroup $\mathcal{IS}_n$ is sometimes called the *rook monoid*.

Exercise 11.1.5 Verify that the natural module is indeed a module, and show that the natural module is simple if and only if $S = \mathcal{IS}_n$ or $S = \mathcal{PT}_n$.

Exercise 11.1.6 Let M be an S-module. Show that M is simple if and only if for any nonzero $m \in M$ the linear span of $s \cdot m$, $s \in S$, coincides with M.

Let V and W be two S-modules. A linear mapping $f : V \to W$ is called a *homomorphism* or an *S-homomorphism* provided that it commutes with the action of all elements from S, that is, for every $s \in S$ the following diagram is commutative:

$$
\begin{array}{ccc}
V & \xrightarrow{\;s\cdot\;} & V \\
{\scriptstyle f}\downarrow & & \downarrow{\scriptstyle f} \\
W & \xrightarrow{\;s\cdot\;} & W.
\end{array}
$$

The trivial example of a homomorphism is the zero mapping. A less trivial example is the identity homomorphism in the case $V = W$. The set of all S-homomorphisms from V to W is denoted by $\mathrm{Hom}_S(V, W)$. This set carries the natural structure of a vector space.

Two S-modules V and W are called *isomorphic* or *equivalent* provided that $\mathrm{Hom}_S(V, W)$ contains an isomorphism. The fact that V and W are isomorphic is denoted by $V \cong W$. As usually, for most of the questions about S-modules, in particular, classification problems, it makes sense to formulate and study them only up to isomorphism. Some basic facts about homomorphisms and isomorphisms are collected in the following statement:

Proposition 11.1.7 *Let V and W be S-modules and $f \in \mathrm{Hom}_S(V, W)$.*

(i) The kernel $\mathrm{Ker}(f)$ of f is a submodule of V.

(ii) The image $\mathrm{im}(f)$ of f is a submodule of W.

(iii) If both V and W are simple module, then f is either an isomorphism or zero.

Proof. Let $v \in V$ be such that $f(v) = 0$. Then $f(s \cdot v) = s \cdot f(v) = s \cdot 0 = 0$ and hence $s \cdot v \in \mathrm{Ker}(f)$. This proves the statement (i).

Let $v \in V$. Then $s \cdot f(v) = f(s \cdot v)$, which implies the statement (ii).

If V is simple, then for the vector space $\mathrm{Ker}(f)$, which is a submodule of V by (i), we have only two possibilities: $\mathrm{Ker}(f) = V$ or $\mathrm{Ker}(f) = 0$. In the first case, f is the zero mapping. In the second case, f is injective. Now $\mathrm{im}(f)$ is a submodule of W by (ii). As W is simple, we have two options: $\mathrm{im}(f) = 0$ or $\mathrm{im}(f) = W$. The first one is not possible as $V \neq 0$ and f is injective. For the second one, we obtain that f is surjective, hence it is an isomorphism. $\qquad\square$

Corollary 11.1.8 *Let V and W be two simple S-modules. Then*

$$
\mathrm{Hom}_S(V, W) \cong
\begin{cases}
\mathbb{C}, & V \cong W; \\
0, & V \not\cong W.
\end{cases}
$$

Proof. If $V \not\cong W$, then the fact that $\mathrm{Hom}_S(V, W) = 0$ follows immediately from Proposition 11.1.7(iii).

If $V \cong W$, we have an obvious isomorphism of vector spaces between $\mathrm{Hom}_S(V, W)$ and $\mathrm{Hom}_S(V, V)$. Let $f \in \mathrm{Hom}_S(V, V)$ be a nonzero homomorphism. As $\mathbb{C}$ is algebraically closed, f has an eigenvalue, $\lambda \in \mathbb{C}$ say. Then $f - \lambda\varepsilon \in \mathrm{Hom}_S(V, V)$ (here ε is the identity transformation of V). As $f - \lambda\varepsilon$ has a nontrivial kernel, we have $f - \lambda\varepsilon = 0$ by Proposition 11.1.7(iii). Thus $f = \lambda\varepsilon$ and we are done. $\square$

Example 11.1.9 Let S be a finite semigroup. Consider the vector space $\mathbb{C}S$ with the formal basis $\{\mathbf{v}_t : t \in S\}$. For every $s \in S$ define a linear operator on $\mathbb{C}S$ by prescribing its action on basis elements in the following way: $s \cdot \mathbf{v}_t = \mathbf{v}_{st}$. We have

$$r \cdot (s \cdot \mathbf{v}_t) = r \cdot \mathbf{v}_{st} = \mathbf{v}_{rst} = rs \cdot \mathbf{v}_t,$$

which implies that the above assignment defines on $\mathbb{C}S$ the structure of an S-module. This module is called the *regular S-module*.

On the one hand, for every $a \in S$ the linear mapping $f_a : \mathbb{C}S \to \mathbb{C}S$ defined via $f_a(\mathbf{v}_t) = \mathbf{v}_{ta}$ is an endomorphism of $\mathbb{C}S$ since for any $s \in S$ we have

$$f_a(s \cdot \mathbf{v}_t) = f_a(\mathbf{v}_{st}) = \mathbf{v}_{sta} = s \cdot \mathbf{v}_{ta} = s \cdot f_a(\mathbf{v}_t).$$

On the other hand, if S is a monoid with identity $\mathbf{1}$, then any homomorphism $f \in \mathrm{Hom}_S(\mathbb{C}S, \mathbb{C}S)$ is uniquely determined by its value on $\mathbf{v_1}$ since

$$f(\mathbf{v}_s) = f(s \cdot \mathbf{v_1}) = s \cdot f(\mathbf{v_1})$$

for all $s \in S$. Hence if $f(\mathbf{v_1}) = \sum_{a \in S} c_a \mathbf{v}_a = \sum_{a \in S} c_a f_a(\mathbf{v_1})$ for some $c_a \in \mathbb{C}$, we have $f = \sum_{a \in S} c_a f_a$. The homomorphisms f_a's are linearly independent as the set $\{\mathbf{v}_a = f_a(\mathbf{v_1}) : a \in S\}$ is a basis of $\mathbb{C}S$. This means that we have an isomorphism of vector spaces

$$\mathrm{Hom}_S(\mathbb{C}S, \mathbb{C}S) \cong \mathbb{C}S.$$

Exercise 11.1.10 Prove that the homomorphisms $\{f_g : g \in S\}$ form a semigroup with respect to the composition. Further, show that this semigroup is isomorphic to $\overleftarrow{S}$.

11.2 $\mathcal{L}$-Induced S-Modules

In this section, we present a construction of some S-modules using modules over maximal subgroup. The idea is very similar to the one described in Sect. 10.4.

Let S be a semigroup for which all regular $\mathcal{L}$-classes are finite. Fix some $e \in \mathcal{E}(S)$ and let M denote some $\mathcal{H}(e)$-module. Let $\mathfrak{a} = \{a_i : i \in I\}$

be a fixed collection of representatives from all $\mathcal{H}$-classes inside $\mathcal{L}(e)$, one for each $\mathcal{H}$-class. In particular, I is finite and indexes $\mathcal{R}$-classes inside the $\mathcal{D}$-class $\mathcal{D}(e)$. Then, analogously to Sect. 10.4, we have that the mapping $x \mapsto a_i x$ is a bijection from $\mathcal{H}(e)$ to $\mathcal{H}(a_i)$ for every $i \in I$, in particular, every $y \in \mathcal{L}(e)$ can be uniquely written in the form $a_i x$ for some $i \in I$ and $x \in \mathcal{H}(e)$. For $i \in I$ let $M^{(i)}$ denote a copy of M and $\eta_i : M \to M^{(i)}$ denote the canonical identification. Consider the vector space

$$V_{\mathfrak{a}}(M) = \bigoplus_{i \in I} M^{(i)}.$$

Note that $\dim(V_{\mathfrak{a}}(M)) = m \cdot \dim(M)$, where m is the number of $\mathcal{H}$-classes inside $\mathcal{L}(e)$ (this number is finite by our assumptions). For $s \in S$, $i \in I$ and $v \in M$ define

$$s \cdot \eta_i(v) = \begin{cases} \eta_j(x \cdot v), & sa_i = a_j x, j \in I, x \in \mathcal{H}(e); \\ 0, & \text{otherwise.} \end{cases}$$

Proposition 11.2.1 *(i) The above assignment defines on the vector space $V_{\mathfrak{a}}(M)$ the structure of an S-module.*

(ii) Let $\mathfrak{a}'$ be another collection of representatives of $\mathcal{H}$-classes inside $\mathcal{L}(e)$. Then $V_{\mathfrak{a}}(M) \cong V_{\mathfrak{a}'}(M)$.

Proof. Let $s, t \in S$, $i \in I$ and $v \in M$. Because of the first paragraph of the proof of Theorem 10.4.1 we have that $sta_i \in \mathcal{L}(e)$ implies that $ta_i \in \mathcal{L}(e)$ as well. Hence it is enough to consider the case $sta_i, ta_i \in \mathcal{L}(e)$. In this case let $ta_i = a_j g$, $sa_j = a_k h$ and $sta_i = a_l z$. Then from (10.3) we have $l = k$ and $z = hg$. Hence

$$st \cdot \eta_i(v) = \eta_k(z \cdot v) = \eta_k(hg \cdot v) = \eta_k(h \cdot (g \cdot v)) = s \cdot \eta_j(g \cdot v) = s \cdot (t \cdot \eta_i(v)).$$

The statement (i) follows.

To prove (ii) we consider $\mathfrak{a}$ and $\mathfrak{a}'$. As $a_i' \in \mathcal{H}(a_i)$ for all $i \in I$, we can find $g_i \in \mathcal{H}(e)$ such that $a_i' = a_i g_i$. Define now the linear mapping $f : V_{\mathfrak{a}}(M) \to V_{\mathfrak{a}'}(M)$ as follows: For $v \in M$ and $i \in I$ let $f_i : M^{(i)} \to M^{(i)}$ be the linear mapping defined via $f_i(\eta_i(v)) = \eta_i(g_i^{-1} \cdot v)$, and set $f = \oplus_{i \in I} f_i$. We consider f as a linear mapping from $V_{\mathfrak{a}}(M)$ to $V_{\mathfrak{a}'}(M)$. As M is an $\mathcal{H}(e)$-module, $\mathcal{H}(e)$ is a group and $g_i \in \mathcal{H}(e)$, we obtain that f_i is an isomorphism. This yields that f is an isomorphism. Now let $s \in S$, $i \in I$ and $v \in M$. If $sa_i \notin \mathcal{L}(e)$, then $s \cdot f(\eta_i(v)) = f(s \cdot \eta_i(v)) = 0$ is obvious. So, we assume $sa_i = a_j x$ for some $j \in I$ and $x \in \mathcal{H}(e)$, in particular,

$$sa_i' = sa_i g_i = a_j x g_i = a_j' g_j^{-1} x g_i. \tag{11.2}$$

To distinguish the actions on $V_\mathfrak{a}(M)$ and $V_{\mathfrak{a}'}(M)$ we for the moment will use the notation $\star$ for the action on $V_{\mathfrak{a}'}(M)$. Using (11.2) we compute

$$
\begin{aligned}
s \star f(\eta_i(v)) &= s \star f_i(\eta_i(v)) \\
&= s \star \eta_i(g_i^{-1} \cdot v) \\
&= \eta_j(g_j^{-1} x g_i \cdot (g_i^{-1} \cdot v)) \\
&= \eta_j(g_j^{-1} x \cdot v) \\
&= \eta_j(g_j^{-1} \cdot (x \cdot v)) \\
&= f_j(\eta_j(x \cdot v)) \\
&= f(\eta_j(x \cdot v)) \\
&= f(s \cdot \eta_i(v)).
\end{aligned}
$$

The statement (ii) follows and the proof is complete.   □

Because of Proposition 11.2.1(ii) in the isomorphism class we can remove the index $\mathfrak{a}$ in the notation $V_\mathfrak{a}(M)$. We will use simply the notation $V(M)$ in the sequel. The module $V(M)$ will be called an S-module $\mathcal{L}$-*induced* from M. Now we would also like to remove the dependence on the choice of e inside a given $\mathcal{D}$-class.

Proposition 11.2.2 *Let* $e, \hat{e} \in \mathcal{E}(S)$ *be such that* $e \mathcal{D} \hat{e}$ *and let* M *be an* $\mathcal{H}(e)$-*module. Then there exists an* $\mathcal{H}(\hat{e})$-*module* $\hat{M}$ *such that* $V(M) \cong V(\hat{M})$.

Proof. Using the above notation, we consider the module $V_\mathfrak{a}(M)$ and assume $e \in \mathfrak{a}$. Let $l \in I$ be such that $a_l \mathcal{R} \hat{e}$. Set $u = a_l$. By Green's Lemma the mapping $x \mapsto ux$ is a bijection from $\mathcal{R}(e)$ to $\mathcal{R}(u)$, which preserves $\mathcal{L}$-classes. Hence there exists $v \in \mathcal{R}(e) \cap \mathcal{L}(\hat{e})$ such that $uv = \hat{e}$.

For all $i \in I$ define $\hat{a}_i = a_i v$, and set $\hat{\mathfrak{a}} = \{\hat{a}_i : i \in I\}$. In particular, $ev = v$ and $uv = \hat{e}$ belong to $\hat{\mathfrak{a}}$. We also have $vu = e$ by the proof of Theorem 4.7.5. Set further $\hat{M} = M^{(l)}$. Let $x \in \mathcal{H}(\hat{e})$ and x' be the inverse to x in $\mathcal{H}(\hat{e})$. Then $(xu)v = x\hat{e} = x$, implying $xu \mathcal{R} x$ by Proposition 4.4.2, in particular, $xu \mathcal{R} u$ as $u \mathcal{R} x$ by our assumptions. At the same time $x'(xu) = \hat{e}u = u$. This implies $xu \mathcal{L} u$ by Proposition 4.4.1 and thus $xu \mathcal{H} u$. Hence $xu = ux'$ for some $x' \in \mathcal{H}(e)$, which by construction means that $\hat{M}$ is stable under the action of $\mathcal{H}(\hat{e})$, in particular, is an $\mathcal{H}(\hat{e})$-module. For $i \in I$ let $\hat{M}^{(i)}$ denote a copy of $\hat{M}$, and $\hat{\eta}_i : \hat{M} \to \hat{M}^{(i)}$ denote the canonical identification. We thus obtain the module $V_{\hat{\mathfrak{a}}}(\hat{M})$.

Define the linear mapping $\varphi : V_\mathfrak{a}(M) \to V_{\hat{\mathfrak{a}}}(\hat{M})$ as follows: $\varphi(\eta_i(m)) = \hat{\eta}_i(\eta_l(m))$, $i \in I$, $m \in M$. By construction, φ is an isomorphism of vector spaces.

To prove our statement for $s \in S$, $i \in I$ and $m \in M$ we have to check the following:

$$s \cdot \varphi(\eta_i(m)) = \varphi(s \cdot \eta_i(m)). \tag{11.3}$$

Note that $sa_i \notin \mathcal{L}(e)$ if and only if $sa_iv = s\hat{a}_i \notin \mathcal{L}(\hat{e})$, in which case both sides of (11.3) are zero. Hence we may assume $sa_i \in \mathcal{L}(e)$. Let $sa_i = a_j x$ for some $j \in I$ and $x \in \mathcal{H}(e)$. Observe that $x = ex = vux$ and hence

$$s\hat{a}_i = sa_iv = a_jxv = a_jvuxv = a_jv(uxv) = \hat{a}_j(uxv).$$

We have $uxv \cdot u = uxe = u \cdot x = a_lx$ and thus the left-hand side of (11.3) equals (we again use the notation $\star$ similarly to the one used in the proof of Proposition 11.2.1)

$$
\begin{aligned}
s \star \varphi(\eta_i(m)) &= s \star \hat{\eta}_i(\eta_l(m)) \\
&= \hat{\eta}_j(uxv \cdot \eta_l(m)) \\
&= \hat{\eta}_j(\eta_l(x \cdot m)).
\end{aligned}
$$

On the other hand, the right-hand side of (11.3) equals

$$
\begin{aligned}
\varphi(s \cdot \eta_i(m)) &= \varphi(\eta_j(x \cdot m)) \\
&= \hat{\eta}_j(\eta_l(x \cdot m)).
\end{aligned}
$$

This proves that φ is an isomorphism, and the statement of the proposition follows. $\qquad\square$

Remark 11.2.3 From the construction used in Proposition 11.2.2 it follows that under the identification of $\mathcal{H}(e)$ and $\mathcal{H}(\hat{e})$, constructed in Theorem 4.7.5, the modules M and $\hat{M}$ are isomorphic.

11.3 Simple Modules over $\mathcal{IS}_n$ and $\mathcal{PT}_n$

In this section, we describe all simple modules for the semigroups $\mathcal{IS}_n$ and $\mathcal{PT}_n$.

For $k \in \{0, 1, 2, \ldots, n\}$ let $\varepsilon_n^{(k)}$ denote the idempotent of $\mathcal{IS}_n$ with domain $\{1, 2, \ldots, k\}$. In particular, $\varepsilon_n^{(n)} = \varepsilon$ and $\varepsilon_n^{(0)} = \mathbf{0}$. Using Theorem 5.1.4 we identify $\mathcal{H}(\varepsilon_n^{(k)})$ with $\mathcal{S}_k$ in the natural way. Our main result in this section is the following:

Theorem 11.3.1 *Let* $S = \mathcal{IS}_n$ *or* $S = \mathcal{PT}_n$.

(i) *For any* $k \in \{0, 1, 2, \ldots, n\}$ *and any simple* $\mathcal{S}_k$-*module* M *the module* $V(M)$ *is a simple* S-*module.*

(ii) *Let* $k, \hat{k} \in \{0, 1, 2, \ldots, n\}$, *and* M *and* $\hat{M}$ *be simple* $\mathcal{S}_k$- *and* $\mathcal{S}_{\hat{k}}$-*modules, respectively. Then* $V(M) \cong V(\hat{M})$ *if and only if* $k = \hat{k}$ *and* $M \cong \hat{M}$.

(iii) *Every simple* S-*module is isomorphic to the module* $V(M)$ *for some* $k \in \{0, 1, 2, \ldots, n\}$ *and some simple* $\mathcal{S}_k$-*module* M.

Proof. The $\mathcal{H}$-classes inside $\mathcal{L}(\varepsilon_n^{(k)})$ are indexed by $I = \{A \subset \mathbf{N} : |A| = k\}$ by Theorem 4.5.1. Set for simplicity $\mathbf{A} = \{1, 2, \ldots, k\}$.

Let M be a simple $\mathcal{S}_k$-module and $v \in V(M)$ be a nonzero element. Then $v = \sum_{A \in I} c_A v_A$, where $c_A \in \mathbb{C}$ and $v_A \in M^{(A)}$. We may assume $v_A \neq 0$ for all $A \in I$. Let N be the linear span of $s \cdot v$, $s \in S$. Obviously N is a submodule of $V(M)$. As $v \neq 0$, there exists $B \in I$ such that $c_B \neq 0$. Assume that $\mathfrak{a} = \{\alpha_A : A \in I\}$ as in Sect. 10.5. Then $\varepsilon_A \alpha_A = \alpha_A$ and $\varepsilon_A \cdot v_A = v_A$ for every $A \in I$. If $A \in I$ is such that $A \neq B$, then $\mathrm{rank}(\varepsilon_B \varepsilon_A) < k$ and hence

$$\varepsilon_B \cdot \sum_{A \in I} c_A v_A = \sum_{A \in I} c_A \varepsilon_B \varepsilon_A \cdot v_A = c_B v_B \neq 0.$$

As $c_B \neq 0$, we have $v_B \in N$, in particular, $v_B = \eta_B(w)$ for some nonzero $w \in M$. As $\alpha_B^{-1} \alpha_B = \varepsilon_n^{(k)}$, we have $\alpha_B^{-1} \cdot v_B = \alpha_B^{-1} \cdot \eta_B(w) = w$. Hence $w \in N$. Since the $\mathcal{S}_k$-module M is simple and $w \neq 0$, the minimal $\mathcal{S}_k$-invariants subspace of $M^{(\mathbf{A})}$ containing w must be the whole $M^{(\mathbf{A})}$, implying $M^{(\mathbf{A})} \subset N$. By definition, $\alpha_A \cdot M^{(\mathbf{A})} = M^{(A)}$, which means $M^{(A)} \subset N$ for all $A \in I$. This finally implies $N = V(M)$. Now the fact that $V(M)$ is simple follows from Exercise 11.1.6. This proves the statement (i).

If $k = \hat{k}$ and $M \cong \hat{M}$, then $V(M) \cong V(\hat{M})$ is obvious. Hence we assume that $k, \hat{k} \in \{0, 1, 2, \ldots, n\}$ and M and $\hat{M}$ are such that there exists an isomorphism $\varphi : V(M) \cong V(\hat{M})$.

Assume first that $k \neq \hat{k}$. Without loss of generality we may assume $k < \hat{k}$. Then $\varepsilon_n^{(k)} \cdot V(\hat{M}) = 0$ by definition, whereas $\varepsilon_n^{(k)} \cdot M^{(\mathbf{A})} = M^{(\mathbf{A})}$. At the same time, the action of $\varepsilon_n^{(k)}$ on $V(\hat{M})$ is just a conjugate of the corresponding action on $V(M)$ by the isomorphism φ, a contradiction. This proves that $k = \hat{k}$.

As $V(M) = \oplus_{A \in I} M^{(A)}$ and $\varepsilon_n^{(k)} \alpha_A \notin \mathcal{L}(\varepsilon_n^{(k)})$ if $A \neq \mathbf{A}$, we have the equality $\varepsilon_n^{(k)} \cdot V(M) = M^{(\mathbf{A})}$. Analogously $\varepsilon_n^{(k)} \cdot V(\hat{M}) = \hat{M}^{(\mathbf{A})}$. As φ commutes with the action of $\varepsilon_n^{(k)}$ we obtain that $\varphi(M^{(\mathbf{A})}) \subset \hat{M}^{(\mathbf{A})}$. Analogously, one shows that $\varphi^{-1}(\hat{M}^{(\mathbf{A})}) \subset M^{(\mathbf{A})}$. As both mappings are injective, we conclude that the restriction of φ to $M^{(\mathbf{A})}$ is an isomorphism with image $\hat{M}^{(\mathbf{A})}$. Since φ commutes with the action of S, it, in particular, commutes with the action of $\mathcal{S}_k$. This proves the statement (ii).

Let L be a simple S-module and $k \in \{0, 1, 2, \ldots, n\}$ be minimal such that there exists an element $\alpha \in S$ of rank k, which acts on L in a nonzero way. As $\alpha \in S \varepsilon_n^{(k)} S$ by Theorem 4.2.8, it follows that $\varepsilon_n^{(k)}$ acts on L in a nonzero way. Set $M = \varepsilon_n^{(k)} \cdot L \neq 0$.

Lemma 11.3.2 *M is a simple $\mathcal{S}_k$-module.*

Proof. Obviously, M is an $\mathcal{S}_k$-module. Assume that this is not the case and let N be a proper $\mathcal{S}_k$-submodule of M. For $A \in I$ set

$$L_A = \alpha_A \alpha_A^{-1} \cdot L = \varepsilon_A \cdot L.$$

By the same arguments as above we have $L_A \neq 0$. Further, as $\varepsilon_A \varepsilon_B \notin \mathcal{L}(\varepsilon_n^{(k)})$ if $A \neq B$, it also follows that the sum of all L_A's inside L is direct.

The minimal subspace of L, which is invariant with respect to the action of S and contains N, is the linear span $\hat{N}$ of $s \cdot v$, $s \in S$ and $v \in N$. For any $s \in S$ and any $v \in N$ we have $s \cdot v = s\varepsilon_n^{(k)} \cdot v$. If $s\varepsilon_n^{(k)} \notin \mathcal{L}(\varepsilon_n^{(k)})$, then $\text{rank}(s\varepsilon_n^{(k)}) < k$ and we have $s \cdot v = 0$ by our assumption on k. If $s\varepsilon_n^{(k)} \in \mathcal{L}(\varepsilon_n^{(k)})$, then there exists $A \in I$ and $\beta \in \mathcal{S}_k$ such that $s\varepsilon_n^{(k)} = \alpha_A \beta$. As $\varepsilon_A \alpha_A = \alpha_A$, we obtain

$$\hat{N} \subset N \oplus \bigoplus_{A \neq \mathbf{A}} L_A \neq L,$$

which contradicts the fact that L is a simple S-module. This completes the proof. □

Define a linear mapping $\varphi : V(M) \to L$ as follows: $\varphi(\eta_A(v)) = \alpha_A \cdot v$ for all $v \in M$ and $A \in I$. Obviously, $\varphi \neq 0$ as $\varphi(v) = v$ for any $v \in M$.

We claim that φ is a homomorphism of S-modules. To prove this we have to check

$$s \cdot \varphi(\eta_A(v)) = \varphi(s \cdot \eta_A(v)) \qquad (11.4)$$

for all $s \in S$, $A \in I$, and $v \in M$. If $\text{rank}(s\alpha_A) < k$, then $s \cdot \eta_A(v) = 0$ by definition and $s \cdot \varphi(\eta_A(v)) = s\alpha_A \cdot v = 0$ by our assumption on k. If $\text{rank}(s\alpha_A) = k$, we can write $s\alpha_A = \alpha_B \beta$ for some $B \in I$ and $\beta \in \mathcal{H}(\varepsilon_n^{(k)})$. We compute

$$
\begin{aligned}
s \cdot \varphi(\eta_A(v)) &= s \cdot (\alpha_A \cdot v) \\
&= s\alpha_A \cdot v \\
&= \alpha_B \beta \cdot v \\
&= \alpha_B \cdot (\beta \cdot v) \\
&= \varphi(\eta_B(\beta \cdot v)) \\
&= \varphi(s \cdot \eta_A(v)).
\end{aligned}
$$

This proves (11.4). So, we have constructed a nonzero homomorphism φ from $V(M)$ to L. The module $V(M)$ is simple by Lemma 11.3.2 and the statement (i), the module L is simple by assumption. Now (iii) follows from Proposition 11.1.7(iii). □

Corollary 11.3.3 *All simple $\mathcal{IS}_n$ modules are obtained from simple $\mathcal{PT}_n$-modules by restriction, in particular, the corresponding simple $\mathcal{IS}_n$- and $\mathcal{PT}_n$-modules have the same dimension.*

Proof. Follows from Theorem 11.3.1, construction of $V(M)$ and the fact that the $\mathcal{L}$-classes of $\varepsilon_n^{(k)}$ in $\mathcal{IS}_n$ and $\mathcal{PT}_n$ coincide. □

11.4 Effective Representations

Let S be a semigroup and $\varphi : S \to \mathrm{End}_{\mathbb{C}}(V)$ be a representation of S. The representation φ is called *effective* provided that it is injective. The same notion is used for the corresponding module.

Theorem 11.4.1 *Let S denote one of the semigroups $\mathcal{T}_n$, $\mathcal{PT}_n$, or $\mathcal{IS}_n$.*

(i) If V is an effective S-module, then $\dim(V) \geq n$.

(ii) The natural S-module is effective.

(iii) If V is an effective S-module of dimension n, then V is isomorphic to the natural S-module.

Proof. The statement (ii) is obvious. To prove the statement (i) let us first assume that $S = \mathcal{PT}_n$ or $S = \mathcal{IS}_n$. Consider the element $\alpha = [1, 2, \ldots, n]$. We have $\alpha^n = \mathbf{0}$ and $\alpha^k \neq \mathbf{0}$ for any $k < n$. Note that $\mathbf{0}$ is an idempotent and $\mathbf{0}\alpha^k = \alpha^k \mathbf{0} = \mathbf{0}$. If V is now an effective S-module, then α defines a linear operator on V, whose Jordan decomposition contains either Jordan cells of size 1 with eigenvalue 1 or nilpotent Jordan cells. Moreover, it must contain at least one nilpotent Jordan cell of dimension n. Hence $\dim(V) \geq n$.

For $S = \mathcal{T}_n$ we consider

$$\beta = \begin{pmatrix} 1 & 2 & 3 & 4 & \ldots & n \\ 1 & 1 & 2 & 3 & \ldots & n-1 \end{pmatrix}.$$

We have $\beta^{n-1} = 0_1$ and $\beta^k \neq 0_1$ for all $k < n - 1$. Hence if now V is an effective $\mathcal{T}_n$-module, then, by the same arguments as above, β defines a linear operator on V, whose Jordan decomposition must have at least one nilpotent Jordan cell of dimension $(n-1)$. Furthermore, the element 0_1 acts on V in a nonzero way, for otherwise all $0_a \in \mathcal{T}_n 0_1 \mathcal{T}_n$ would act as zeros and V could not be effective. This implies that β also contains at least one Jordan cell with eigenvalue 1. Thus $\dim(V) \geq n$ again and the statement (i) is proved.

To prove (iii) we again first consider the case $S = \mathcal{PT}_n$ or $S = \mathcal{IS}_n$. Our first observation is:

Lemma 11.4.2 *Let $S = \mathcal{PT}_n$ or $S = \mathcal{IS}_n$ and V be an effective module of dimension n. Then $\mathbf{0} \cdot V = 0$.*

Proof. Let $W = \mathbf{0} \cdot V$ and assume $W \neq 0$. As $\mathbf{0}$ is an idempotent, for any $s \in S$ and $w \in W$ we have

$$s \cdot w = s \cdot (\mathbf{0} \cdot w) = (s\mathbf{0}) \cdot w = \mathbf{0} \cdot w = w.$$

In particular, W is a submodule of V. As we have seen above, $s \cdot w = w$ for all $w \in W$. As V is effective, we get that for any $s \neq t \in S$ there should exist $v \in V$ such that $s \cdot v \neq t \cdot v$. Assume $s \cdot v = t \cdot v + w$ for some $w \in W$. Applying $\mathbf{0}$ we obtain $\mathbf{0} \cdot (s \cdot v) = \mathbf{0} \cdot (t \cdot v + w)$, which yields $\mathbf{0} \cdot v = \mathbf{0} \cdot v + w$. Hence $w = 0$. However, $w = 0$ is impossible since $s \cdot v \neq t \cdot v$. This proves that $s \cdot v + W \neq t \cdot v + W$. The latter implies that the module V/W is an effective module of dimension strictly smaller than n, which contradicts Theorem 11.4.1(i).   □

As V is effective and $\mathbf{0} \cdot V = 0$ by Lemma 11.4.2, we have $\varepsilon_{\{1\}} \cdot V \neq 0$. Fix any nonzero element $v \in \varepsilon_{\{1\}} \cdot V$. Let M be the trivial $\mathcal{H}(\varepsilon_{\{1\}})$-module and $m \in M$ be any nonzero element.

Lemma 11.4.3 *The assignment* $s \cdot m \mapsto s \cdot v$, $s \in S$, *defines a nonzero homomorphism* $f : V(M) \to V$.

Proof. As $\varepsilon_{\{1\}} \cdot m = m$, for $s \in S$ we have

$$s \cdot m = s \cdot (\varepsilon_{\{1\}} \cdot m) = (s\varepsilon_{\{1\}}) \cdot m.$$

We have $S\varepsilon_{\{1\}} = \{0, \alpha_{\{i\}} : i \in \mathbf{N}\}$ in the notation of Sect. 10.5. We further have $\mathbf{0} \cdot v = 0$ by Lemma 11.4.2 and $\mathbf{0} \cdot m = 0$ by definition. Observe that for $s, t \in S$ the equality $s \cdot m = t \cdot m$ implies $s\varepsilon_{\{1\}} = t\varepsilon_{\{1\}}$ and thus $s \cdot v = t \cdot v$. As $\alpha_{\{i\}} \cdot m$ are linearly independent in $V(M)$, the mapping f is a well-defined linear mapping. It commutes with the action of S by definition, and hence it is a homomorphism of S-modules. Finally, f is obviously nonzero.   □

The homomorphism f is nonzero and hence injective since the module $V(M)$ is simple by Theorem 11.3.1(i). As $\dim(V(M)) = \dim(V) = n$, we obtain that f is an isomorphism, implying that an effective module of dimension n is unique up to isomorphism. For $S = \mathcal{PT}_n$ and $S = \mathcal{IS}_n$ the statement (iii) now follows from the statement (ii).

Consider now the case $S = \mathcal{T}_n$. First we note that $0_1 \cdot V \neq 0$ as V is effective. Indeed, as $0_i \in \mathcal{T}_n 0_1$, $i \in \mathbf{N}$, the equality $0_1 \cdot V = 0$ would imply $0_i \cdot V = 0$ for all i, a contradiction. Now we note that $(1, 2)0_1 = 0_2$, in particular, the idempotent linear transformations 0_1 and 0_2 of V have the same rank and the same kernel. Since they must be different as V is effective, we conclude that their images are different. As these images have the same dimension, they cannot be included into each other and hence there exists a nonzero $v \in 0_1 \cdot V$ such that $v \notin 0_2 \cdot V$.

Let M be the trivial representation of $\mathcal{H}(0_1)$ and $m \in M$ be a nonzero element. Analogously to Lemma 11.4.3 one shows that the assignment $s \cdot m \mapsto s \cdot v$, $s \in \mathcal{T}_n$, defines a nonzero homomorphism $f : V(M) \to V$. To proceed we have to analyze the module $V(M)$. Let $\{v_1, \ldots, v_n\}$ denote the standard basis of $\mathbb{C}^n$.

Lemma 11.4.4 (i) *For* $i \in \mathbf{N}$ *the assignment* $0_i \cdot m \mapsto v_i$ *defines an isomorphism from* $V(M)$ *to the natural* $\mathcal{T}_n$-*module* $\mathbb{C}^n$.

(ii) The natural $\mathcal{T}_n$-module $\mathbb{C}^n$ has exactly three submodules, namely, 0, $\mathbb{C}^n$ and the submodule

$$N = \left\{ \sum_{i=1}^{n} a_i v_i \ : \ \sum_{i=1}^{n} a_i = 0 \right\}.$$

Proof. The statemet (i) follows from the relation $\alpha 0_i = 0_{\alpha(i)}$ in $\mathcal{T}_n$. To prove (ii) let W be a proper $\mathcal{T}_n$-submodule of $\mathbb{C}^n$, and $w = (w_1, \ldots, w_n) \in W$ be a nonzero element.

Assume first that $w_1 + \cdots + w_n \neq 0$. In this case, we have $0_1 \cdot w = (w_1 + \cdots + w_n, 0, \ldots, 0)$, implying that W contains v_1. Then W contains $(1, i) \cdot v_1 = v_i$ for every $i \in \mathbf{N}$ and hence $W = \mathbb{C}^n$, a contradiction.

Assume now that $w_1 + \cdots + w_n = 0$. As the action of $\mathcal{S}_n$ simply permutes the entries of w, without loss of generality we may assume $w_1 \neq 0$. Then we have

$$\begin{pmatrix} 1 & 2 & 3 & 4 & \cdots & n \\ 1 & 2 & 2 & 2 & \cdots & 2 \end{pmatrix} \cdot w = (w_1, w_2 + \cdots + w_n, 0, \ldots, 0),$$

which implies that W contains the element $z = (1, -1, 0, \ldots, 0)$. Applying the elements from $\mathcal{S}_n$ we get that W contains all elements of the form $(0, \ldots, 0, 1, -1, 0, \ldots, 0)$. As such elements form a basis in N, we obtain $N \subset W$. As the codimension of N in $\mathbb{C}^n$ is 1 and $W \neq \mathbb{C}^n$, it follows that $W = N$. This completes the proof. $\qquad\square$

The mapping f is nonzero and hence from Lemma 11.4.4(ii) we have that either $\mathrm{Ker}(f) = 0$ or $\mathrm{Ker}(f) \cong N$. Assume $\mathrm{Ker}(f) \cong N$. As $\mathrm{im}(f) \cong V(M)/\mathrm{Ker}(f)$, we have that $\mathrm{im}(f)$ has dimension 1. At the same time, it contains both v and the element $0_2 \cdot v$, which is linearly independent with v because of our assumptions. This is a contradiction, which proves $\mathrm{Ker}(f) = 0$. As both $V(M)$ and V have dimension n, it follows that f is an isomorphism. Now the proof is completed by applying Lemma 11.4.4(i). $\qquad\square$

11.5 Arbitrary $\mathcal{IS}_n$-Modules

Let S be a semigroup and V and W be S-modules. Then the vector space $V \oplus W = \{(v, w) \ : \ v \in V, w \in W\}$ carries the natural structure of an S-module via $s \cdot (v, w) = (s \cdot v, s \cdot w)$, $s \in S$, $v \in V$, $w \in W$. The module $V \oplus W$ is called the *direct sum* of the modules V and W. The direct sum is called *trivial* if $V = 0$ or $W = 0$.

An S-module U is called *indecomposable* provided that U cannot be decomposed into a nontrivial direct sum of two modules. In other words, U is indecomposable if $U \cong V \oplus W$ implies $V = 0$ or $W = 0$. Every simple module is obviously indecomposable. The converse is not true in general.

Exercise 11.5.1 Show that for $n > 1$ the natural $\mathcal{T}_n$-module is indecomposable but not simple.

An S-module U is called *semisimple* provided that it is isomorphic to a direct sum of simple modules. The corresponding representation is called *completely reducible*. Obviously every simple module is semisimple. If U is a semisimple module, then there exists essentially a unique way to write it as a direct sum of simple modules as shown in the following statement.

Proposition 11.5.2 *Let U be a semisimple S-module such that*

$$U \cong V_1 \oplus V_2 \oplus \cdots \oplus V_k \cong W_1 \oplus W_2 \oplus \cdots \oplus W_m,$$

where all V_i's and W_j's are simple modules. Then $k = m$ and there exists $\alpha \in \mathcal{S}_k$ such that $V_i \cong W_{\alpha(i)}$ for all i.

Proof. Let L be some simple S-module. We have

$$\mathrm{Hom}_S(L, U) = \mathrm{Hom}_S(L, V_1 \oplus \cdots \oplus V_n) = \mathrm{Hom}_S(L, V_1) \oplus \cdots \oplus \mathrm{Hom}(L, V_n)$$

(see Exercise 11.7.2). As both L and V_i are simple, from Corollary 11.1.8 we obtain that

$$|\{i \, : \, V_i \cong L\}| = \dim \mathrm{Hom}(L, U).$$

Analogously one obtains

$$|\{j \, : \, W_j \cong L\}| = \dim \mathrm{Hom}(L, U).$$

The statement of the proposition follows. ∎

The main aim of the present section is to prove the following theorem, which, together with Proposition 11.5.2, gives a complete description of all $\mathcal{IS}_n$-modules:

Theorem 11.5.3 *Every $\mathcal{IS}_n$-module is semisimple.*

Proof. Let V be an $\mathcal{IS}_n$-module. We prove the statement by induction on $\dim(V)$. If $\dim(V) = 1$, the module V is simple (see Example 11.1.3) and hence semisimple.

Let k be such that $\varepsilon_n^{(k)} \cdot V \neq 0$ while $\varepsilon_n^{(i)} \cdot V = 0$ for all $i < k$. In particular, we have $\alpha \cdot V = \alpha \alpha^{-1} \alpha \cdot V = \alpha \cdot (\alpha^{-1}\alpha \cdot V) = 0$ for any $\alpha \in \mathcal{IS}_n$ such that $\mathrm{rank}(\alpha) < k$. Let $I = \{A \subset \mathbf{N} \, : \, |A| = k\}$ and for $A \in I$ set $U_A = \varepsilon_A \cdot V$. Define

$$W = \bigcap_{A \in I} \{v \in V \, : \, \varepsilon_A \cdot v = 0\}, \qquad U = \sum_{A \in I} U_A.$$

Lemma 11.5.4   *(i) U is a direct sum of U_A's.*

(ii) $V = U \oplus W$.

(iii) Both U and W are $\mathcal{IS}_n$-submodules.

Proof. Let $v_A \in U_A$ be nonzero and $c_A \in \mathbb{C}$ be such that $\sum_{A \in I} c_A v_A = 0$. As for $A \neq B \in I$ we have $\operatorname{rank}(\varepsilon_A \varepsilon_B) < k$, we compute

$$\varepsilon_B \cdot \sum_{A \in I} c_A v_A = \sum_{A \in I} c_A \varepsilon_B \cdot (\varepsilon_A \cdot v_A) = \sum_{A \in I} c_A \varepsilon_B \varepsilon_A \cdot v_A = c_B v_B.$$

This yields $c_B = 0$ for any $B \in I$, and the statement (i) follows.

As all ε_A's are idempotents, we have $U \cap W = 0$ from the definition. Let $v \in V$. Then $u = \sum_{A \in I} \varepsilon_A \cdot v \in U$ and for any $B \in I$, using the same arguments as in the previous paragraph, we have

$$
\begin{aligned}
\varepsilon_B \cdot (v - u) &= \varepsilon_B \cdot \left(v - \sum_{A \in I} \varepsilon_A \cdot v \right) \\
&= \varepsilon_B \cdot v - \sum_{A \in I} \varepsilon_B \varepsilon_A \cdot v \\
&= \varepsilon_B \cdot v - \varepsilon_B \cdot v \\
&= 0.
\end{aligned}
$$

Hence $v - u \in W$, implying $V = U + W$. The statement (ii) follows.

Let $A \in I$, $v_A \in U_A$ and $\alpha \in \mathcal{IS}_n$. If $\operatorname{rank}(\alpha \varepsilon_A) < k$, then

$$\alpha \cdot v_A = \alpha \varepsilon_A \cdot v_A = 0 \in U.$$

If $\operatorname{rank}(\alpha \varepsilon_A) = k$, then

$$\alpha \cdot v_A = \alpha \varepsilon_A \cdot v_A = \varepsilon_{\alpha(A)} \alpha \varepsilon_A \cdot v_A = \varepsilon_{\alpha(A)} \cdot (\alpha \cdot v_A).$$

and hence $\alpha \cdot v_A \in U_{\alpha(A)}$. This implies that U is an $\mathcal{IS}_n$-submodule of V.

Let $w \in W$, $\alpha \in \mathcal{IS}_n$, and $A \in I$. If $\operatorname{rank}(\varepsilon_A \alpha) < k$, then

$$\varepsilon_A \cdot (\alpha \cdot w) = \varepsilon_A \alpha \cdot w = 0$$

by our assumption on k. If $\operatorname{rank}(\varepsilon_A \alpha) = k$, then $\varepsilon_A \alpha = \alpha \varepsilon_B$, where $B = \alpha^{-1}(A)$. Hence

$$\varepsilon_A \cdot (\alpha \cdot w) = \varepsilon_A \alpha \cdot w = \alpha \varepsilon_B \cdot w = \alpha \cdot (\varepsilon_B \cdot w) = 0.$$

This shows that W is an $\mathcal{IS}_n$-submodule of V and completes the proof. $\square$

By our assumptions, we have $U \neq 0$. If $W \neq 0$, then $\dim(U) < \dim(V)$ and $\dim(W) < \dim(V)$ and hence applying the inductive assumption to both U and W we obtain a decomposition of V into a direct sum of simple modules, as required. Hence in what follows we may assume $W = 0$.

Let $\mathbf{A} = \{1, 2, \ldots, k\}$ and for $B \in I$ let α_B denote the unique increasing bijection from $\mathbf{A}$ to B.

Lemma 11.5.5 *For $B \in I$ the action of α_B induces a bijection from $U_\mathbf{A}$ to U_B.*

Proof. This follows from the obvious equalities $\alpha_B^{-1}\alpha_B = \varepsilon_\mathbf{A}$ and $\alpha_B\alpha_B^{-1} = \varepsilon_B$. $\qquad\square$

Let $(\cdot, \cdot)_1$ be any Hermitian scalar product on $U_\mathbf{A}$. For $v, w \in U_\mathbf{A}$ set

$$(v, w) = \sum_{g \in \mathcal{H}(\varepsilon_n^{(k)})} (g \cdot v, g \cdot w)_1.$$

For $v \in U_C$ and $w \in U_B$, $C \neq B \in I$ we set

$$(v, w) = (w, v) = 0.$$

Finally, for $v, w \in U_\mathbf{A}$ and $B \in I$ we set

$$(\alpha_B \cdot v, \alpha_B \cdot w) = (v, w)$$

and extend the product $(\cdot, \cdot)$ to the whole V by skew-linearity.

Lemma 11.5.6 *(i) $(\cdot, \cdot)$ is a Hermitian scalar product on V.*

(ii) For all $v, w \in V$ and $\alpha \in \mathcal{IS}_n$ we have $(\alpha \cdot v, w) = (v, \alpha^{-1} \cdot w)$.

Proof. The product $(\cdot, \cdot)$ is bilinear and skew-symmetric by construction. The restriction of $(\cdot, \cdot)$ to $U_\mathbf{A}$ is positive definite since $(\cdot, \cdot)_1$ is positive definite. Now the fact that $(\cdot, \cdot)$ is positive definite follows from the construction and Lemma 11.5.4(i). This proves (i).

To prove (ii) let $v \in U_C$, $w \in U_B$ and $\beta \in \mathcal{IS}_n$ be chosen such that we have $(\beta \cdot v, w) \neq 0$. This, in particular, means $0 \neq \beta \cdot v \in U_B$ and thus $\text{rank}(\varepsilon_B \beta \varepsilon_C) = k$. Hence $\text{rank}(\varepsilon_C \beta^{-1} \varepsilon_B) = k$ as well and thus $\varepsilon_C \beta^{-1} \varepsilon_B \cdot w \in U_C$. By definition we have

$$(\beta \cdot v, w) = \left(\alpha_B^{-1} \cdot (\beta \cdot v), \alpha_B^{-1} \cdot w\right) = \left((\alpha_B^{-1}\beta\alpha_C) \cdot (\alpha_C^{-1} \cdot v), \alpha_B^{-1} \cdot w\right) \quad (11.5)$$

and analogously

$$(v, \beta^{-1} \cdot w) = \left(\alpha_C^{-1} \cdot v, (\alpha_C^{-1}\beta^{-1}\alpha_B) \cdot (\alpha_B^{-1} \cdot w)\right). \quad (11.6)$$

The element $\gamma = \alpha_B^{-1}\beta\alpha_C$ belongs to $G = \mathcal{H}(\varepsilon_\mathbf{A})$, $\gamma^{-1} = \alpha_C^{-1}\beta^{-1}\alpha_B$ and both $\alpha_C^{-1} \cdot v$ and $\alpha_B^{-1} \cdot w$ belong to $U_\mathbf{A}$. At the same time for any $x, y \in U_\mathbf{A}$, using the fact that G is a group, we have

$$
\begin{aligned}
(\gamma \cdot x, y) &= \sum_{g \in G} \left(g \cdot (\gamma \cdot x), g \cdot y \right)_1 \\
&= \sum_{g \in G} \left(g\gamma \cdot x, g \cdot y \right)_1 \\
&= \sum_{g \in G} \left(g \cdot x, g\gamma^{-1} \cdot y \right)_1 \\
&= \sum_{g \in G} \left(g \cdot x, g \cdot (\gamma^{-1} \cdot y) \right)_1 \\
&= (x, \gamma^{-1} \cdot y).
\end{aligned}
$$

The latter together with (11.5) and (11.6) implies the statement (ii). The proof is complete. $\square$

Let now X be some proper submodule of V. Then for the orthogonal complement

$$
X^{\perp} = \{ y \in V \ : \ (y, x) = 0 \text{ for all } x \in X \}
$$

using Lemma 11.5.6(ii) we have

$$
(\alpha \cdot y, x) = (y, \alpha^{-1} \cdot x) = 0
$$

for any $y \in X^{\perp}$, $x \in X$ and $\alpha \in \mathcal{IS}_n$. Hence $X^{\perp}$ is a proper submodule of V and we have $V = X \oplus X^{\perp}$. Applying now the inductive assumption to both X and $X^{\perp}$ we produce a decomposition of V into a direct sum of simple modules. This completes the proof. $\square$

11.6   Addenda and Comments

11.6.1 The results, presented in this chapter, mostly follow from more general results for representation of semigroups, obtained by Munn in [Mu1, Mu2, Mu3, Mu4] and Ponizovskiy [Po3]. These more general results, which also are more technically complicated and require more advanced knowledge about algebras, can be found in [CP1, Chap. 5]. The only results which we did not manage to find in the literature are those presented in Sect. 11.4.

There is also a well-developed theory of characters of semigroups. Characters of commutative semigroups were studied by Schwarz in [Sw1, Sw2, Sw3] and Hewitt and Zuckerman in [HZ1]. This theory is also presented in [CP1, Chap. 5].

Some recent developments in the character theory for modules, especially over transformation semigroups, can be found in the works of Putcha [Pu1, Pu2, Pu3]. For inverse semigroups some new approaches appear in [Ste2]. Character tables for $\mathcal{IS}_n$ were studied in [So]. A modern point of view

on classification of simple modules over finite semigroups is presented in
Exercises 11.7.19 and 11.7.20. It also recently appeared in [GMS].

11.6.2 All simple $\mathcal{T}_n$-modules were explicitly constructed by Hewitt and
Zuckerman in [HZ2], and their characters were described in [Pu1, Section 2].
This description is more complicated than the corresponding description for
$\mathcal{PT}_n$ and $\mathcal{IS}_n$ presented in Theorem 11.3.1, and the proof is technically
much more complicated. To state the result we will need some new notation
for the symmetric group.

Let $\sigma \in \mathcal{S}_n$ be a permutation. A pair $(i,j) \in \mathbf{N} \times \mathbf{N}$ is called an *inversion*
for σ provided that $i < j$ and $\sigma(i) > \sigma(j)$. The number of inversions for σ
is denoted inv_σ. The permutation σ is called *odd* or *even* provided that inv_σ
is odd or even, respectively. For example, the identity permutation is even,
while any transposition is odd.

For $c \in \mathbb{C}$ the assignment $\sigma \cdot c = (-1)^{\mathrm{inv}_\sigma} c$ defines on $\mathbb{C}$ the structure
of an $\mathcal{S}_n$-module, called the *sign module*. This module is simple as it is one-
dimensional. For $n > 1$ the trivial module and the sign module are the only
one-dimensional $\mathcal{S}_n$-modules. They are obviously nonisomorphic. For $n = 1$
the definition is awkward, however, it is convenient to use the convention
that the trivial and the sign $\mathcal{S}_1$-modules coincide. Now we can formulate the
statement describing all simple $\mathcal{T}_n$-modules. We refer the reader to [HZ2]
and [Pu1] for the proofs.

Theorem 11.6.1   *(i) The $\mathcal{T}_n$-module $V(M)$ is simple if and only if $k = n$
or M is not isomorphic to the sign $\mathcal{S}_k$-module.*

(ii) *If M is isomorphic to the sign $\mathcal{S}_k$-module, then the module $V(M)$ has
a unique simple quotient $\overline{V(M)}$ of dimension $\binom{n-1}{k-1}$.*

(iii) *Every simple $\mathcal{T}_n$-module is isomorphic either to $\overline{V(M)}$, where M is the
sign $\mathcal{S}_k$-module, or to $V(M)$, where M is an $\mathcal{S}_k$-module, which is not
isomorphic to the sign module.*

The statement in Theorem 11.6.1(iii) is proved similarly to the corresponding statement in Theorem 11.3.1. The really hard part here is to prove
the first two statements of Theorem 11.6.1.

A naive straightforward generalization of the arguments from Sect. 11.3
leads to some interesting combinatorial questions. Let us follow the arguments from Sect. 11.3 and try to prove that the $\mathcal{T}_n$-module $V(M)$ is simple
in the case when M is the trivial module and $k > 1$. To prove this we have
to show that for any nonzero vector $v \in V(M)$ the minimal submodule of
$V(M)$, containing v, coincides with $V(M)$. This is equivalent to showing
that for any nonzero vector $v \in V(M)$ there exists some idempotent $e \in \mathcal{T}_n$
of rank k such that $e \cdot v \neq 0$. The latter statement has the following combinatorial formulation: Let X denote the matrix whose rows are indexed by

k-element subsets of $\mathbf{N}$ and columns are indexed by unordered partitions of $\mathbf{N}$ into k disjoint nonempty blocks. Note that the same objects index the $\mathcal{L}$-classes and the $\mathcal{R}$-classes of $\mathcal{T}_n$ for elements of rank k. In particular, the elements of X correspond naturally to $\mathcal{H}$-classes inside $\mathcal{D}_k$. Let the element of X be 1 if the corresponding $\mathcal{H}$-class contains an idempotent and 0 otherwise. Then, following the above arguments, the irreducibility of $V(M)$ is equivalent to the statement that the rank of X equals the number of rows in X, that is, $\binom{n}{k}$. This result follows from Theorem 11.6.1(i). However, we do not know any combinatorial proof.

The above matrix X contains a natural square submatrix defined using the following natural injection from the set of all k-element subsets of $\mathbf{N}$ to the set of all partitions of $\mathbf{N}$ into k-blocks. With each k-element subset $\{i_1 < i_2 < \cdots < i_k\}$ one associates the partition

$$\{i_k + 1, \ldots, n, 1, \ldots, i_1\} \cup \{i_1 + 1, \ldots, i_2\} \cup \cdots \cup \{i_{k-1} + 1, \ldots, i_k\}.$$

We suspect that the determinant of the corresponding square submatrix of X is always a power of 2 (up to sign), but we cannot prove that.

11.6.3 There is a beautiful combinatorial description of all simple $\mathcal{S}_n$-modules, see for example [Sa]. A *partition* of n is a vector $\lambda = (\lambda_1, \lambda_2, \ldots, \lambda_l)$ with positive integer coefficients such that $\lambda_1 \geq \lambda_2 \geq \cdots \geq \lambda_l$ and $n = \lambda_1 + \lambda_2 + \cdots + \lambda_l$. That λ is a partition of n is usually denoted $\lambda \vdash n$. The *Ferrers diagram* or *Young diagram* of *shape* λ is an array of n boxes having l left-justified rows with λ_i boxes in the i-th row for $1 \leq i \leq l$. For example, the Young diagram of shape $\lambda = (4, 2, 2, 1) \vdash 9$ is the following:

With every $\lambda \vdash n$ one associates a simple $\mathcal{S}_n$-module S_λ, called the *Specht module*.

Let $\lambda \vdash n$ and $a_1 < a_2 < \cdots < a_n$ be a sequence of integers. A *standard λ-tableau* of content $\{a_1, \ldots, a_n\}$ is obtained if we write the numbers $a_1, \ldots, a_n$ in the boxes of the Young tableau so that the entries in all rows increase left-to-right and entries in all columns increase top-to-bottom. Here is an example of a standard $(4, 2, 2, 1)$-tableau of content $\mathbf{N}_9$:

1	2	6	7
3	5		
4	8		
9			

Theorem 11.6.2 (i) Each simple S_n-module is isomorphic to S_λ for some $\lambda \vdash n$.

(ii) For $\lambda, \mu \vdash n$ we have $S_\lambda \cong S_\mu$ if and only if $\lambda = \mu$.

(iii) Let $\lambda \vdash n$. Then the Specht module S_λ has a natural basis, which is indexed by standard λ-tableaux of content $\mathbf{N}$.

The combinatorial description from Theorem 11.6.2(iii) was generalized to $\mathcal{IS}_n$ by Grood in [Grd]. As simple $\mathcal{IS}_n$-modules are obtained from simple $\mathcal{PT}_n$-modules by restriction (Corollary 11.3.3), the same description applies to $\mathcal{PT}_n$ as well. As usually, for $k \leq n$ we identify S_k with the group $\mathcal{H}(\varepsilon_n^{(k)})$ in the natural way. Combining the results of Grood with the results from Sect. 11.3 we have:

Theorem 11.6.3 Let $S = \mathcal{IS}_n$ or $S = \mathcal{PT}_n$.

(i) Each simple S-module is isomorphic to $M(S_\lambda)$ for some $\lambda \vdash k$, $k \leq n$.

(ii) For $\lambda \vdash k_1$ and $\mu \vdash k_2$, $k_1, k_2 \leq n$, we have $M(S_\lambda) \cong M(S_\mu)$ if and only if $\lambda = \mu$.

(iii) Let $\lambda \vdash k$, $k \leq n$. Then the module $M(S_\lambda)$ has a natural basis, which is indexed by standard λ-tableaux, whose contents is a k-element subset of $\mathbf{N}$.

Let $k \leq n$ and I denote the set of all k-element subsets of $\mathbf{N}$. If $\lambda \vdash k$, then we have a decomposition

$$M(S_\lambda) \cong \bigoplus_{A \in I} \varepsilon_A \cdot M(S_\lambda)$$

into a direct sum of vector spaces. The space $\varepsilon_A \cdot M(S_\lambda)$ is a simple $\mathcal{H}(\varepsilon_A)$-module, which is isomorphic to S_λ after an appropriate identification of $\mathcal{H}(\varepsilon_A)$ with S_k. The bases of Theorems 11.6.2(iii) and 11.6.3(iii) are naturally compatible with this identification.

Tableaux combinatorics can also be used to describe the decomposition of the restriction of simple $\mathcal{IS}_n$-modules to $\mathcal{IS}_{n-1}$ (the so-called *Branching rule*), see [MR]. It is very similar to the corresponding result for the symmetric group, see for example [Sa] for the latter one.

11.6.4 Let S be a finite semigroup. The *semigroup algebra* $\mathbb{C}[S]$ is the set of all formal linear combinations $\sum_{s \in S} c_s s$ with complex coefficients, endowed with the natural bilinear multiplication induced from the multiplication in the semigroup S.

Each S-module naturally extends to a $\mathbb{C}[S]$-module, and each $\mathbb{C}[S]$-module restricts to an S-module. This produces an isomorphism of categories of all finite-dimensional S-modules and all finite-dimensional $\mathbb{C}[S]$-modules.

Denote by $\mathrm{Mat}_n(\mathbb{C})$ the algebra of all complex $n \times n$ matrices. In the language of algebras Theorem 11.5.3 says that the algebra $\mathbb{C}[\mathcal{IS}_n]$ is semisimple, in particular, it is isomorphic to a direct sum of some $\mathrm{Mat}_{n_i}(\mathbb{C})$.

One can also show that $\mathbb{C}[\mathcal{IS}_n]$ is isomorphic to a direct sum of matrix algebras over symmetric groups as follows: For $k \le n$ the algebra

$$\mathcal{A}_k = \mathrm{Mat}_{\binom{n}{k}}(\mathbb{C}) \otimes_{\mathbb{C}} \mathbb{C}[\mathcal{S}_k]$$

can be realized as the algebra of all $\binom{n}{k} \times \binom{n}{k}$ matrices with coefficients from $\mathbb{C}[\mathcal{S}_k]$.

Theorem 11.6.4 ([Mu3]) $\mathbb{C}[\mathcal{IS}_n] \cong \displaystyle\bigoplus_{k=0}^{n} \mathcal{A}_k$.

The most beautiful part of the proof of this fact, presented for example in [So, Section 2], is the construction of pairwise orthogonal idempotents, which determine the components $\mathcal{A}_k$'s. This is done as follows: For $A \subset \mathbf{N}$ define

$$\eta_A = \sum_{B \subset A} (-1)^{|A \setminus B|} \varepsilon_B. \tag{11.7}$$

By the inclusion–exclusion formula we have

$$\varepsilon_A = \sum_{B \subset A} \eta_B. \tag{11.8}$$

Lemma 11.6.5 *For $A, B \subset \mathbf{N}$ we have*

$$\eta_A \eta_B = \begin{cases} \eta_B, & A = B; \\ 0, & A \ne B. \end{cases}$$

Proof. We first claim that for $A, B \subset \mathbf{N}$ we have

$$\varepsilon_A \eta_B = \begin{cases} \eta_B, & B \subset A; \\ 0, & \text{otherwise.} \end{cases} \tag{11.9}$$

This is obvious if $B \subset A$. Assume now that $B \setminus A \ne \varnothing$. Then

$$\varepsilon_A \eta_B = \sum_{C \subset B} (-1)^{|B \setminus C|} \varepsilon_{A \cap C}.$$

The latter sum is a linear combination of the ε_Y's, where $Y \subset A \cap B$. Write $C = Y \cup Z$, where $Z \subset B \setminus A \ne \varnothing$. Then the coefficient at ε_Y is

$$\sum_{Z \subset B \setminus A} (-1)^{|B \setminus (Y \cup Z)|} = 0.$$

This proves the formula (11.9). From (11.7) and (11.9) we have

$$\eta_A \eta_B = \sum_{X \subseteq A} (-1)^{|A \backslash X|} \varepsilon_X \eta_B = \left(\sum_{B \subseteq X \subseteq A} (-1)^{|A \backslash X|} \right) \eta_B$$

The coefficient on the right-hand side is zero unless $A = B$, when it is equal to 1. This completes the proof. $\square$

For $k \leq n$ now set $\eta_k = \sum_{|A|=k} \eta_A$. Then Lemma 11.6.5 implies

$$\eta_k \eta_r = \begin{cases} \eta_k, & k = r; \\ 0, & k \neq r. \end{cases}$$

Moreover, (11.8) even says $\varepsilon = \sum_{k=0}^{n} \eta_k$, which gives us a decomposition of the identity element in $\mathbb{C}[\mathcal{IS}_n]$ into a sum of pairwise orthogonal idempotents.

Exercise 11.6.6 Show that $\eta_n \alpha = \alpha \eta_n$ for any $\alpha \in \mathcal{IS}_n$.

From Exercise 11.6.6 it follows that

$$\mathbb{C}[\mathcal{IS}_n] \cong \bigoplus_{k=0}^{n} \mathbb{C}[\mathcal{IS}_n] \eta_k = \bigoplus_{k=0}^{n} \eta_k \mathbb{C}[\mathcal{IS}_n] \eta_k$$

is a decomposition into a direct sum of subalgebras. It is now not difficult to check that $\eta_k \mathbb{C}[\mathcal{IS}_n] \eta_k \cong \mathcal{A}_k$. Using Möbius function, these arguments generalize to arbitrary finite inverse semigroups, see [Ste1].

11.6.5 Description of all finite-dimensional modules for semigroups $\mathcal{T}_n$ and $\mathcal{PT}_n$ is a very hard problem. It is for example known (see [Po4, Pu2, Ri]) that the semigroup $\mathcal{T}_n$ has finitely many isomorphism classes of indecomposable modules if and only if $n < 5$.

11.6.6 Let S be a semigroup and V and W be two S-modules. Then the vector space $V \otimes_{\mathbb{C}} W$ carries the natural structure of an S-module via

$$s \cdot (v \otimes w) = (s \cdot v) \otimes (s \cdot w).$$

The module $V \otimes_{\mathbb{C}} W$ is called the *tensor product* of the modules V and W. Set $V^{\otimes k} = V \otimes_{\mathbb{C}} V \otimes_{\mathbb{C}} \cdots \otimes_{\mathbb{C}} V$, where the right-hand side contains exactly k factors.

Let now V be the natural representation of $\mathcal{IS}_n$. Then the symmetric group $\mathcal{S}_k$ acts on the space $V^{\otimes k}$ permuting components of the tensor product. This action obviously commutes with the action of $\mathcal{IS}_n$. However, the linear closure of this action of $\mathcal{S}_k$ does not give all linear operators which commute with the $\mathcal{IS}_n$-action. The set of all operators, which commute with some action, is usually called the *centralizer* of the action. The centralizer of the $\mathcal{IS}_n$-action on $V^{\otimes k}$ can be explicitly described in terms of the *dual symmetric inverse semigroup* $\mathcal{I}_k^*$, see [KuMa5]. Furthermore, the original $\mathbb{C}[\mathcal{IS}_n]$-action turns out to give the full centralizer of the $\mathbb{C}[\mathcal{I}_k^*]$-action, which gives rise to a *Schur-Weyl duality* connecting $\mathcal{IS}_n$ and $\mathcal{I}_k^*$. Moreover, one can also show that the kernel of the action of $\mathbb{C}[\mathcal{IS}_n]$ on $V^{\otimes n}$ coincides with $\mathbb{C}0$. In particular, the module $V^{\otimes n}$ contains as submodules, except for the trivial module, representatives from all isomorphism classes of simple $\mathcal{IS}_n$-modules.

There is another version of the Schur-Weyl duality for $\mathcal{IS}_n$ in which $\mathcal{IS}_n$ occurs on the other side. It was discovered in [So, Sect. 5]. Let V be the natural n-dimensional representation of the group $\mathbf{GL}_n$ of all nondegenerate complex $n \times n$ matrices. Let $\mathbb{C}$ be the trivial representation of $\mathbf{GL}_n$. Set $U = V \oplus \mathbb{C}$. Then $\mathcal{IS}_k$ acts on $U^{\otimes k}$ in the following way: Let $\pi : U \to \mathbb{C}$ be the projection with kernel V. For $\alpha \in \mathcal{IS}_k$ and $u_i \in U$, $i = 1, \ldots, k$, set

$$\alpha \cdot (u_1 \otimes u_2 \otimes \cdots \otimes u_k) = v_1 \otimes v_2 \otimes \cdots \otimes v_k,$$

where

$$v_j = \begin{cases} u_{\alpha(j)}, & j \in \mathrm{dom}(\alpha); \\ \pi(u_j), & j \notin \mathrm{dom}(\alpha). \end{cases}$$

It turns out that the action of $\mathbb{C}[\mathcal{IS}_k]$ gives the full centralizer of the $\mathbb{C}[\mathbf{GL}_n]$-action and vice versa.

11.6.7 Let S be a semigroup. An element $w \in S$ is called an *involution* provided that $w \in G_e$ for some $e \in \mathcal{E}(S)$ and $w^2 = e$. Let $\mathcal{B}$ denote the set of all involutions in $\mathcal{IS}_n$ and let $\mathbb{C}\mathcal{B}$ denote the $\mathbb{C}$-linear span of $\mathbf{v}_w$, $w \in \mathcal{B}$. For $\alpha \in \mathcal{IS}_n$ and an involution $w \in \mathcal{H}(e)$, $e \in \mathcal{E}(\mathcal{IS}_n)$, such that $\alpha e \in \mathcal{D}(e)$ set

$$\mathrm{inv}_w(\alpha) = |\{(i,j) \ : \ i,j \in \mathrm{dom}(w), i < j, w(i) = j \text{ and } \alpha(i) > \alpha(j)\}|.$$

Theorem 11.6.7 ([KuMa6])   *(i) For $\alpha \in \mathcal{IS}_n$ and an involution $w \in \mathcal{H}(e)$, $e \in \mathcal{E}(\mathcal{IS}_n)$, the assignment*

$$\alpha \cdot \mathbf{v}_w = \begin{cases} (-1)^{\mathrm{inv}_w(\alpha)} \mathbf{v}_{(\alpha e)w(\alpha e)^{-1}}, & \alpha e \in \mathcal{D}(e); \\ 0, & \text{otherwise.} \end{cases}$$

defines on $\mathbb{C}\mathcal{B}$ the structure of an $\mathcal{IS}_n$-module.

(ii) $\mathbb{C}\mathcal{B}$ is a multiplicity-free direct sum of all simple $\mathcal{IS}_n$-modules.

Theorem 11.6.7(ii) says that $\mathbb{C}\mathcal{B}$ is a *Gelfand model* for $\mathbb{C}[\mathcal{IS}_n]$. Theorem 11.6.7 is based on the corresponding result for $\mathcal{S}_n$, proved in [APR].

11.7 Additional Exercises

11.7.1 Let S be a finite monoid with the unit $\mathbf{1}$, V be any S-module and $v \in V$.

(a) Show that the assignment $\mathbf{v_1} \mapsto v$ extends to a homomorphism from the regular S-module $\mathbb{C}S$ to V.

(b) Use (a) to show that V is a quotient of some $\mathbb{C}S \oplus \cdots \oplus \mathbb{C}S$.

11.7.2 Let S be a semigroup and U, V, W be S-modules. Show that

(a) $\mathrm{Hom}_S(U \oplus V, W) \cong \mathrm{Hom}_S(U, W) \oplus \mathrm{Hom}_S(V, W)$.

(b) $\mathrm{Hom}_S(U, V \oplus W) \cong \mathrm{Hom}_S(U, V) \oplus \mathrm{Hom}_S(U, W)$.

11.7.3 Describe all modules, in particular, all simple modules for the semigroup $\mathcal{B}(\mathbf{N})$ with respect to the operation $\cap$.

11.7.4 Determine the minimal dimension of an effective representation of the semigroup $\mathcal{B}(\mathbf{N})$ with respect to the operation $\cap$.

11.7.5 Construct examples of indecomposable but not simple $\mathcal{PT}_n$-modules.

11.7.6 Generalize the arguments from Sect. 11.5 and prove an analogue of Theorem 11.5.3 for arbitrary finite inverse semigroup.

11.7.7 Let S be a finite semigroup. Show that for every $s \in S$ there exists some simple S-module V such that $s \cdot V \neq 0$, while $t \cdot V = 0$ for any $t \in S$ such that $s \notin StS$.

11.7.8 Let $k \leq n$. Show that the subalgebra of $\mathbb{C}[\mathcal{IS}_n]$, generated by the ideal $\mathcal{I}_k$, is isomorphic to $\bigoplus\limits_{i=0}^{k} \mathcal{A}_i$ in the notation of 11.6.4.

11.7.9 (a) Show that every semigroup has at least two nonisomorphic simple modules: the trivial module, and the one-dimensional module with the zero action of S.

(b) Show that every finite nilpotent semigroup has exactly two nonisomorphic simple modules.

11.7.10 By Theorem 11.5.3, the regular module $\mathbb{C}\mathcal{IS}_n$ is semisimple. Prove that for every simple $\mathcal{IS}_n$-module V we have

$$\dim \mathrm{Hom}_{\mathcal{IS}_n}(V, \mathbb{C}\mathcal{IS}_n) = \dim(V).$$

11.7.11 Let S be a finite monoid and G its maximal subgroup of invertible elements. Let V denote the one-dimensional S-module on which the elements from G act trivially, and all other elements act as zero. Let now M be any simple S-module. Show that the module $M \otimes_{\mathbb{C}} V$ is semisimple.

11.7.12 Let S be a semigroup acting on some set M. Let $\mathbb{C}M$ denote the vector space with the basis $\mathbf{v}_m$, $m \in M$. Show that $\mathbb{C}M$ becomes an S-module via $s \cdot \mathbf{v}_m = \mathbf{v}_{s \cdot m}$, $s \in S$, $m \in M$.

11.7.13 ([Ste2]) Let S be a finite inverse semigroup. Show that the regular S-module is isomorphic to the module, obtained from the Preston–Wagner representation of S via the procedure described in Exercise 11.7.12.

11.7.14 ([KuMa5]) Let V denote the natural $\mathcal{IS}_n$-module. Show that the module $V^{\otimes k}$ is effective for any k.

11.7.15 ([So]) Let V denote the natural $\mathcal{IS}_n$-module with the standard basis $v_1, \ldots, v_n$. For $0 \le k \le n$ consider the exterior power $\Lambda^k(V)$.

(a) Show that for $\alpha \in \mathcal{IS}_n$ the assignment

$$\alpha \cdot (v_{i_1} \wedge \cdots \wedge v_{i_k}) = \begin{cases} v_{\alpha(i_1)} \wedge \cdots \wedge v_{\alpha(i_k)}, & \{i_1, \ldots, i_k\} \subset \operatorname{dom}(\alpha); \\ 0, & \text{otherwise}, \end{cases}$$

where $i_1, \ldots, i_k \in \mathbf{N}$ (and $\alpha \cdot 1 = 1$ for $k = 0$) defines on $\Lambda^k(V)$ the structure of an $\mathcal{IS}_n$-module.

(b) Show that the module $\Lambda^k(V)$ is simple for every k.

(c) Show that the module $\Lambda^k(V)$ is isomorphic to $V(M)$, where M is the sign $\mathcal{S}_k$-module.

11.7.16 ([KuMa4]) Let S denote one of the semigroups $\mathcal{T}_n$, $\mathcal{PT}_n$, or $\mathcal{IS}_n$, and let $\alpha, \beta \in S$. Show that $\alpha \sim_S \beta$ if and only if for every representation φ of S the traces of the linear operators $\varphi(\alpha)$ and $\varphi(\beta)$ coincide.

11.7.17 ([KuMa4]) Show that the statement of Exercise 11.7.16 is false for finite semigroups in general.

11.7.18 Classify all simple modules over a finite rectangular band.

11.7.19 ([GMS]) Let S be a finite semigroup, $e \in \mathcal{E}(S)$, and M be a simple $\mathcal{H}(e)$-module.

(a) Show that the set

$$N(M) = \{v \in V(M) : s \cdot v = 0 \text{ for all } s \in \mathcal{D}(e)\}$$

is a submodule of $V(M)$.

(b) Prove that any submodule $X \subsetneq V(M)$ is contained in $N(M)$.

(c) Use (b) to show that $V(M)$ has the unique simple quotient $\overline{V(M)} = V(M)/N(M)$.

11.7.20 ([GMS]) Let S be a finite monoid.

(a) Prove that every simple S-module is isomorphic to $\overline{V(M)}$ for appropriate e and M as defined in Exercise 11.7.19.

(b) Use (a) to classify all simple S-modules.

Chapter 12

Cross-Sections

12.1 Cross-Sections

Let X be a set and ρ an equivalence relation on X. A subset Y of X is called a *ρ-cross-section* provided that Y contains exactly one representative from each equivalence class.

In the study of semigroups, it is natural to expect that prospective cross-sections capture some semigroup theoretical properties. This can be understood in several different ways: one may expect semigroup-theoretical properties of either the equivalence relation or the cross-section or both. We will concentrate on the most restrictive case, and require that both the equivalence relation and its cross-section have some semigroup theoretical meaning.

Let S be a semigroup. We basically know the following natural equivalence relations on S: Green's relations, conjugacy, and congruences. By a *cross-section* with respect to any of these equivalence relations we will mean a subsemigroup of S which contains exactly one representative from every equivalence class. The main problem of course is that there is absolutely no guarantee that given some of these equivalence relations one can find in S some cross-section for this relation. We will see that in some cases cross-sections do not exist. The first negative result is the following:

Theorem 12.1.1 *Let S denote one of the semigroups $\mathcal{T}_n$, $\mathcal{PT}_n$ or $\mathcal{IS}_n$, and ρ be either the relation of $\mathcal{S}_n$-conjugation or the relation of S-conjugation.*

(i) If $n = 1$, then S contains exactly one cross-section with respect to ρ, namely, S itself.

(ii) If $n > 1$, then S does not contain any cross-section with respect to ρ.

Proof. The statement (i) is obvious. To prove (ii) we will consider two cases.

First we assume $n \geq 3$. In this case we observe that conjugacy classes of $\mathcal{S}_n$ form separate ρ-classes regardless whether ρ is the $\mathcal{S}_n$-conjugation or

O. Ganyushkin, V. Mazorchuk, *Classical Finite Transformation Semigroups*, Algebra and Applications 9, DOI: 10.1007/978-1-84800-281-4_12,
© Springer-Verlag London Limited 2009

the S-conjugation, see Sect. 6.4. Assume that T is a ρ-cross-section. Then T must contain some cycle $\alpha = (a, b, c)$ of length three. As T is a subsemigroup, $\alpha^2 = (a, c, b)$ belongs to T as well. However, $\alpha^2 = (b, c)\alpha(b, c)$, implying that α and α^2 are $\mathcal{S}_n$-conjugate (and thus also S-conjugate). This contradicts our assumption that T is a ρ-cross-section.

For $n = 2$ consider first the case $S = \mathcal{IS}_2$ or $S = \mathcal{T}_2$. In this case $\mathcal{S}_2$ is commutative and thus must be contained in any ρ-cross-section T. From Proposition 6.4.3 and Theorem 6.4.13 we obtain that T must also contain one of the two idempotents of rank one. Hence $T = S$ by Theorem 3.1.4 or Theorem 3.1.3, respectively. However, S is not a ρ-cross-section since all idempotents of rank one are $\mathcal{S}_n$-conjugate (and hence S-conjugate as well).

Finally, let $S = \mathcal{PT}_2$ and T be a ρ-cross-section. As above, T contains $\mathcal{S}_2$ and at least one idempotent e of rank 1. But then T must also contain the $\mathcal{S}_2$-conjugated (and hence also S-conjugated) idempotent $(1, 2)e(1, 2)$, which contradicts our assumption that T is a ρ-cross-section. This completes the proof. $\square$

12.2   Retracts

Let S be a semigroup and ρ be a congruence on S. A cross-section with respect to ρ is called a *retract* or ρ-*retract*. Description of retracts for S being one of the semigroups $\mathcal{T}_n$, $\mathcal{PT}_n$, or $\mathcal{IS}_n$ is an easy application of Theorem 6.3.10, where all congruences on S were described.

Lemma 12.2.1 *Let S be one of the semigroups $\mathcal{T}_n$, $\mathcal{PT}_n$, or $\mathcal{IS}_n$ and ρ be a congruence on S. Then a ρ-retract exists if and only if ρ is one of the following:*

(i) *The identity congruence*

(ii) *The uniform congruence*

(iii) *The congruence $\equiv_R$ from Sect. 6.3, where R is a subgroup of $\mathcal{S}_n$, moreover, if $S = \mathcal{T}_n$, then, additionally, $R \neq \{e\}$*

Proof. Let $1 < k < n$ and R be a normal subgroup of $\mathcal{S}_k$. Let further $\equiv_R$ be the corresponding congruence from Sect. 6.3 and T be an $\equiv_R$-retract. Then, by the construction of $\equiv_R$, T contains $\mathcal{S}_n$ and all idempotents of rank $(n - 1)$. Hence $T = S$ by Theorems 3.1.3–3.1.5, respectively. Since for $k > 1$ the relation $\equiv_R$ is not the identity congruence by construction, we get a contradiction with the assumption that T is a cross-section for $\equiv_R$.

Assume that $S = \mathcal{T}_n$, $k = n$, $R = \{e\}$ and T is an $\equiv_R$-retract. Then, by construction, T contains $\mathcal{S}_n$ and, apart from that, T must contain a unique noninvertible element e. Let $x, y \in \mathbf{N}$ be such that $x \in \text{im}(e)$ while $y \notin \text{im}(e)$. Then $(x, y)e \neq e$, which means that such T is not closed under the multiplication, a contradiction.

Now the statement of the lemma follows from Theorem 6.3.10, which says that any ρ which is not of the form (i)–(iii), is equal to $\rho ==_R$ for k and R as in the previous paragraphs. $\square$

Using Lemma 12.2.1 we just describe the retracts for each of the semigroups $\mathcal{T}_n$, $\mathcal{PT}_n$, and $\mathcal{IS}_n$ on a case-by-case basis. Recall from 6.5.5 that normal subgroups of $\mathcal{S}_n$ are: the $\mathcal{S}_n$ itself, the subgroup $\mathcal{A}_n$ of even permutations and the identity subgroup $\{\varepsilon\}$ (for $n \neq 4$). The group $\mathcal{S}_4$ has an additional normal subgroup $\mathcal{V}_4$.

Proposition 12.2.2   *(i) The identity congruence on $\mathcal{IS}_n$ has the unique retract, namely, $\mathcal{IS}_n$ itself.*

(ii) If ρ is the uniform congruence on $\mathcal{IS}_n$, then there exist 2^n retracts for ρ, all of the form $\{e\}$, where $e \in \mathcal{E}(\mathcal{IS}_n)$.

(iii) If $\rho ==_{\mathcal{S}_n}$, then $\mathcal{IS}_n$ contains $2^n - 1$ retracts ρ, all of the form $\{\varepsilon, e\}$, where $e \in \mathcal{E}(\mathcal{IS}_n)\backslash\{\varepsilon\}$.

(iv) If $\rho ==_{\mathcal{A}_n}$, then $\mathcal{IS}_n$ contains

$$\sum_{k=2}^{n} \binom{n}{k} \sum_{s=1}^{\lfloor \frac{k+2}{4} \rfloor} \frac{k!}{(k - 2(2s-1))! \cdot (2s-1)! \cdot 2^{2s-1}}$$

ρ-retracts, all of the form $\{\varepsilon, \alpha, e\}$, where $e \in \mathcal{E}(\mathcal{IS}_n)\backslash\{\varepsilon\}$ and the element $\alpha \in \mathcal{S}_n\backslash\mathcal{A}_n$ has order two and is such that $\alpha(x) = x$ for any $x \in \mathrm{im}(e)$.

(v) If $\rho ==_{\{\varepsilon\}}$, then $\mathcal{IS}_n$ contains the unique ρ-retract $\mathcal{S}_n \cup \{\mathbf{0}\}$.

(vi) If $\rho ==_{\mathcal{V}_4}$, then $\mathcal{IS}_4$ contains eight ρ-retracts of the form

$$Q_0^{\{a,b,c\}} = \{\varepsilon, (a,b), (a,c), (b,c), (a,b,c), (a,c,b), \mathbf{0}\};$$
$$Q_1^{\{a,b,c\}} = \{\varepsilon, (a,b), (a,c), (b,c), (a,b,c), (a,c,b), \varepsilon_{\{d\}}\},$$

where $\{a, b, c\} \subset \mathbf{N}_4$, and $\{d\} = \mathbf{N}_4\backslash\{a, b, c\}$.

Proof. The statement (i) is obvious. If ρ is the uniform congruence, then T must contain exactly one element, which is automatically idempotent. Any idempotent is a retract for the uniform congruence. Hence (ii) follows from Corollary 2.7.3.

If $\rho ==_{\mathcal{S}_n}$, then ρ has two equivalence classes: $\mathcal{S}_n$ and $\mathcal{IS}_n\backslash\mathcal{S}_n$. Each of them is a subsemigroup of $\mathcal{IS}_n$, and hence any retract of ρ must consist of idempotents. There is a unique idempotent ε in $\mathcal{S}_n$. On the other hand, for any $e \in \mathcal{E}(\mathcal{IS}_n)\backslash\{\varepsilon\}$, the set $\{\varepsilon, e\}$ is a retract for ρ. Hence (iii) follows again from Corollary 2.7.3.

If $\rho ==_{\mathcal{S}_n}$, then ρ has three equivalence classes: $\mathcal{A}_n$, $\mathcal{S}_n\backslash\mathcal{A}_n$, and $\mathcal{IS}_n\backslash\mathcal{S}_n$. As both $\mathcal{A}_n$ and $\mathcal{IS}_n\backslash\mathcal{S}_n$ are subsemigroups, any retract T must contain an idempotent from each of these two classes. Hence $\mathcal{A}_n$ must be represented in T by ε and $\mathcal{IS}_n\backslash\mathcal{S}_n$ by some $e \in \mathcal{E}(\mathcal{IS}_n)\backslash\{\varepsilon\}$. As $\alpha^2 \in \mathcal{A}_n$ for any $\alpha \in \mathcal{S}_n\backslash\mathcal{A}_n$, the latter class must be represented by some α of order two. As $\alpha \in \mathcal{S}_n\backslash\mathcal{A}_n$, α must be a product of an odd number $r = 2s - 1$ of pairwise commuting transpositions. The condition which would guarantee that T is closed under multiplication is $\alpha e = e\alpha = e$. This is obviously equivalent to $\alpha(x) = x$ for any $x \in \mathrm{dom}(e)$. Now if we let k be the cardinality of $\overline{\mathrm{dom}(e)}$, the statement (iv) follows easily from the formula (7.7).

If $\rho ==_{\{\varepsilon\}}$, then any retract T must contain $\mathcal{S}_n$ and one additional noninvertible element x. To be closed under multiplication, such x must satisfy $\pi x = x\pi = x$ for all $\pi \in \mathcal{S}_n$. The only element of $\mathcal{IS}_n$ which satisfies this condition is the element $\mathbf{0}$. Indeed, if $\alpha \in \mathcal{IS}_n$ is noninvertible and such that $\alpha \neq \mathbf{0}$, then $n > 1$, $\mathrm{im}(\alpha) \neq \varnothing$ and taking $a \in \mathrm{im}(\alpha)$ and $b \in \mathbf{N}\backslash\{a\}$ we obtain $(a,b)\alpha \neq \alpha$. The statement (v) follows. The statement (vi) is left as an exercise to the reader.   □

Proposition 12.2.3   *(i) The identity congruence on $\mathcal{T}_n$ has the unique retract, namely, $\mathcal{T}_n$ itself.*

(ii) If ρ is the uniform congruence on $\mathcal{T}_n$, then there exist $\sum_{k=1}^{n} \binom{n}{k}k^{n-k}$ retracts for ρ, all of the form $\{e\}$, where $e \in \mathcal{E}(\mathcal{T}_n)$.

(iii) If $\rho ==_{\mathcal{S}_n}$, then $\mathcal{T}_n$ contains $\sum_{k=1}^{n-1} \binom{n}{k}k^{n-k}$ retracts for ρ, all of the form $\{\varepsilon, e\}$, where $e \in \mathcal{E}(\mathcal{T}_n)\backslash\{\varepsilon\}$.

(iv) If $\rho ==_{\mathcal{A}_n}$, then $\mathcal{T}_n$ contains

$$\sum_{k=2}^{n}\binom{n}{k}\sum_{s=1}^{\lfloor \frac{k+2}{4}\rfloor} \frac{k!(n-k)^{k-(2s-1)}}{(k-2(2s-1))!\cdot(2s-1)!\cdot 2^{2s-1}}$$

ρ-retracts, all of the form $\{\varepsilon, \alpha, e\}$, where $e \in \mathcal{E}(\mathcal{T}_n)\backslash\{\varepsilon\}$ and $\alpha \in \mathcal{S}_n\backslash\mathcal{A}_n$ is an element of order two, which preserves the blocks of ρ_e and satisfies $\alpha(x) = x$ for any $x \in \mathrm{im}(e)$.

(v) If $\rho ==_{V_4}$, then $\mathcal{T}_4$ contains four ρ-retracts of the form

$$Q^{\{a,b,c\}} = \{\varepsilon, (a,b), (a,c), (b,c), (a,b,c), (a,c,b), 0_d\};$$

where $\{a,b,c\} \subset \mathbf{N}_4$ and $\{d\} = \mathbf{N}_4\backslash\{a,b,c\}$.

Proof. Analogous to that of Proposition 12.2.2 and is left to the reader.   □

Proposition 12.2.4   *(i) The identity congruence on $\mathcal{PT}_n$ has the unique retract, namely, $\mathcal{PT}_n$ itself.*

(ii) If ρ is the uniform congruence on $\mathcal{PT}_n$, then there exist $\sum_{k=0}^{n} \binom{n}{k}(k+1)^{n-k}$ retracts for ρ, all of the form $\{e\}$, where $e \in \mathcal{E}(\mathcal{PT}_n)$.

(iii) If $\rho ==_{\mathcal{S}_n}$, then $\mathcal{PT}_n$ contains $\sum_{k=0}^{n-1} \binom{n}{k}(k+1)^{n-k}$ retracts ρ, all of the form $\{\varepsilon, e\}$, where $e \in \mathcal{E}(\mathcal{PT}_n)\backslash\{\varepsilon\}$.

(iv) If $\rho ==_{\mathcal{A}_n}$, then $\mathcal{PT}_n$ contains

$$\sum_{k=2}^{n} \binom{n}{k} \sum_{s=1}^{\lfloor \frac{k+2}{4} \rfloor} \frac{k!(n-k+1)^{k-(2s-1)}}{(k-2(2s-1))! \cdot (2s-1)! \cdot 2^{2s-1}}$$

ρ-retracts, all of the form $\{\varepsilon, \alpha, e\}$, where $e \in \mathcal{E}(\mathcal{IS}_n)\backslash\{\varepsilon\}$ and $\alpha \in \mathcal{S}_n\backslash\mathcal{A}_n$ is an element of order two, which preserves the blocks of ρ_e and satisfies $\alpha(x) = x$ for any $x \in \mathrm{dom}(e)$.

(v) If $\rho ==_{\{\varepsilon\}}$, then $\mathcal{PT}_n$ contains the unique ρ-retract $\mathcal{S}_n \cup \{0\}$.

(vi) If $\rho ==_{V_4}$, then $\mathcal{PT}_4$ contains twelve ρ-retracts of the form $Q_0^{\{a,b,c\}}$, $Q_1^{\{a,b,c\}}$, $Q^{\{a,b,c\}}$, as defined in Propositions 12.2.2(vi) and 12.2.3(v).

Proof. Analogous to that of Proposition 12.2.2 and is left to the reader. □

12.3   *H*-Cross-Sections in $\mathcal{T}_n$, $\mathcal{PT}_n$, and $\mathcal{IS}_n$

In this section, we describe cross-sections in $\mathcal{T}_n$, $\mathcal{PT}_n$, and $\mathcal{IS}_n$ with respect to the relation $\mathcal{H}$. We will call such cross-sections simply $\mathcal{H}$-*cross-sections*. We start with the following negative result:

Theorem 12.3.1 *Let* $S = \mathcal{T}_n$ *or* $S = \mathcal{PT}_n$.

(i) If $n = 1$, then S contains the unique $\mathcal{H}$-cross-section, namely, S itself.

(ii) If $n = 2$, then S contains the unique $\mathcal{H}$-cross-section, namely, the subsemigroup $S\backslash\{(1,2)\}$.

(iii) If $n > 2$, then S does not contain any $\mathcal{H}$-cross-sections.

Proof. The statement (i) is obvious. To prove the statement (ii) one just observes that in this case the only $\mathcal{H}$-class of S containing more than one element is $\mathcal{S}_2$. As $\mathcal{S}_2$ is a subsemigroup of S, any $\mathcal{H}$-cross-section must contain an idempotent from $\mathcal{S}_2$, which is the identity transformation ε. We have $\mathcal{S}_2\backslash\{\varepsilon\} = \{(1,2)\}$. As $S\backslash\{(1,2)\}$ is a subsemigroup, it is an $\mathcal{H}$-cross-section. This proves the statement (ii).

To prove (iii) we consider the following elements from S:

$$\alpha_1 = \begin{pmatrix} 1 & 2 & 3 & \cdots & n \\ 1 & 3 & 3 & \cdots & 3 \end{pmatrix}, \quad \alpha_2 = \begin{pmatrix} 1 & 2 & 3 & \cdots & n \\ 3 & 1 & 1 & \cdots & 1 \end{pmatrix}.$$

$$\beta_1 = \begin{pmatrix} 1 & 2 & 3 & \cdots & n \\ 2 & 2 & 3 & \cdots & 3 \end{pmatrix}, \quad \beta_2 = \begin{pmatrix} 1 & 2 & 3 & \cdots & n \\ 3 & 3 & 2 & \cdots & 2 \end{pmatrix}.$$

$$\gamma_1 = \begin{pmatrix} 1 & 2 & 3 & 4 & \cdots & n \\ 1 & 2 & 1 & 2 & \cdots & 2 \end{pmatrix}, \quad \gamma_2 = \begin{pmatrix} 1 & 2 & 3 & 4 & \cdots & n \\ 2 & 1 & 2 & 1 & \cdots & 1 \end{pmatrix}.$$

From Theorem 4.5.1 we have $\mathcal{H}(\alpha_1) = \{\alpha_1, \alpha_2\}$, $\mathcal{H}(\beta_1) = \{\beta_1, \beta_2\}$ and $\mathcal{H}(\gamma_1) = \{\gamma_1, \gamma_2\}$. Assume that T is an $\mathcal{H}$-cross-section of S. Then the intersection of T with each of $\mathcal{H}(\alpha_1)$, $\mathcal{H}(\beta_1)$ and $\mathcal{H}(\gamma_1)$ consists of exactly one element. Since all these $\mathcal{H}$-classes are subgroups of S, the intersection of T with each of them must contain the identity elements of the corresponding subgroup; namely, α_1, β_1, and γ_1 respectively. Then T contains the element

$$\pi = \gamma_1 \beta_1 \alpha_1 = \begin{pmatrix} 1 & 2 & 3 & \cdots & n \\ 2 & 1 & 1 & \cdots & 1 \end{pmatrix}.$$

As $\pi \neq \pi^2$ and $\pi^2 \in \mathcal{H}(\pi)$, we obtain that T must contain two elements from $\mathcal{H}(\pi)$. This contradicts our assumption that T is an $\mathcal{H}$-cross-section, which completes the proof. $\square$

Let $\prec$ be a linear order on $\mathbf{N}$. A partial permutation $\alpha \in \mathcal{IS}_n$ is said to be $\prec$-*order-preserving* provided that $\alpha(i) \prec \alpha(j)$ for all $i, j \in \mathrm{dom}(\alpha)$ such that $i \prec j$. Denote by $\mathcal{IO}_n^{\prec}$ the set of all $\prec$-order-preserving partial permutations from $\mathcal{IS}_n$. If $\prec$ is the natural order, then one simply says *order-preserving* instead of $\prec$-order-preserving, and uses the notation $\mathcal{IO}_n$ for the corresponding set $\mathcal{IO}_n^{\prec}$. We denote by $\overleftarrow{\prec}$ the linear order opposite to $\prec$. We will also denote by $<$ the natural order on $\mathbf{N}$.

Exercise 12.3.2 Check that $\mathcal{IO}_n^{\prec}$ is a subsemigroup of $\mathcal{IS}_n$.

Theorem 12.3.3 *(i) For every linear order $\prec$ on $\mathbf{N}$ the subsemigroup $\mathcal{IO}_n^{\prec}$ is an $\mathcal{H}$-cross-section of $\mathcal{IS}_n$.*

(ii) For two linear orders $\prec_1$ and $\prec_2$ we have $\mathcal{IO}_n^{\prec_1} = \mathcal{IO}_n^{\prec_2}$ if and only if $\prec_1 = \prec_2$ or $\prec_1 = \overleftarrow{\prec}_2$.

(iii) If $n \neq 3$, then every $\mathcal{H}$-cross-section of $\mathcal{IS}_n$ has the form $\mathcal{IO}_n^{\prec}$ for some linear order $\prec$ on $\mathbf{N}$.

Proof. From Exercise 12.3.2 we know that $\mathcal{IO}_n^{\prec}$ is a subsemigroup of $\mathcal{IS}_n$. By Theorem 4.5.1(iii), each $\mathcal{H}$-class of $\mathcal{IS}_n$ is determined by some k, $0 \leq k \leq n$, and a pair A and B of k-element subsets of $\mathbf{N}$, and consists of all bijections $\alpha : A \to B$. Write $A = \{a_1 \prec a_2 \prec \cdots \prec a_k\}$ and $B = \{b_1 \prec b_2 \prec \cdots \prec b_k\}$. Then the element

$$\begin{pmatrix} a_1 & a_2 & \cdots & a_k \\ b_1 & b_2 & \cdots & b_k \end{pmatrix}$$

is the only representative of the corresponding $\mathcal{H}$-class in $\mathcal{IO}_n^{\prec}$. Hence $\mathcal{IO}_n^{\prec}$ is an $\mathcal{H}$-cross-section, which proves the statement (i).

It follows directly from the definitions that $\mathcal{IO}_n^{\prec_1} = \mathcal{IO}_n^{\prec_2}$ provided that $\prec_1 = \prec_2$ or $\prec_1 = \overleftarrow{\prec_2}$. Assume now that $\prec_1 \neq \prec_2$ and $\prec_1 \neq \overleftarrow{\prec_2}$. Without loss of generality we may even assume that $\prec_1$ is the natural order $<$ and set $\prec = \prec_2$. Since $\prec$ is neither the natural order nor the opposite to it, there exists $i \in \mathbf{N}$ such that either $i \pm 1 \prec i$ or $i \prec i \pm 1$. Then for the element

$$\alpha = \begin{pmatrix} i-1 & i \\ i & i+1 \end{pmatrix} \in \mathcal{IO}_n^{\leq}$$

of rank 2 we have $\alpha \notin \mathcal{IO}_n^{\prec}$. The statement (ii) follows.

The really interesting part here is to prove the statement (iii). Note that for $n = 1$ the statement is obvious. If $n = 2$, then we have a unique $\mathcal{H}$-class, containing more than one element, namely, $\mathcal{S}_2$. As in Theorem 12.3.1(ii) we obtain that $\mathcal{IS}_2 \backslash \{(1,2)\}$ is the unique $\mathcal{H}$-cross-section, which obviously has the necessary form. Hence from now on we may assume $n \geq 4$.

Let T be an $\mathcal{H}$-cross-section of $\mathcal{IS}_n$, $n \geq 4$. Define the binary relation $\prec = \prec_T$ on $\mathbf{N}$ as follows: $a \prec b$ if and only if there exists an element $\alpha \in T$ of rank two such that $\alpha(1) = a$ and $\alpha(2) = b$. Note that for $a \neq b$ we always have that exactly one of the inequalities $a \prec b$ or $b \prec a$ holds as there always exists a unique $\alpha \in T$ of rank two such that $\alpha(\{1, 2\}) = \{a, b\}$. In other words, $\prec$ is antisymmetric.

Lemma 12.3.4 T preserves $\prec$ in the sense that for any $a \prec b$ and any $\beta \in T$ such that $\{a, b\} \subset \mathrm{dom}(\beta)$ we have $\beta(a) \prec \beta(b)$.

Proof. Let $\alpha \in T$ be an element of rank two such that $\alpha(1) = a$ and $\alpha(2) = b$. Then $\mathrm{im}(\alpha) \subset \mathrm{dom}(\beta)$, which implies that $\beta\alpha$ has rank two as well. We have $\beta\alpha(1) = \beta(\alpha(1)) = \beta(a)$ and $\beta\alpha(2) = \beta(\alpha(2)) = \beta(b)$. This implies $\beta(a) \prec \beta(b)$ by definition. $\square$

Lemma 12.3.5 The relation $\prec$ is transitive in the sense that $a \prec b$ and $b \prec c$ implies $a \prec c$.

Proof. Let $a, b, c \in \mathbf{N}$ be such that $a \prec b$ and $b \prec c$. Assume that $c \prec a$. As $n \geq 4$, we can take some $d \in \mathbf{N} \backslash \{a, b, c\}$. Let $\alpha \in T$ be such that $\alpha(\{a, b, c\}) = \{b, c, d\}$.

If $\alpha(a) = d$, then $\alpha(\{b, c\}) = \{b, c\}$ and as α preserves $\prec$ by Lemma 12.3.5, we have $\alpha(b) = b$ and $\alpha(c) = c$. This implies $c \prec d$.

If $\alpha(a) = c$, then $\alpha(b) = b$ is impossible as α must preserve $\prec$. Hence $\alpha(c) = b$ and $\alpha(b) = d$. As α preserves $\prec$, we again obtain $c \prec d$. Analogously one shows that $c \prec d$ in the last case $\alpha(a) = b$.

On the other hand, let $\beta \in T$ be such that $\beta(\{a, b, c\}) = \{a, c, d\}$. If $\beta(a) = d$, then $\beta(b) = c$ and $\beta(c) = a$ as α preserves $\prec$. This implies $d \prec c$. Analogous arguments lead to the same conclusion in the case $\beta(a) = a$ and $\beta(a) = c$. Thus $d \prec c$, a contradiction.

Therefore, $c \prec a$ is not possible. As we already know that either $c \prec a$ or $a \prec c$, we conclude that $a \prec c$, which completes our proof. $\square$

The relation $\prec$ is antireflexive by definition. We already observed that $\prec$ is also antisymmetric. By Lemma 12.3.5, it is transitive. Hence $\prec$ is a partial order. As we already saw, any two elements of $\mathbf{N}$ are comparable with respect to $\prec$. This implies that $\prec$ is a linear order. Thus we have $T \subset \mathcal{IO}_n^{\prec}$ by Lemma 12.3.4. On the other hand, $\mathcal{IO}_n^{\prec}$ is an $\mathcal{H}$-cross-section of $\mathcal{IS}_n$ by (i). Hence T and $\mathcal{IO}_n^{\prec}$ have the same cardinalities (the total number of $\mathcal{H}$-classes in $\mathcal{IS}_n$) and we obtain $T = \mathcal{IO}_n^{\prec}$. This proves (iii) and completes the proof of the theorem. $\square$

Corollary 12.3.6 *(i) For $n \neq 1,3$ the semigroup $\mathcal{IS}_n$ contains exactly $\frac{n!}{2}$ cross-sections with respect to the relation $\mathcal{H}$.*

(ii) An element $\alpha \in \mathcal{IS}_n$ belongs to some $\mathcal{H}$-cross-section if and only if α does not have cycles of length greater than one.

Proof. By Theorem 12.3.3(iii), each $\mathcal{H}$-cross-section of $\mathcal{IS}_n$, $n \neq 3$, is of the form $\mathcal{IO}_n^{\prec}$. For $n \neq 1$, the $n!$ linear orders $\prec$ on $\mathbf{N}$ are divided into $n!/2$ pairs of the type $\{\prec, \overleftarrow{\prec}\}$. Now the statement (i) follows from Theorem 12.3.3(ii).

If $\alpha \in \mathcal{IS}_n$ has a cycle of length greater than one, then it is obvious that α cannot preserve any linear order on $\mathbf{N}$. On the other hand, if α does not have such cycles, we can write

$$\alpha = (a_1) \cdots (a_k)[b_1, b_2, \ldots, b_l] \cdots$$

and α preserves the order $a_1 \prec \cdots \prec a_k \prec b_1 \prec b_2 \prec \cdots \prec b_l \prec \cdots$. This completes the proof. $\square$

Exercise 12.3.7 Prove that for $n \neq 3$ all $\mathcal{H}$-cross-sections of $\mathcal{IS}_n$ are isomorphic.

12.4 $\mathcal{L}$-Cross-Sections in $\mathcal{T}_n$ and $\mathcal{PT}_n$

Let S denote one of the semigroups $\mathcal{T}_n$ or $\mathcal{PT}_n$. Fix a linear order $\prec$ on $\mathbf{N}$. Let

$$\mathbf{N} = \{u_1 \prec u_2 \prec \cdots \prec u_n\}. \tag{12.1}$$

For a nonempty $A \subset \mathbf{N}$ denote by $\min(A, \prec)$ the minimal element of A with respect to the order $\prec$. If $A, B \subset \mathbf{N}$ are nonempty and disjoint, we will write $A \prec B$ provided that $\min(A, \prec) \prec \min(B, \prec)$. Let $L(\prec)$ denote the set of all elements of S, which have the form

$$\alpha = \begin{pmatrix} A_1 & A_2 & \cdots & A_k \\ u_1 & u_2 & \cdots & u_k \end{pmatrix}, \tag{12.2}$$

where $A_1 \prec A_2 \prec \cdots \prec A_k$.

Theorem 12.4.1   (i) *For every linear order* $\prec$ *on* **N** *the set* $L(\prec)$ *is an L-cross-section of* S.

(ii) *For two orders* $\prec_1$ *and* $\prec_2$ *we have* $L(\prec_1) = L(\prec_2)$ *if and only if* $\prec_1 = \prec_2$.

(iii) *Every L-cross-section of* S *has the form* $L(\prec)$ *for some linear order* $\prec$ *on* **N**.

To prove this theorem we will need several lemmas.

Lemma 12.4.2 *For every linear order* $\prec$ *on* **N** *the set* $L(\prec)$ *is a subsemigroup of* S.

Proof. Let $\alpha, \beta \in L(\prec)$ be such that the element α is given by (12.2), while

$$\beta = \left(\begin{array}{cccc} B_1 & B_2 & \cdots & B_m \\ u_1 & u_2 & \cdots & u_m \end{array} \right),$$

where $B_1 \prec B_2 \prec \cdots \prec B_m$. We have to prove that $\alpha\beta \in L(\prec)$.

Let $A_i \prec A_j$ and assume that $u_j \in \mathrm{im}(\alpha\beta)$. This yields that for some $a \in A_j$ we have $a \in \mathrm{im}(\beta)$. As $\min(A_i, \prec) \prec \min(A_j, \prec) \prec a$, we conclude that $\min(A_i, \prec) \in \mathrm{im}(\beta)$ as well. Thus $u_i \in \mathrm{im}(\alpha\beta)$. Therefore, $\mathrm{im}(\alpha\beta) = \{u_1, \ldots, u_l\}$ for some $l \leq k$.

Let $u_i \prec u_j$ be elements from $\mathrm{im}(\alpha\beta)$ and $u_s = \min(A_i, \prec)$. Then by the definition we have

$$\{x \in \mathbf{N} \, : \, \alpha\beta(x) = u_i\} = \bigcup_{u_z \in A_i} B_z \qquad (12.3)$$

and

$$\{x \in \mathbf{N} \, : \, \alpha\beta(x) = u_j\} = \bigcup_{u_z \in A_j} B_z,$$

where we assume that $B_z = \varnothing$ if $z > m$. Note that $u_s \prec v$ for any $v \in A_j$ because $A_i \prec A_i$ and $\alpha \in L(\prec)$. Hence for any $u_z \in A_j$ we have the inequality $\min(B_s, \prec) \prec \min(B_z, \prec)$ because $\beta \in L(\prec)$. As the set from (12.3) contains B_s, we get

$$\{x \in \mathbf{N} \, : \, \alpha\beta(x) = u_i\} \prec \{x \in \mathbf{N} \, : \, \alpha\beta(x) = u_j\},$$

which finally yields $\alpha\beta \in L(\prec)$. $\square$

Lemma 12.4.3 *For every linear order* $\prec$ *on* **N** *the set* $L(\prec)$ *contains exactly one element from every L-class of* S.

Proof. Let $X \subset \mathbf{N}$ if $S = \mathcal{PT}_n$ or $X = \mathbf{N}$ if $S = \mathcal{T}_n$. Let further $X = X_1 \cup X_2 \cup \cdots \cup X_k$ be an unordered partition of X into a disjoint union of nonempty subsets. Without loss of generality we may assume that we have $X_1 \prec X_2 \prec \cdots \prec X_k$. By Theorem 4.5.1(ii), such partitions bijectively correspond to $\mathcal{L}$-classes of the semigroup S inside the $\mathcal{D}$-class $\mathcal{D}_k$. Then the element

$$\begin{pmatrix} X_1 & X_2 & \cdots & X_k \\ u_1 & u_2 & \cdots & u_k \end{pmatrix}$$

is the unique representative of $L(\prec)$ in the $\mathcal{L}$-class, defined by the partition $X_1 \cup X_2 \cup \cdots \cup X_k$. This completes the proof. $\qquad\square$

Lemma 12.4.4 *Let T be an $\mathcal{L}$-cross-section of S and $\alpha, \beta \in \mathcal{D}_k \cap T$. Then $\operatorname{im}(\alpha) = \operatorname{im}(\beta)$.*

Proof. Assume that this is not the case. Set

$$C = \operatorname{im}(\alpha) \cap \operatorname{im}(\beta), \quad A = \operatorname{im}(\alpha)\backslash C, \quad B = \operatorname{im}(\beta)\backslash C.$$

As $|\operatorname{im}(\alpha)| = |\operatorname{im}(\beta)| = k$, we have $|A| = |B| = m$. Let us write

$$A = \{a_1, \ldots, a_m\}, \quad B = \{b_1, \ldots, b_m\}, \quad C = \{c_1, \ldots, c_{k-m}\}.$$

Let $\gamma \in T$ be the element which belongs to the $\mathcal{L}$-class given by the following partition of $\mathbf{N}$:

$$\{a_1, b_1\} \cup \cdots \cup \{a_m, b_m\} \cup \{c_1\} \cup \cdots \cup \{c_{k-m-1}\} \cup (\{c_{k-m}\} \cup (\mathbf{N}\backslash(A \cup B \cup C))).$$

We have that all the elements α, β, γ, $\gamma\alpha$ and $\gamma\beta$ have rank k. This yields

$$\operatorname{im}(\gamma\alpha) = \operatorname{im}(\gamma) = \operatorname{im}(\gamma\beta). \qquad (12.4)$$

Moreover, we also have $\rho_\alpha = \rho_{\gamma\alpha}$, implying $\alpha = \gamma\alpha$ as T is an $\mathcal{L}$-cross-section. Analogously one shows that $\beta = \gamma\beta$. The statement of the lemma now follows from (12.4). $\qquad\square$

Lemma 12.4.5 *Let T be some $\mathcal{L}$-cross-section of S, $\alpha \in \mathcal{D}_k \cap T$ and $\beta \in \mathcal{D}_{k+1} \cap T$. Then $\operatorname{im}(\alpha) \subset \operatorname{im}(\beta)$.*

Proof. Let $A = \operatorname{im}(\alpha) = \{a_1, \ldots, a_k\}$ and γ denote the representative of T in the $\mathcal{L}$-class, which corresponds to the partition

$$\{a_1\} \cup \cdots \cup \{a_k\} \cup (\mathbf{N}\backslash A).$$

Then we have $\rho_\alpha = \rho_{\gamma\alpha}$ and thus $\gamma\alpha = \alpha$ since T is an $\mathcal{L}$-cross-section. At the same time $\operatorname{im}(\gamma\alpha) \subset \operatorname{im}(\gamma)$, which yields $\operatorname{im}(\alpha) \subset \operatorname{im}(\gamma)$. As $\gamma \in \mathcal{D}_{k+1}$, we have $\operatorname{im}(\gamma) = \operatorname{im}(\beta)$ by Lemma 12.4.4. This completes the proof. $\qquad\square$

Proof of Theorem 12.4.1. The statement (i) follows from Lemmas 12.4.2 and 12.4.3.

To prove (ii) let $\prec_1$ and $\prec_2$ be two different linear orders on **N**. Assume that

$$\begin{aligned} \mathbf{N} &= \{a_1 \prec_1 a_2 \prec_1 \cdots \prec_1 a_n\}, \\ \mathbf{N} &= \{b_1 \prec_2 b_2 \prec_2 \cdots \prec_2 b_n\}, \end{aligned} \quad (12.5)$$

and that k is minimal such that $a_k \neq b_k$. Consider some fixed L-class inside D_k. Let α and β be the representatives from $L(\prec_1)$ and $L(\prec_2)$ inside this L-class, respectively. Then by the definition we have

$$\mathrm{im}(\alpha) = \{a_1, \ldots, a_k\} \neq \{b_1, \ldots, b_k\} = \mathrm{im}(\beta).$$

Hence $\alpha \neq \beta$ and thus $L(\prec_1) \neq L(\prec_2)$. The statement (ii) is proved.

It remains to prove the statement (iii). Let T be an L-cross-section of S. From Lemma 12.4.5 it follows that there exists a linear order $\prec$ on **N** given by (12.1), such that for any $\alpha \in T$ of rank k we have $\mathrm{im}(\alpha) = \{u_1, \ldots, u_k\}$. We claim that $T \subset L(\prec)$.

Let $\alpha \in T$ be an element of rank k such that

$$\alpha = \begin{pmatrix} A_1 & A_2 & \cdots & A_k \\ v_1 & v_2 & \cdots & v_k \end{pmatrix},$$

where $A_1 \prec A_2 \prec \cdots \prec A_k$ and $\{v_1, \ldots, v_k\} = \{u_1, \ldots, u_k\}$. Assume that $\alpha \notin L(\prec)$ and let m be the minimal possible index such that $v_m \neq u_m$, in particular $u_m \prec v_m$. Let $u_l = \min(A_m, \prec)$. Then

$$u_l \prec x \quad \text{for any } x \in A_j \text{ such that } j > m. \quad (12.6)$$

Let $\beta \in T$ be an element of rank l. Then $\mathrm{im}(\beta) = \{u_1, \ldots, u_l\}$ by our assumptions. Consider the element $\alpha\beta \in T$. The image of $\alpha\beta$ is the union of all the $\alpha(A_i)$'s such that there exists $1 \leq j \leq l$ for which $u_j \in A_i$. If $j < l$, we have $u_j \prec u_l$ and thus $u_j \in A_i$ for some $i < m$ by (12.6) and our choice of u_l. At the same time $\alpha(A_m) = v_m \in \mathrm{im}(\alpha\beta)$. Hence $\mathrm{im}(\alpha\beta)$ has the form $\{v_m\} \cup N$ for some subset $N \subset \{u_1, \ldots, u_{m-1}\}$. As $v_m \neq u_m$, the latter set is not of the form $\{u_1, \ldots, u_i\}$ for any i. Hence $\alpha\beta \notin T$ by our assumptions, a contradiction. This shows that $\alpha \in L(\prec)$ and hence $T \subset L(\prec)$.

As both T and $L(\prec)$ are L-cross-sections, they have the same cardinality. Hence $T \subset L(\prec)$ implies $T = L(\prec)$. This completes the proof of (iii) and of the theorem. $\qquad\square$

Corollary 12.4.6 *(i) S contains exactly $n!$ different L-cross-sections.*

(ii) All L-cross-sections of S are isomorphic.

Proof. The statement (i) follows immediately from Theorem 12.4.1 as there are exactly $n!$ different linear orders on **N**.

To prove the statement (ii) assume that $\prec_1$ and $\prec_2$ are two different linear orders on $\mathbf{N}$ given by (12.5). Let $\pi \in \mathcal{S}_n$ be such that $\pi(a_i) = b_i$, $i = 1, \ldots, n$. Then from the definition we have $L(\prec_1) = \pi^{-1}L(\prec_2)\pi$. This implies that $L(\prec_2)$ is mapped to $L(\prec_1)$ by an inner automorphism of S. Hence these two semigroups are isomorphic and the proof is complete. □

12.5 $\mathcal{L}$-Cross-Sections in $\mathcal{IS}_n$

In this section we classify all $\mathcal{L}$-cross-sections in $\mathcal{IS}_n$. This description turns out to be more complicated than the descriptions presented in the previous sections. Hence we will start with some preparation and some auxiliary lemmas.

We first note that, by Theorem 4.5.1(ii), for $\alpha, \beta \in \mathcal{IS}_n$ the condition $\alpha \mathcal{L} \beta$ is equivalent to the condition $\mathrm{dom}(\alpha) = \mathrm{dom}(\beta)$. Hence the equalities $\alpha = \beta$ and $\mathrm{dom}(\alpha) = \mathrm{dom}(\beta)$ are equivalent for elements α, β from an arbitrary $\mathcal{L}$-cross-section T of $\mathcal{IS}_n$. We will frequently use this fact in this section.

Lemma 12.5.1 *Let T be an $\mathcal{L}$-cross-section of $\mathcal{IS}_n$. Then $\alpha \notin \langle T \backslash \{\alpha\}\rangle$ for every $\alpha \in T \cap \mathcal{D}_{n-1}$.*

Proof. Let $\alpha \in T \cap \mathcal{D}_{n-1}$ and $\beta, \gamma \in T$ be such that $\alpha = \beta\gamma$. Assume that $\beta \neq \alpha$ and $\gamma \neq \alpha$. Then, in particular, $\beta, \gamma \neq \varepsilon$. As T is an $\mathcal{L}$-cross-section, we thus get $\beta, \gamma \in \mathcal{I}_{n-1}$. The equality $\alpha = \beta\gamma$ then implies that both β and γ have rank $(n-1)$. Thus $\mathrm{dom}(\alpha) = \mathrm{dom}(\beta\gamma) = \mathrm{dom}(\gamma)$, which yields $\alpha = \gamma$, a contradiction. Thus we either have $\beta = \alpha$ or $\gamma = \alpha$, and the statement of the lemma follows. □

From Lemma 12.5.1 it follows that every generating system of each $\mathcal{L}$–cross-section T of $\mathcal{IS}_n$ must contain the set $T \cap \mathcal{D}_{n-1}$. Now we would like to study the latter set in more details.

Lemma 12.5.2 *Let T be an $\mathcal{L}$-cross-section of $\mathcal{IS}_n$ and $\alpha \in T \cap \mathcal{D}_{n-1}$. Then the chain decomposition of α contains exactly one chain, $[a_1, \ldots, a_k]$ say, and $\alpha(x) = x$ for all $x \in \mathbf{N} \backslash \{a_1, \ldots, a_k\}$.*

Proof. The element α contains exactly one chain as $\mathrm{def}(\alpha) = 1$. If this chain is $[a_1, \ldots, a_k]$, then $\mathrm{dom}(\alpha^k) = \mathrm{dom}(\alpha^{k+1})$, and hence $\alpha^k = \alpha^{k+1}$ as T is an $\mathcal{L}$-cross-section. By our assumptions, the element α induces a bijection on the set $\mathbf{N} \backslash \{a_1, \ldots, a_k\}$, hence α^k induces a bijection on this set as well. Let $x, y \in \mathbf{N} \backslash \{a_1, \ldots, a_k\}$ be such that $\alpha^k(y) = x$. Then, using $\alpha^k = \alpha^{k+1}$, we have

$$\alpha(x) = \alpha(\alpha^k(y)) = \alpha^{k+1}(y) = \alpha^k(y) = x,$$

which completes the proof. □

For $k = 1, \ldots, n$ set $\alpha_k = [1, 2, \ldots, k](k+1) \cdots (n)$ and denote by $\overline{K}_n$ the subsemigroup of $\mathcal{IS}_n$, generated by $\alpha_1, \ldots, \alpha_n$.

Exercise 12.5.3 ([GM3]) (a) Show that $\alpha_k \cdot \alpha_l = \alpha_l \cdot \alpha_{k-1}$ for every l and k such that $k \leq l$.

(b) Show that $\alpha_k^m = \alpha_k^{m-1} \cdot \alpha_{k-1}$ for all $m > 1$ and $k > 1$.

Corollary 12.5.4 *Every element $\alpha \in \overline{K}_n$ can be written in the form*

$$\alpha = \alpha_n^{a_n} \alpha_{n-1}^{a_{n-1}} \cdots \alpha_1^{a_1},$$

where $a_i \in \{0, 1\}$ for all $i = 1, 2, \ldots, n$, and $a_1 + \cdots + a_n > 0$.

Proof. From Exercise 12.5.3(a) it follows easily that every element $\alpha \in \overline{K}_n$ can be written in the form $\alpha = \alpha_n^{a_n} \alpha_{n-1}^{a_{n-1}} \cdots \alpha_1^{a_1}$, where $a_1, a_2, \ldots, a_n \in \{0, 1, 2, \ldots\}$. Taking this into account, the statement of the corollary follows from Exercise 12.5.3(b) and the observation that $\alpha_1^2 = \alpha_1$. $\square$

Exercise 12.5.5 ([GM3]) Let $1 \leq i_1 < i_2 < \cdots < i_k \leq n$ and define $\alpha = \alpha_{i_k} \alpha_{i_{k-1}} \cdots \alpha_{i_1}$. Show that

(a) $\mathrm{dom}(\alpha) = \mathbf{N} \backslash \{i_1, \ldots, i_k\}$.

(b) $\mathrm{im}(\alpha) = \mathbf{N} \backslash \{1, \ldots, k\}$.

(c) α coincides with the unique increasing bijection from $\mathbf{N} \backslash \{i_1, \ldots, i_k\}$ to $\mathbf{N} \backslash \{1, \ldots, k\}$.

Corollary 12.5.6 *The semigroup $K_n = \overline{K}_n \cup \{\varepsilon\}$ is an $\mathcal{L}$-cross-section of the semigroup $\mathcal{IS}_n$.*

Proof. From Exercise 12.5.5(a) and Corollary 12.5.4 it follows that for every $A \subset \mathbf{N}$ the semigroup K_n contains exactly one element α with domain A. Hence K_n contains exactly one element from every $\mathcal{L}$-class of $\mathcal{IS}_n$ by Theorem 4.5.1. $\square$

Lemma 12.5.7 (i) *In the semigroup K_n the inequality $x^{n-1} \neq x^n$ has the unique solution α_n.*

(ii) *The element α_n satisfies $\alpha_n^n = \alpha_n^{n+1}$.*

Proof. The statement (ii) is proved by a direct calculation.

We prove the statement (i) by induction on n. If $n = 1$, then the statement (i) is obvious. Let $A \subset \mathbf{N}$ be such that $|A| = k$ and assume that $\alpha : A \to \{n - k + 1, \ldots, n\}$ be an increasing bijection. Note that, by Exercise 12.5.5, in this way we get all elements from K_n. If $n \in A$, then, obviously, $\alpha(n) = n$. Applying the inductive assumption and the statement (ii) to the restriction of α to the invariant subset $\mathbf{N}_{n-1} \subset \mathbf{N}$, we obtain $\alpha^{n-1} = \alpha^n$.

Assume now that $n \notin A$ and $\text{rank}(\alpha) < n-1$. Then $\text{im}(\alpha) = \{i, \ldots, n\} \subset \{3, \ldots, n\}$ by Exercise 12.5.5(b). As α is an increasing function, $\text{im}(\alpha)$ is invariant with respect to α. Let β be the restriction of α to its image. Because of Exercise 12.5.5, we can apply the inductive assumption and the statement (ii) to β and get $\beta^{n-2} = \beta^{n-1}$. Multiplying with α from the right yields $\alpha^{n-1} = \alpha^n$.

If $n \notin A$ and $\text{rank}(\alpha) = n-1$, then $\text{dom}(\alpha) = \mathbf{N}_{n-1}$ and $\alpha = \alpha_n$ follows from Corollary 12.5.6. The inequality $\alpha_n^{n-1} \neq \alpha_n^n$ is checked by a direct calculation. This completes the proof.   □

Lemma 12.5.8 *Let T be an $\mathcal{L}$-cross-section of $\mathcal{IS}_n$. Assume that T contains the element $\alpha = \alpha_k$. Let $\beta \in T \cap \mathcal{D}_{n-1}$ be such that $\beta(x) = x$ for some $x \in \{1, \ldots, k\}$. Then $\beta(y) = y$ for all y, $x \leq y \leq k$.*

Proof. Without loss of generality we may assume that x is the minimal element such that $\beta(x) = x$. Consider the set

$$M = \{y : x \leq y \leq k \text{ and } \beta(y) \neq y\}$$

and assume that it is not empty. Let p be the maximal element in this set.

Assume first that $p \notin \text{dom}(\beta)$. Then $\text{dom}(\beta) = \mathbf{N} \setminus \{p\}$, and, taking into account $\beta(z) = z$ for all $p < z \leq k$, we obtain

$$\text{dom}(\alpha^{k-p}\beta) = \text{dom}(\alpha^{k-p+1}) = N \setminus \{p, p+1, \ldots, k\}.$$

Thus $\alpha^{k-p}\beta = \alpha^{k-p+1}$. At the same time

$$\alpha^{k-p}\beta(x) = x + k - p \neq x + k - p + 1 = \alpha^{k-p+1}(x).$$

Therefore, $p \notin \text{dom}(\beta)$ is not possible.

Assume now that $p \in \text{dom}(\beta)$. Set $\beta(p) = q \neq p$ and assume that the length of the unique chain in the chain decomposition of β equals m. Since $\beta(p) \neq p$, from Lemma 12.5.2 we deduce that p belongs to the chain of β. This implies that for the idempotent β^m we have $p, q \notin \text{dom}(\beta^m)$. Set $A = \{k+1, \ldots, n\}$, $B = \text{dom}(\beta^m)$, $C = A \cap B$ and $A_1 = A \backslash C$.

All elements from the set A_1 occur in the chain of β and hence we have $\text{im}(\beta^m) \cap A_1 = \varnothing$. Moreover, because of our choice of x we have that all elements from $\{1, 2, \ldots, x-1\}$ belong to the chain of the element β as well. Thus $\text{im}(\beta^m) \subset \{x, x+1, \ldots, k\} \cup C$. This implies that

$$\text{im}(\alpha^{p-x}\beta^m) \subset \{p, p+1, \ldots, k\} \cup C \subset \text{dom}(\beta).$$

The latter yields $\text{dom}(\alpha^{p-x}\beta^m) = \text{dom}(\beta\alpha^{p-x}\beta^m)$ and hence $\alpha^{p-x}\beta^m = \beta\alpha^{p-x}\beta^m$ as T is an $\mathcal{L}$-cross-section. But

$$(\alpha^{p-x}\beta^m)(x) = \alpha^{p-x}(\beta^m(x)) = \alpha^{p-x}(x) = p \neq q = \beta(p) = (\beta\alpha^{p-x}\beta^m)(x).$$

Hence $p \in \text{dom}(\beta)$ is not possible either. Thus, we obtain a contradiction with our assumption that M is not empty, and the statement of the lemma follows.   □

Let $[a_1, a_2, \ldots, a_k]$ be some chain. For every $l \in \{1, 2, \ldots, k\}$ the chain $[a_1, a_2, \ldots, a_l]$ will be called a *prefix* of $[a_1, a_2, \ldots, a_k]$.

Lemma 12.5.9 *Let T be an $\mathcal{L}$-cross-section of $\mathcal{IS}_n$ and $\alpha, \beta \in T \cap \mathcal{D}_{n-1}$. Assume that the chains of α and β have at least one common element. Then one of these chains is a prefix of the other one.*

Proof. Without loss of generality we can assume that $\alpha = \alpha_k$. Let $A = \{1, 2, \ldots, k\}$, B be the set of all elements from the chain of the element β and $C = A \cap B$. From Lemma 12.5.8 it follows that $C = \{1, 2, \ldots, m\}$, where $m \leq k$.

Assume that $x \in \mathbf{N} \backslash A$ is such that $x \neq \beta^a(x) = y \in C$ for some a. Then from $\alpha(x) = x$ and Lemma 12.5.8 we obtain that $\alpha(y) = y$, which contradicts the fact that $y \in C$. Hence $\beta = [i_1, i_2, \ldots, i_m, j_1, \ldots, j_l](f) \cdots (g)$, where $i_1, i_2, \ldots, i_m$ is a permutation of $1, 2, \ldots, m$.

Suppose that $(i_1, i_2, \ldots, i_m) \neq (1, 2, \ldots, m)$ (this is an inequality of vectors, not cycles). Then there exist elements $u, v, p, q \in \{1, 2, \ldots, m\}$ such that $i_p = u$, $i_q = v$, $u < v$, $p > q$. This implies $\alpha^{v-u}(u) = v$, $\beta^{p-q}(v) = u$ and $(\beta^{p-q} \alpha^{v-u})(u) = u$. For the element $\gamma = \beta^{p-q} \alpha^{v-u}$ there exists t such that γ^t is an idempotent. Clearly, $u \in \text{dom}(\gamma^t)$ and $\gamma^t(u) = u$. Moreover, $\text{dom}(\gamma^t) \subset \text{dom}(\gamma) \subset \text{dom}(\alpha)$. Now $\gamma^{2t} = \gamma^t$ implies that $\text{dom}(\gamma^t) = \text{im}(\gamma^t)$. Hence $\text{dom}(\gamma^t) = \text{dom}(\alpha \gamma^t)$ and thus $\gamma^t = \alpha \gamma^t$. But the last equality is impossible for $\gamma^t(u) = u \neq u + 1 = (\alpha \gamma^t)(u)$.

From the previous paragraph we have $(i_1, i_2, \ldots, i_m) = (1, 2, \ldots, m)$ and thus $\beta = [1, 2, \ldots, m, j_1, \ldots, j_l](f) \cdots (g)$. To complete the proof it is now enough to show that either $k = m$ or $l = 0$. Assume that this is not the case, that is, $k > m$ and $l > 0$. Note that $\{m+1, \ldots, k\} \cap \{j_1, \ldots, j_l\} = \varnothing$, α acts as the identity on all elements from $B \backslash C = \{j_1, \ldots, j_l\}$ and β acts as the identity on all elements from $A \backslash C = \{m+1, \ldots, k\}$. This implies that $k, j_l \notin \text{dom}(\beta \alpha)$ and $k, j_l \notin \text{dom}(\alpha \beta)$. Since $\text{def}(\beta \alpha) \leq 2$ and $\text{def}(\alpha \beta) \leq 2$, we have that $\text{dom}(\beta \alpha) = \text{dom}(\alpha \beta) = \mathbf{N} \backslash \{k, j_l\}$ and $\alpha \beta = \beta \alpha$ as T is an $\mathcal{L}$-cross-section. But $(\beta \alpha)(m) = m + 1 \neq j_1 = (\alpha \beta)(m)$. This contradiction completes the proof of the lemma. $\qquad \square$

Let now $\mathbf{N} = M_1 \cup M_2 \cup \cdots \cup M_k$ be an arbitrary partition of $\mathbf{N}$ into an unordered disjoint union of nonempty blocks. For $i = 1, 2, \ldots, k$ we set $m_i = |M_i|$ and choose some linear order $\prec_i$ on M_i. The collection $\{\prec_1, \ldots, \prec_k\}$ uniquely determines the partition $M_1 \cup M_2 \cup \cdots \cup M_k$. Assume that

$$M_i = \{u_1^i \prec_i u_2^i \prec_i \cdots \prec_i u_{m_i}^i\}.$$

For every $i \in \{1, \ldots, k\}$ and every $j \in \{1, \ldots, m_i\}$ let α_j^i denote the unique element of rank $(n-1)$, which can be written as follows:

$$\alpha_j^i = [u_1^i, u_2^i, \ldots, u_j^i](a)(b) \cdots (c).$$

Let $L(\prec_1, \ldots, \prec_k)$ denote the subsemigroup of $\mathcal{IS}_n$, which is generated by ε and all α_j^i, $i \in \{1, \ldots, k\}$, $j \in \{1, \ldots, m_i\}$.

Theorem 12.5.10   *(i) For any partition* $\mathbf{N} = M_1 \cup M_2 \cup \cdots \cup M_k$ *and any choice of a linear order* $\prec_i$ *for each block of the partition the semigroup* $L(\prec_1, \ldots, \prec_k)$ *is an* $\mathcal{L}$-*cross-section of* $\mathcal{IS}_n$.

(ii) $L(\prec_1, \ldots, \prec_k) = L(\prec_1', \ldots, \prec_m')$ *if and only if we have the equality* $\{\prec_1, \ldots, \prec_k\} = \{\prec_1', \ldots, \prec_m'\}$.

(iii) Every $\mathcal{L}$-*cross-section of* $\mathcal{IS}_n$ *has the form* $L(\prec_1, \ldots, \prec_k)$ *for appropriate partition* $\mathbf{N} = M_1 \cup M_2 \cup \cdots \cup M_k$ *and some choice* $\prec_i$ *of a linear order for every block of the partition.*

Proof. Set $L = L(\prec_1, \ldots, \prec_k)$. For $X \subset \mathbf{N}$ and $i = 1, 2, \ldots, k$ set $X_i = X \cap M_i$ and denote by L_i the subsemigroup of L, generated by ε and all the α_j^i, $j \in \{1, \ldots, m_i\}$. Then for all $\alpha \in L_i$ and $x \notin M_i$ we have $\alpha(x) = x$, so we can consider L_i as a subsemigroup of $\mathcal{IS}(M_i)$. Let us identify M_i with $\mathbf{N}_{m_i}$ such that the order $\prec_i$ corresponds to the natural order on $\mathbf{N}_{m_i}$. Then the semigroup L_i is identified with the semigroup K_{m_i} from Corollary 12.5.6. The latter corollary also says that L_i is an $\mathcal{L}$-cross-section of $\mathbf{N}_{m_i}$. This means that there exists a unique element $\beta_i \in L_i$ such that $\mathrm{dom}(\beta_i) = \mathbf{N} \backslash X_i$.

From the definition we have that every element from L_i commutes with every element from L_j if $i \neq j$. As every β_i acts as the identity outside X_i, we conclude that

$$\mathrm{dom}(\beta_1 \beta_2 \cdots \beta_k) = \bigcap_{i=1}^{k}(\mathbf{N} \backslash X_i) = \mathbf{N} \backslash X.$$

Hence L contains a representative in every $\mathcal{L}$-class of $\mathcal{IS}_n$.

From Corollary 12.5.6 we have $|L_i| = 2^{m_i}$. As every element from L_i commutes with every element from L_j if $i \neq j$, from the definition we have that every $\beta \in L$ can be written as the product $\beta = \beta_1 \beta_2 \cdots \beta_k$, where $\beta_i \in L_i$ for every i. Hence

$$|L| \leq |L_1| \cdot |L_2| \cdots |L_k| = 2^{m_1} \cdot 2^{m_2} \cdot \cdots \cdot 2^{m_k} = 2^{m_1 + m_2 + \cdots + m_k} = 2^n.$$

As L contains a representative in every $\mathcal{L}$-class of $\mathcal{IS}_n$, we have $|L| \geq 2^n$. This yields $|L| = 2^n$ and thus L must contain a unique representative in every $\mathcal{L}$-class of $\mathcal{IS}_n$. This proves the statement (i).

To prove the statement (ii) we consider two collections $\prec = \{\prec_1, \ldots, \prec_k\}$ and $\prec' = \{\prec_1', \ldots, \prec_k'\}$. Set $L = L(\prec_1, \ldots, \prec_k)$ and $L' = L(\prec_1', \ldots, \prec_m')$. For every $x \in \mathbf{N}$ consider the orbit of x with respect to L:

$$\mathfrak{o}_{\prec}(x) = \{\alpha(x) : \alpha \in L \text{ and } x \in \mathrm{dom}(\alpha)\}.$$

It is easy to see that the blocks M_i's are maximal orbits with respect to inclusions. Hence the equality $L = L'$ implies $k = m$ and uniquely determines the partition $\mathbf{N} = M_1 \cup \cdots \cup M_k$. Let L_i and L'_i denote the subsemigroups of L and L', defined as in the first part of the proof. By Lemma 12.5.7(i), the inequality $x^{m_i-1} \neq x^{m_i}$ has unique solutions in both L and L', which must coincide provided that $L = L'$. The solution has the form $[u_1^i, \ldots, u_{m_i}^i](a) \cdots (b)$ and uniquely determines the order $\prec_i$. Hence $\prec_i = \prec'_i$ and the statement (ii) follows.

It remains to prove the statement (iii). Let L be an $\mathcal{L}$-cross-section of $\mathcal{IS}_n$. By Lemma 12.5.9 the chains of two arbitrarily chosen elements from $L \cap \mathcal{D}_{n-1}$ are either disjoint or one of these chains is a prefix of the other one. The chains of the elements from $L \cap \mathcal{D}_{n-1}$, which are not proper prefixes for chains of any other element from $L \cap \mathcal{D}_{n-1}$, define a disjoint family $M_1, \ldots, M_k$ of subsets of $\mathbf{N}$. Moreover, each such maximal chain defines a natural linear order on the corresponding M_i in the following way: the chain $[x_1, \ldots, x_m]$ defines $M_i = \{x_1, \ldots, x_m\}$ with the order $x_1 \prec_i x_2 \prec_i \cdots \prec_i x_m$. For every $x \in \mathbf{N}$ there exists $\alpha \in L \cap \mathcal{D}_{n-1}$ such that $\mathrm{dom}(\alpha) = \mathbf{N} \backslash \{x\}$. This means that x belongs to the unique chain of α and hence x belongs to some M_i. Therefore L defines a partition $\mathbf{N} = M_1 \cup \cdots \cup M_k$.

From the construction of this partition and Lemma 12.5.9 it follows that

$$L \cap \mathcal{D}_{n-1} = L(\prec_1, \ldots, \prec_k) \cap \mathcal{D}_{n-1}$$

and hence L contains the subsemigroup of $L(\prec_1, \ldots, \prec_k)$, generated by all elements of rank n and $(n-1)$. However, $L(\prec_1, \ldots, \prec_k)$ is generated by elements of rank n and $(n-1)$ by definition. Hence $L(\prec_1, \ldots, \prec_k) \subset L$.

Recall that $L(\prec_1, \ldots, \prec_k)$ is an $\mathcal{L}$-cross-section of $\mathcal{IS}_n$ by (i), while L is an $\mathcal{L}$-cross-section of $\mathcal{IS}_n$ by our assumption. Hence $|L(\prec_1, \ldots, \prec_k)| = |L|$, implying $L(\prec_1, \ldots, \prec_k) = L$. This completes the proof of the theorem. $\qquad \square$

12.6 $\mathcal{R}$-Cross-Sections in $\mathcal{IS}_n$

Let $\mathbf{N} = M_1 \cup M_2 \cup \cdots \cup M_k$ be an arbitrary partition of $\mathbf{N}$ into an unordered disjoint union of nonempty blocks. For $i = 1, \ldots, k$ set $m_i = |M_i|$ and choose some linear order $\prec_i$ on M_i. Assume that

$$M_i = \{u_1^i \prec_i u_2^i \prec_i \cdots \prec_i u_{m_i}^i\}.$$

For every $i \in \{1, \ldots, k\}$ and every $j \in \{1, \ldots, m_i\}$ let β_j^i denote the unique element of rank $(n-1)$, which can be written as follows:

$$\beta_j^i = [u_j^i, u_{j+1}^i, \ldots, u_{m_i}^i](a)(b) \cdots (c).$$

Let $R(\prec_1, \ldots, \prec_k)$ denote the subsemigroup of $\mathcal{IS}_n$, which is generated by ε and all β_j^i, $i \in \{1, \ldots, k\}$, $j \in \{1, \ldots, m_i\}$.

Theorem 12.6.1   *(i) For any partition* $\mathbf{N} = M_1 \cup M_2 \cup \cdots \cup M_k$ *and any choice of a linear order* $\prec_i$ *for each block of the partition the semigroup* $R(\prec_1, \ldots, \prec_k)$ *is an* $\mathcal{R}$-*cross-section of* $\mathcal{IS}_n$.

(ii) $R(\prec_1, \ldots, \prec_k) = R(\prec_1', \ldots, \prec_m')$ *if and only if we have the equality* $\{\prec_1, \ldots, \prec_k\} = \{\prec_1', \ldots, \prec_m'\}$.

(iii) Every $\mathcal{R}$-*cross-section of* $\mathcal{IS}_n$ *has the form* $R(\prec_1, \ldots, \prec_k)$ *for appropriate partition* $\mathbf{N} = M_1 \cup M_2 \cup \cdots \cup M_k$ *and some choice* $\prec_i$ *of a linear order for every block of the partition.*

Proof. The mapping $\alpha \mapsto \alpha^{-1}$ is an antiinvolution on $\mathcal{IS}_n$. It swaps $\mathcal{L}$- and $\mathcal{R}$-classes of $\mathcal{IS}_n$. It also swaps formulations of Theorems 12.5.10 and 12.6.1. Hence Theorem 12.6.1 follows from Theorem 12.5.10. $\square$

Corollary 12.6.2 *The semigroup* $\mathcal{IS}_n$ *contains exactly* $\displaystyle\sum_{k=1}^{n} \frac{n!}{k!}\binom{n-1}{k-1}$ *different* $\mathcal{L}$-*cross-sections, and the same number of different* $\mathcal{R}$-*cross-sections.*

Proof. It is enough to prove the statement for $\mathcal{L}$-cross-sections. Let us count the number of $\mathcal{L}$-cross-sections, which correspond to partitions of $\mathbf{N}$ into k blocks.

For this fixed k consider an arbitrary permutation $i_1, \ldots, i_n$ of $1, 2, \ldots, n$, and a $(k-1)$-element subset, $\{j_1, \ldots, j_{k-1}\}$, of $\{1, 2, \ldots, n-1\}$. Assume that $j_1 < j_2 < \cdots < j_{k-1}$. This defines a decomposition of $\mathbf{N}$ into k blocks $M_1 = \{i_1, \ldots, i_{j_1}\}$, $M_2 = \{i_{j_1+1}, \ldots, i_{j_2}\}, \ldots, M_k = \{i_{j_{k-1}+1}, \ldots, i_n\}$ together with linear orders on these blocks. Since the order of the blocks is not important for $L(\prec_1, \ldots, \prec_k)$ we get that each $\mathcal{L}$-cross-section is counted exactly $k!$ times. Therefore we have exactly $n!\binom{n-1}{k-1}\frac{1}{k!}$ different $\mathcal{L}$-cross-sections for our fixed k. Summing this over all k we get the necessary statement. $\square$

Corollary 12.6.3 *The only* $\mathcal{L}$-*cross-section of* $\mathcal{IS}_n$, *which is an* $\mathcal{R}$-*cross-section at the same time, is the semigroup* $\mathcal{E}(\mathcal{IS}_n)$ *of all idempotents of* $\mathcal{IS}_n$.

Proof. Let $L = L(\prec_1, \ldots, \prec_k)$ be an $\mathcal{L}$-cross-section and $\mathbf{N} = M_1 \cup \cdots \cup M_k$ be the corresponding partition of $\mathbf{N}$.

Assume that $m_i > 1$ for some i. Let X and Y be two different subsets of M_i of the same cardinality. Set $X' = \mathbf{N} \backslash X$ and $Y' = \mathbf{N} \backslash Y$. Let $\alpha, \beta \in L$ be such that $\mathrm{dom}(\alpha) = X'$ and $\mathrm{dom}(\beta) = Y'$. From $|M_i \backslash X| = |M_i \backslash Y|$ and Exercise 12.5.5 we obtain $\mathrm{im}(\alpha) = \mathrm{im}(\beta)$. This means that L contains two different elements from the same $\mathcal{R}$-class and hence L cannot be an $\mathcal{R}$-cross-section.

On the other hand, if $m_i = 1$ for all i, we obtain that $k = n$, all $\prec_i$'s are trivial and $\alpha_1^i = \varepsilon_{\mathbf{N}\backslash M_i}$. Hence $L = \mathcal{E}(\mathcal{IS}_n)$. As $\mathcal{E}(\mathcal{IS}_n)$ is stable under the mapping $\alpha \mapsto \alpha^{-1}$, $\mathcal{E}(\mathcal{IS}_n)$ is an $\mathcal{R}$-cross-section as well. This completes the proof. $\square$

12.7 Addenda and Comments

12.7.1 Propositions 12.2.2–12.2.4 are proved in [ST1], [ST2], and [ST3], respectively. Proposition 12.2.3 can also be found in [Pek1]. Theorems 12.3.1 and 12.4.1 are taken from [Pek2]. Theorem 12.3.3 is proved in [CR]. The results of Sects. 12.5 and 12.6 are obtained in [GM3]. These results were also independently obtained in [YY] by completely different methods.

12.7.2 Cross-sections of other semigroups, in particular some infinite transformation semigroups, were studied in [Pek3, Pek4, KMM].

12.7.3 If G is an abelian group and ρ is the conjugacy relation on G, then G contains a unique ρ-cross-section, namely, G itself. In the non-abelian case we have the following:

Proposition 12.7.1 *Let G be a finite non-abelian group and ρ be the conjugacy relation on G. Then G does not contain any ρ-cross-section.*

Proof. Assume that T is a ρ-cross-section. Let H be a conjugacy class in G of maximal cardinality k. Note that $k > 1$ as G is not abelian. At the same time G contains a one-element conjugacy class consisting of the identity element. Since T is a ρ-cross-section, we get

$$|T| > \frac{|G|}{|H|}. \tag{12.7}$$

The group G acts on H by conjugation. Let $a \in H \cap T$. As the action of G on H is transitive, from Theorem 10.2.4(i) we have $|G| = |H| \cdot |\mathrm{St}_G(a)|$. In particular, $|\mathrm{St}_G(a)| = |G|/|H|$. Now we claim that $T \subset \mathrm{St}_G(a)$. Indeed, if $g \in T$, then $g^{-1}ag \in H \cap T$ and thus $g^{-1}ag = a$ as T is a ρ-cross-section. This yields $g \in \mathrm{St}_G(a)$. Hence

$$|T| \leq |\mathrm{St}_G(a)| = \frac{|G|}{|H|}. \tag{12.8}$$

Comparing (12.7) and (12.8) we get a contradiction. This proves the statement of our proposition. □

After Proposition 12.7.1 a natural question is: Does Proposition 12.7.1 generalize to infinite groups?

12.7.4 Description of $\mathcal{D}$-cross-section for semigroups $\mathcal{T}_n$, $\mathcal{PT}_n$ and $\mathcal{IS}_n$ seems to be a very difficult problem, even if one considers some special classes of cross-sections.

For example, consider the easiest case of the semigroup $\mathcal{IS}_n$. A natural problem is to try to classify all $\mathcal{D}$-cross-sections of $\mathcal{IS}_n$, consisting of idempotents. As $\mathcal{E}(\mathcal{IS}_n) \cong (\mathcal{B}(\mathbf{N}), \cap)$, the problem can be reformulated as follows: Classify all collections $(X_0, X_1, \ldots, X_n)$ of subsets of $\mathbf{N}$ such that the following two conditions are satisfied:

- $|X_i| = i$ for every $i = 0, 1, \ldots, n$.

- For any $i, j \in \{0, 1, \ldots, n\}$ there exists $l \in \{0, 1, \ldots, n\}$ such that $X_i \cap X_j = X_l$.

We do not know how to solve this problem (which appears already in [GM3]). We even do not know how to count the number of such collections. Classification of $\mathcal{D}$-cross-section for both $\mathcal{T}_n$ and $\mathcal{PT}_n$ contains this problem as a subproblem. For $\mathcal{PT}_n$ the latter statement is obvious as $\mathcal{IS}_n \subset \mathcal{PT}_n$ and this inclusion induces a bijection on $\mathcal{D}$-classes. For $\mathcal{T}_n$ it follows from the embedding of $\mathcal{IS}_n$ into $\mathcal{T}_{n+1}$ via

$$
\begin{pmatrix} a_1 & a_2 & \cdots & a_k \\ b_1 & b_2 & \cdots & b_k \end{pmatrix} \mapsto \begin{pmatrix} a_1 & a_2 & \cdots & a_k & \mathbf{N}_{n+1} \backslash \{a_1, \ldots, a_k\} \\ b_1 & b_2 & \cdots & b_k & n+1 \end{pmatrix}.
$$

12.7.5 At the moment there is no description/classification of $\mathcal{R}$-cross-sections for semigroups $\mathcal{T}_n$ and $\mathcal{PT}_n$. As the inclusion $\mathcal{T}_n \subset \mathcal{PT}_n$ induces a bijection on $\mathcal{R}$-classes, the problem of description of $\mathcal{R}$-cross-sections for $\mathcal{T}_n$ is a subproblem of the corresponding problem for $\mathcal{PT}_n$. From the inclusion $\mathcal{IS}_n \subset \mathcal{PT}_n$ we have that every $\mathcal{R}$-cross-section of $\mathcal{IS}_n$ is an $\mathcal{R}$-cross-section of $\mathcal{PT}_n$ as well.

A natural family of $\mathcal{R}$-cross-sections for $\mathcal{T}_n$ is the following: For a nonempty subset $X \subset \mathbf{N}$ such that $X = \{i_1 < i_2 < \cdots < i_k\}$ define the element

$$
\gamma_X = \begin{pmatrix} 1 & 2 & \cdots & k-2 & k-1 & k & k+1 & \cdots & n \\ i_1 & i_2 & \cdots & i_{k-2} & i_{k-1} & i_k & i_k & \cdots & i_k \end{pmatrix}.
$$

Let $R = \{\gamma_X : \varnothing \neq X \subset \mathbf{N}\}$.

Lemma 12.7.2 R is an $\mathcal{R}$-cross-section for $\mathcal{T}_n$.

Proof. As $\operatorname{im}(\gamma_X) = X$, the set R contains exactly one representative from each $\mathcal{R}$-class of $\mathcal{T}_n$. A direct calculation shows that R is closed under the multiplication. Hence R is an $\mathcal{R}$-cross-section for $\mathcal{T}_n$. $\square$

Conjugating R with elements from $\mathcal{S}_n$ one obtains $n!$ different $\mathcal{R}$-cross-sections for $\mathcal{T}_n$. However, $\mathcal{T}_n$ contains many other $\mathcal{R}$-cross-sections, see for example Exercise 12.8.10. We suspect that R is the only $\mathcal{R}$-cross-section for $\mathcal{T}_n$, which contains the element $\gamma_{\{2,3,\ldots,n\}}$, and that this fact could be used to classify $\mathcal{R}$-cross-sections for $\mathcal{T}_n$ along the arguments, dual to those of Sect. 12.5.

12.8 Additional Exercises

12.8.1 Prove that the only element $x \in \mathcal{PT}_n$ which satisfies $\pi x = x\pi = x$ for all $\pi \in \mathcal{S}_n$ is the element $\mathbf{0}$.

12.8.2 Prove the statement of Proposition 12.2.2(vi).

12.8.3 Prove Proposition 12.2.3.

12.8.4 Prove Proposition 12.2.4.

12.8.5 ([CR]) Show that the set

$$\mathcal{I}_1 \cup \{\varepsilon, \varepsilon_{\{1,2\}}, \varepsilon_{\{1,3\}}, \varepsilon_{\{2,3\}}, [2,1,3], [1,2,3], [3,1,2], [1,3,2], [3,2,1], [2,3,1]\}$$

is an $\mathcal{H}$-cross-section of $\mathcal{IS}_3$, which is not equal to $\pi \mathcal{IO}_3\pi^{-1}$ for any $\pi \in \mathcal{S}_3$.

12.8.6 Describe all $\mathcal{H}$-cross-sections of $\mathcal{IS}_3$.

12.8.7 ([GM3]) Show that the semigroup $\overline{K}_n$ has the following presentation:

$$\overline{K}_n = \langle \alpha_1, \ldots, \alpha_n | \alpha_1^2 = \alpha_1; \alpha_k^2 = \alpha_k\alpha_{k-1}, k = 2, \ldots, n;$$
$$\alpha_k\alpha_l = \alpha_l\alpha_{k-1}, 1 \le k < l \le n \rangle.$$

12.8.8 Let S be a semigroup and T be an $\mathcal{L}$-cross-section of S. Assume that T contains some element b. Show that T contains all idempotents $e \in \mathcal{E}(S)$ such that $b\mathcal{R}e$.

12.8.9 Classify all $\mathcal{L}$-cross-sections of a rectangular band.

12.8.10 Show that the following set is an $\mathcal{R}$-cross-section of $\mathcal{T}_4$, which is not conjugate to the $\mathcal{R}$-cross-section R from Lemma 12.7.2:

$$\{\varepsilon, \varepsilon_{1,2}, \varepsilon_{2,1}, \varepsilon_{3,4}, \varepsilon_{4,3}, \varepsilon_{1,2}\varepsilon_{3,4}, \varepsilon_{1,2}\varepsilon_{4,3}, \varepsilon_{2,1}\varepsilon_{3,4}, \varepsilon_{2,1}\varepsilon_{4,3},$$
$$(2,3)\varepsilon_{1,2}\varepsilon_{3,4}, (2,3)\varepsilon_{2,1}\varepsilon_{4,3}, 0_1, 0_2, 0_3, 0_4, \}.$$

12.8.11 ([GM3]) Let $n > 3$ and assume that the unordered partition $\mathbf{N} = M_1 \cup \cdots \cup M_k$ contains exactly m blocks of cardinality at least two. Show that for arbitrary linear orders $\prec_i$ on M_i, $i = 1, \ldots, k$, the $\mathcal{L}$-cross-section $L(\prec_1, \ldots, \prec_k)$ of $\mathcal{IS}_n$ is contained in exactly $k! \cdot 2^{m-1}$ different $\mathcal{H}$-cross-sections.

12.8.12 ([GM3]) Show that for $n > 3$ every $\mathcal{H}$-cross-section of $\mathcal{IS}_n$ contains exactly

$$1 + \sum_{l=1}^{\lfloor \frac{n}{2} \rfloor} \left(2^l \cdot \sum_{m=0}^{n-2l} \binom{m+l}{l} \binom{n-m-l-1}{l-1} \right)$$

different $\mathcal{L}$-cross-sections (or $\mathcal{R}$-cross-sections).

12.8.13 ([GM3]) Show that the $\mathcal{L}$-cross-sections

$$L(\prec_1, \ldots, \prec_k) \text{ and } L(\prec_1', \ldots, \prec_k'),$$

corresponding to some choices of linear orders on blocks of the partitions
$\mathbf{N} = M_1 \cup \cdots \cup M_k$ and $\mathbf{N} = M_1' \cup \cdots \cup M_k'$, respectively, are isomorphic if
there exists a bijection $\alpha \in \mathcal{S}_k$ such that $|M_i| = |M_{\alpha(i)}'|$.

12.8.14 ([CR]) Let $n > 3$. Show that an element $\alpha \in \mathcal{IS}_n$ is contained in a
unique $\mathcal{H}$-cross-section if and only if α is a nilpotent element of nilpotency
degree n.

12.8.15 Construct a natural bijection between the set of all nilpotent elements of $\mathcal{IS}_n$ and the set of all $\mathcal{L}$-cross-sections of $\mathcal{IS}_n$.

Chapter 13

Variants

13.1 Variants of Semigroups

Let $S = (S, \cdot)$ be a semigroup and $a \in S$. For $x, y \in S$ set $x \circ_a y = x \cdot a \cdot y$. Then $\circ_a$ is an associative binary operation on S and hence $(S, \circ_a)$ is a semigroup. The semigroup $(S, \circ_a)$ is called a *variant* of S, or, alternatively, the *sandwich semigroup* of S with respect to the *sandwich element* a. The operation $\circ_a$ is usually called the *sandwich operation*. To simplify notation we will denote $(S, \circ_a)$ simply by S^a. If a is the identity element of S, then we obviously have $\circ_a = \cdot$. However, for other a the sandwich operations normally give rise to new semigroups:

Proposition 13.1.1 *(i) We have $S^a \cong S$ provided that $a \in S^*$.*

(ii) If S is a monoid, then $S^a \cong S$ if and only if $a \in S^*$.

Proof. To prove the statement (i) we consider the mapping $\varphi : S \to S$, defined as follows: $x \mapsto xa^{-1}$. For $x, y \in S$ we have

$$
\begin{aligned}
\varphi(xy) &= (xy)a^{-1} \\
&= xa^{-1}axa^{-1} \\
&= \varphi(x)a\varphi(y) \\
&= \varphi(x) \circ_a \varphi(y).
\end{aligned}
$$

Hence $\varphi : S \to S^a$ is a homomorphism. Since a is invertible, the mapping $y \mapsto ya$ is inverse to φ. Therefore, φ is bijective and hence an isomorphism. This proves the statement (i).

To prove (ii) we assume that $S^a \cong S$. In particular, S^a is a monoid as well. Let $b \in S^a$ be the unit element of this monoid. Then for any $s \in S^a$ we have $s \circ_a b = b \circ_a s = s$, which is equivalent to $s(ab) = s$ and $(ba)s = s$. The first equality means that ab is a right identity in S, and the second equality means that ba is a left identity in S. As S is a monoid with the identity

O. Ganyushkin, V. Mazorchuk, *Classical Finite Transformation Semigroups*, Algebra and Applications 9, DOI: 10.1007/978-1-84800-281-4_13, © Springer-Verlag London Limited 2009

element **1**, we have that $ab = ba = \mathbf{1}$. This implies that $a \in S^*$ and proves the statement (ii). $\square$

Corollary 13.1.2 *Let G be a group. Then every variant of G is isomorphic to G.*

Corollary 13.1.2 says that study of variants is a purely semigroup theoretic phenomenon, which simply does not appear in group theory. The next natural question to ask would be: How many nonisomorphic semigroups can one obtain as variants of a given semigroup? Later on we will answer this question for the semigroups $\mathcal{T}_n$, $\mathcal{PT}_n$, and $\mathcal{IS}_n$ completely. For now we just give the following general sufficient condition:

Proposition 13.1.3 *Let S be a monoid, $a \in S$ and $u, v \in S^*$. Then $S^a \cong S^{uav}$.*

Proof. Consider the mapping $\varphi : S \to S$, $x \mapsto v^{-1}xu^{-1}$. As $u, v \in S^*$, the mapping φ is bijective (with the inverse $y \mapsto vyu$). For $x, y \in S$ we have

$$
\begin{aligned}
\varphi(x \circ_a y) &= \varphi(xay) \\
&= v^{-1}xayu^{-1} \\
&= (v^{-1}xu^{-1})(uav)(v^{-1}yu^{-1}) \\
&= \varphi(x) \circ_{uav} \varphi(y).
\end{aligned}
$$

This shows that φ is a homomorphism, and hence an isomorphism. The statement of the proposition follows. $\square$

Proposition 13.1.4 *Let S be a semigroup and $a \in S$. Then $(S^a)^2 = SaS$.*

Proof. For $x, y \in S^a$ we have $x \circ_a y = xay \in SaS$. Conversely, for any $xay \in SaS$ we have $xay = x \circ_a y$. The statement follows. $\square$

A variant of a semigroup is usually a special construction over some subsemigroup, called *inflation*. This point of view gives some feeling about the structure of a variant. To introduce it we need some preparation.

Proposition 13.1.5 *Let S be a semigroup and $a \in S$.*

(i) *The mapping $\varphi_l : S \to aS$, $x \mapsto ax$, is an epimorphism from S^a to aS.*

(ii) *The mapping $\varphi_r : S \to Sa$, $x \mapsto xa$, is an epimorphism from S^a to Sa.*

(iii) *Assume that $a \in \mathcal{E}(S)$, then $(aS, \circ_a) = aS$ (note the equality and not the isomorphism sign!).*

(iv) *Assume that $a \in \mathcal{E}(S)$, then $(Sa, \circ_a) = Sa$ (note the equality and not the isomorphism sign!).*

Proof. The mapping φ_l is surjective by definition. For $x, y \in S$ we have

$$\begin{aligned} \varphi_l(x \circ_a y) &= \varphi_l(xay) \\ &= axay \\ &= \varphi_l(x)\varphi_l(y) \end{aligned}$$

and the statement (i) follows. The statement (ii) is proved similarly.
  If $a \in \mathcal{E}(S)$, then for any $x = ay \in aS$ we have

$$ax = a(ay) = a^2y = ay = x$$

and hence the restriction of the mapping φ_l from (i) to aS is the identity mapping. Now the statement (iii) follows from the statement (i). Analogously, the statement (iv) follows from the statement (ii). $\qquad\square$

A semigroup S is called an *inflation* of its subsemigroup T provided that there exists an epimorphism $\theta : S \to T$ such that the following two conditions are satisfied:

- $\theta^2 = \theta$

- $\theta(x)\theta(y) = xy$ for all $x, y \in S$

Example 13.1.6 A semigroup with zero multiplication is an inflation of its subsemigroup containing the zero element via the obvious projection mapping.

Using Proposition 13.1.5 we define the equivalence relation $\mathfrak{h}$ on S as follows:

$$\mathfrak{h} = \big(\mathrm{Ker}(\varphi_l) \cap \mathrm{Ker}(\varphi_r) \cap \big((S^a \backslash (S^a)^2) \times (S^a \backslash (S^a)^2))\big)\big) \cup$$
$$\cup \{(a, a) : a \in (S^a)^2\}.$$

Proposition 13.1.7 *Let S be a semigroup and $a \in S$.*

(i) $\mathfrak{h}$ is a congruence on S^a.

(ii) Any cross-section T of $\mathfrak{h}$ is a subsemigroup of S^a, moreover, S^a is an inflation of T via the projection mapping π_T, defined as follows:

$$\pi_T(x) = t \in T \text{ if and only if}(x, t) \in \mathfrak{h}.$$

Proof. Let $x, y \in S^a \setminus (S^a)^2$ be such that $(x, y) \in \mathfrak{h}$ and $s \in S^a$. Then $(x, y) \in \mathrm{Ker}(\varphi_l)$ by definition and hence

$$s \circ_a x = sax = say = s \circ_a y.$$

Thus $(s \circ_a x, s \circ_a y) \in \mathfrak{h}$. Analogously, one shows that $(x \circ_a s, x \circ_a s) \in \mathfrak{h}$. The statement (i) follows.

To prove (ii) we first observe that, by definition, $(S^a)^2 \subset T$. If $x, y \in T$, then $x \circ_a y \in (S^a)^2$ and hence $x \circ_a y \in T$, in particular, T is a subsemigroup of S^a. If $x, y \in S^a$, we have

$$
\begin{aligned}
\pi_T(x) \circ_a \pi_T(y) &= \pi_T(x) a \pi_T(y) \\
(\text{as } (x, \pi_T(x)) \in \mathrm{Ker}(\varphi_r)) &= xa\pi_T(y) \\
(\text{as } (y, \pi_T(y)) \in \mathrm{Ker}(\varphi_l)) &= xay \\
&= x \circ_a y.
\end{aligned}
$$

At the same time $\pi_T(x \circ_a y) = x \circ_a y$ as $x \circ_a y \in (S^a)^2$. Hence π_T is an idempotent epimorphism. This proves the statement (ii). $\qquad \square$

13.2   Classification of Variants for $\mathcal{IS}_n$, $\mathcal{T}_n$, and $\mathcal{PT}_n$

In this section, we classify variants of $\mathcal{IS}_n$, $\mathcal{T}_n$, and $\mathcal{PT}_n$ up to isomorphism. Our main result in this section is the following theorem:

Theorem 13.2.1 *Let S denote one of the semigroups $\mathcal{IS}_n$, $\mathcal{T}_n$, or $\mathcal{PT}_n$, and $\alpha, \beta \in S$. Then the following statements are equivalent:*

(a) $S^\alpha \cong S^\beta$.

(b) *There exist $\tau, \sigma \in \mathcal{S}_n$ such that $\tau \alpha \sigma = \beta$.*

(c) $\mathfrak{t}(\alpha) = \mathfrak{t}(\beta)$.

Proof. We first observe that the equivalence (b)$\Leftrightarrow$(c) is just Exercises 2.10.24 and 2.10.25. The implication (b)$\Rightarrow$(a) follows immediately from Proposition 13.1.3. Hence we are left to prove either the implication (a)$\Rightarrow$(b) or the implication (a)$\Rightarrow$(c). We will do this for each of the semigroups separately. We start with the easiest case $S = \mathcal{IS}_n$.

Let $S = \mathcal{IS}_n$ and $\alpha, \beta \in \mathcal{IS}_n$ be such that $\mathcal{IS}_n^\alpha \cong \mathcal{IS}_n^\beta$. Then the sets $(\mathcal{IS}_n^\alpha)^2$ and $(\mathcal{IS}_n^\beta)^2$ must have the same cardinality. As $\mathcal{IS}_n$ is a monoid, from Proposition 13.1.4 we obtain that the principal two-sided ideals, generated by α and β, respectively, must have the same cardinality. From Theorem 4.2.8 we therefore obtain $\mathrm{rank}(\alpha) = \mathrm{rank}(\beta)$. Further, Theorem 4.5.1(iv) implies $\alpha \mathcal{D} \beta$. Finally, Proposition 4.5.4(ii) implies that the condition (b) is satisfied and we obtain (a)$\Rightarrow$(b). This completes the proof.

We proceed now with the case $S = \mathcal{T}_n$. Let $\alpha \in \mathcal{T}_n$. We will need some more detailed information about $\mathcal{T}_n^\alpha$.

Lemma 13.2.2 *The set* $\mathcal{I}_1$ *is the minimum two-sided ideal of* $\mathcal{T}_n^\alpha$ *in the sense that every two-sided ideal of* $\mathcal{T}_n^\alpha$ *contains* $\mathcal{I}_1$.

Proof. The fact that $\mathcal{I}_1$ is an ideal is obvious. Let I be an ideal of $\mathcal{T}_n^\alpha$ and $\beta \in I$. Then $0_i = 0_i \circ_\alpha \beta \in I$ for any $i \in \mathbf{N}$ and the statement follows. $\qquad\square$

Lemma 13.2.2 says that $\mathcal{T}_n^\alpha$ contains the unique minimum two-sided ideal. Define the relation $\sim_\alpha$ on $\mathcal{I}_1$ as follows: for $i, j \in \mathbf{N}$ set $0_i \sim_\alpha 0_j$ if and only if $\beta \circ_\alpha 0_i = \beta \circ_\alpha 0_j$ for any $\beta \in \mathcal{T}_n^\alpha$. Assume that

$$\alpha = \left(\begin{array}{cccc} A_1 & A_2 & \cdots & A_k \\ a_1 & a_2 & \cdots & a_k \end{array} \right)$$

is an element of rank k.

Lemma 13.2.3 *For* $i, j \in \mathbf{N}$ *we have* $0_i \sim_\alpha 0_j$ *if and only if there exists* $l \in \{1, \dots, k\}$ *such that* $i, j \in A_l$.

Proof. If $i, j \in A_l$, then $\alpha 0_i = \alpha 0_j = 0_{a_l}$ and thus $\beta \circ_\alpha 0_i = \beta \circ_\alpha 0_j = \beta(a_l)$. On the other hand, if $i \in A_l$ and $j \in A_m$ for some $l \neq m$, then

$$\varepsilon \circ_\alpha 0_i = 0_{a_l} \neq 0_{a_m} = \varepsilon \circ_\alpha 0_j.$$

$\qquad\square$

Now let $\alpha, \beta \in \mathcal{T}_n$ and assume that $\psi : \mathcal{T}_n^\alpha \to \mathcal{T}_n^\beta$ is an isomorphism. From Lemma 13.2.2 it follows that ψ induces a bijection from $\mathcal{I}_1 \subset \mathcal{T}_n^\alpha$ to $\mathcal{I}_1 \subset \mathcal{T}_n^\beta$.

Lemma 13.2.4 *For* $\gamma_1, \gamma_2 \in \mathcal{I}_1 \subset \mathcal{T}_n^\alpha$ *we have*

$$\gamma_1 \sim_\alpha \gamma_2 \quad \text{if and only if} \quad \psi(\gamma_1) \sim_\beta \psi(\gamma_2).$$

Proof. If $\gamma_1 \sim_\alpha \gamma_2$ and $\sigma \in \mathcal{T}_n^\alpha$ is arbitrary, then

$$\begin{aligned} \psi(\sigma) \circ_\beta \psi(\gamma_1) &= \psi(\sigma \circ_\alpha \gamma_1) \\ &= \psi(\sigma \circ_\alpha \gamma_2) \\ &= \psi(\sigma) \circ_\beta \psi(\gamma_2). \end{aligned}$$

As $\psi(\sigma)$ is then an arbitrary element of $\mathcal{T}_n^\beta$, we get $\psi(\gamma_1) \sim_\beta \psi(\gamma_2)$. The statement of the lemma now follows for the same argument applies to the isomorphism ψ^{-1}. $\qquad\square$

By Lemma 13.2.4 the restriction of ψ to $\mathcal{I}_1$ induces an isomorphism be-tween the equivalence relations $\sim_\alpha$ and $\sim_\beta$. In particular, for every $i \in \mathbf{N}$ these relations have the same number of equivalence classes of cardinality i. By Lemma 13.2.3 and the definition, we obtain that $\mathsf{t}(\alpha) = \mathsf{t}(\beta)$. This com-pletes the proof.

Finally, in the case $S = \mathcal{PT}_n$ the proof is rather similar to that from the case $S = \mathcal{T}_n$. First, we observe that the semigroup $\mathcal{PT}_n^0$ is the semigroup with zero multiplication. At the same time for any element $\alpha \neq \mathbf{0}$ let $i \in \mathrm{dom}(\alpha)$. Then

$$0_i \circ_\alpha 0_i = 0_i \alpha 0_i = 0_i \neq \mathbf{0}$$

and hence $\mathcal{PT}_n^\alpha \not\cong \mathcal{PT}_n^0$. So, it is enough to consider the case when $\alpha, \beta \neq \mathbf{0}$.

For $\alpha \neq \mathbf{0}$ in the same way as in Lemma 13.2.2 one shows that $\mathcal{I}_1$ is the unique minimum element in the set of all nonzero ideals contained in the set $\left(\mathcal{PT}_n^\alpha\right)^2$. As the zero element is obviously preserved by any isomorphism, we derive that any isomorphism from $\mathcal{PT}_n^\alpha$ to $\mathcal{PT}_n^\beta$ (where both α and β are nonzero) must leave the set $\mathcal{I}_1$ invariant. The rest is absolutely analogous to the case of $\mathcal{T}_n$. On $\mathcal{I}_1$ we consider the relations $\sim_\alpha$ and $\sim_\beta$ and it follows that any isomorphism induces an isomorphism of these relations. A computation of the sizes of the equivalence classes implies $\mathsf{t}(\alpha) = \mathsf{t}(\beta)$. This completes the proof of the whole theorem. $\square$

Let $\mathfrak{p}(n)$ denote the *partition function*, that is, the number of nonnegative integer solutions to the equation

$$x_1 + 2x_2 + 3x_3 + \cdots + nx_n = n. \tag{13.1}$$

We also set $\mathfrak{p}(0) = 1$.

Corollary 13.2.5 *(i) The semigroup $\mathcal{IS}_n$ has $n + 1$ pairwise nonisomor-phic variants.*

(ii) The semigroup $\mathcal{T}_n$ has $\mathfrak{p}(n)$ pairwise nonisomorphic variants.

(iii) The semigroup $\mathcal{PT}_n$ has $\displaystyle\sum_{k=0}^{n} \mathfrak{p}(k)$ pairwise nonisomorphic variants.

Proof. From Theorem 13.2.1 we have that the isomorphism classes of vari-ants of $\mathcal{IS}_n$ are classified by ranks of elements in $\mathcal{IS}_n$. These ranks are numbers from $\{0, 1, \ldots, n\}$ and the statement (i) follows.

Analogously, from Theorem 13.2.1 and Exercises 2.10.24 and 1.5.12 it follows that variants of $\mathcal{T}_n$ are classified by solutions to (13.1). The statement (ii) follows. To prove the statement (iii) is left as an exercise to the reader. $\square$

Let $\Delta = (t_0, t_1, \ldots, t_n)$ be a vector with nonnegative integer coefficients such that

$$\sum_{i=0}^{n} t_i = n \quad \text{and} \quad \sum_{i=0}^{n} it_i = k \leq n. \tag{13.2}$$

Exercise 13.2.6 Show that for $\alpha \in \mathcal{PT}_n$ the vector $t(\alpha)$ satisfies (13.2) for $k = |\text{dom}(\alpha)|$.

Let $\pi_\Delta \in \mathcal{PT}_n$ be the idempotent, given by (9.9) (with the convention $\pi_\Delta(x) = \varnothing$ for all $x > k$). In particular, $\pi_\Delta \in \mathcal{T}_n$ if and only if $\sum_{i=0}^{n} it_i = n$; and $\pi_\Delta \in \mathcal{IS}_n$ if and only if $t_i = 0$ for all $i > 1$.

Let S denote one of the semigroups $\mathcal{T}_n$, $\mathcal{PT}_n$, or $\mathcal{IS}_n$. For $\alpha \in S$ from Exercise 13.2.6 we have that $\Delta = t(\alpha)$ satisfies (13.2). From the proof of Lemma 9.3.5 we obtain $t(\pi_\Delta) = \Delta$ and hence by Theorem 13.2.1 we have $S^\alpha \cong S^{\pi_\Delta}$. This means that, up to isomorphism, we may always assume that the sandwich element α is an idempotent.

13.3 Idempotents and Maximal Subgroups

Let S denote one of the semigroups $\mathcal{T}_n$, $\mathcal{PT}_n$, or $\mathcal{IS}_n$. As we have already noted, in the case $\alpha = \mathbf{0}$, the semigroup S^0 is the semigroup with zero multiplication and hence is not really interesting. So, from now on we assume that the sandwich element α is given by

$$\alpha = \begin{pmatrix} A_1 & A_2 & \cdots & A_k \\ a_1 & a_2 & \cdots & a_k \end{pmatrix}, \tag{13.3}$$

has rank $k > 0$ and is an idempotent, that is, $a_i \in A_i$ for all $i = 1, \ldots, k$. We also set $|A_i| = l_i$.

Lemma 13.3.1 *Let*

$$\beta = \begin{pmatrix} B_1 & B_2 & \cdots & B_m \\ b_1 & b_2 & \cdots & b_m \end{pmatrix} \in S \tag{13.4}$$

be an element of rank m. Then β is an idempotent of S^α if and only if there exists an injection $f : \{1, \ldots, m\} \to \{1, \ldots, k\}$ such that $b_i \in A_{f(i)}$ and $a_{f(i)} \in B_i$ for all $i = 1, \ldots, m$.

Proof. Direct computation. $\qquad\square$

Proposition 13.3.2 *(i) If $\alpha \in \mathcal{IS}_n$, then the semigroup $\mathcal{IS}_n^\alpha$ has 2^k idempotents.*

(ii) If $\alpha \in \mathcal{T}_n$, then the number of idempotents in the semigroup $\mathcal{T}_n^\alpha$ equals

$$\sum_{\varnothing \neq X \subset \{1,\ldots,k\}} |X|^{n-|X|} \cdot \prod_{i \in X} l_i.$$

(iii) If $\alpha \in \mathcal{PT}_n$, then the number of idempotents in the semigroup $\mathcal{PT}_n^\alpha$ equals

$$1 + \sum_{\varnothing \neq X \subset \{1,\ldots,k\}} (|X|+1)^{n-|X|} \cdot \prod_{i \in X} l_i.$$

Proof. If $\beta \in \mathcal{IS}_n$ is given by (13.4), then $|B_i| = 1$ for all i and we also have $A_i = \{a_i\}$ for all i. Hence the assertion of Lemma 13.3.1 reads as follows: $\{b_1,\ldots,b_m\} \subset \{a_1,\ldots,a_k\}$ and $B_i = \{b_i\}$ for all i. The statement (i) follows.

To prove (ii) we count the number of idempotents β of $\mathcal{T}_n^\alpha$, whose image intersects exactly the A_i's for which $i \in X \subset \{1,\ldots,k\}$, $X \neq \varnothing$. By Lemma 13.3.1, to determine such β we should choose an image for each a_i, $i \in X$, in the corresponding block A_i, which can be done in l_i different ways; this determines $\mathrm{im}(\beta)$ and after that we should arbitrarily choose the image of all elements from $\mathbf{N} \backslash \{a_i : i \in X\}$ inside $\mathrm{im}(\beta)$. Since all our choices are independent, the statement (ii) now follows applying the product rule.

The statement (iii) is proved similarly to the proof of the statement (ii) with two differences: we should take the zero idempotent into account; and when mapping the elements which are left arbitrarily to $\mathrm{im}(\beta)$, we should leave for all of them the option to be mapped to $\varnothing$. The statement follows.   □

Corollary 13.3.3 *Let* $\beta \in S^\alpha$ *be an idempotent given by (13.4). Then the corresponding maximal subgroup* G_β *of* S^α *is isomorphic to* $\mathcal{S}_m$ *and coincides, as a set, with the* $\mathcal{H}$-*class* $\mathcal{H}(\beta)$ *of the element* β, *considered as an element from* S.

Proof. By Lemma 5.1.2, G_β coincides with the group of units in $\beta \circ_\alpha S^\alpha \circ_\alpha \beta$. In particular, every $\gamma \in G_\beta$ has the form

$$\gamma = \beta \circ_\alpha \gamma' \circ_\alpha \beta = \beta \alpha \gamma' \alpha \beta \tag{13.5}$$

for some $\gamma' \in S$. As S is finite, we also have that β must belong to the cyclic subgroup of S^α, generated by γ. This implies $\mathrm{rank}(\beta) = \mathrm{rank}(\gamma)$, which, together with (13.5), yields $\rho_\beta = \rho_\gamma$ and $\mathrm{im}(\beta) = \mathrm{im}(\gamma)$, that is, $\gamma \in \mathcal{H}(\beta)$ by Theorem 4.5.1.

On the other hand, without loss of generality we may assume that the mapping f from Lemma 13.3.1 is the natural inclusion $f(i) = i$. For $\gamma \in \mathcal{H}(\beta)$, because of the previous paragraph, we have $\gamma(B_i) = b_{\tau(i)}$ for some $\tau \in \mathcal{S}_k$. Then a direct calculation shows that the mapping $\gamma \mapsto \tau$ is an isomorphism from $\mathcal{H}(\beta) \subset S^\alpha$ to $\mathcal{S}_k$. This completes the proof.   □

13.4 Principal Ideals and Green's Relations

For the study of ideals in variants, it is worth mentioning one more time that a variant of a semigroup is not a monoid in general. In this section, S denotes one of the semigroups $\mathcal{T}_n$, $\mathcal{PT}_n$, or $\mathcal{IS}_n$, and α is given by (13.3). The proposition that follows reduces the study of principal ideals in S^α to that of principal ideals in S, see Sect. 4.2.

Proposition 13.4.1 *Assume that $\beta \in S$ is given by (13.4). We have*

(i) *The principal left ideal of S, generated by β, equals $\{\beta\} \cup S\alpha\beta$.*

(ii) *The principal right ideal of S, generated by β, equals $\{\beta\} \cup \beta\alpha S$.*

(iii) *The principal two-sided ideal of S, generated by β, equals*

$$\{\beta\} \cup S\alpha\beta \cup \beta\alpha S \cup S\alpha\beta\alpha S.$$

Proof. This follows directly from the definitions. □

Theorem 13.4.2 *Let $\beta \in S$ be given by (13.4) and $\gamma \in S$ be the following element of rank p:*

$$\gamma = \begin{pmatrix} C_1 & C_2 & \cdots & C_p \\ c_1 & c_2 & \cdots & c_p \end{pmatrix}. \tag{13.6}$$

(i) *$\beta \mathcal{L} \gamma$ if and only if $\beta = \gamma$ or the following conditions are satisfied:*

(a) *$\rho_\beta = \rho_\gamma$ (in particular $p = m$).*

(b) *There exists an injective function $f : \{1,\ldots,m\} \to \{1,\ldots,k\}$ such that $b_i \in A_{f(i)}$, $i = 1,\ldots,m$.*

(c) *There exists an injective function $g : \{1,\ldots,p\} \to \{1,\ldots,k\}$ such that $c_i \in A_{g(i)}$, $i = 1,\ldots,p$.*

(ii) *$\beta \mathcal{R} \gamma$ if and only if $\beta = \gamma$ or the following conditions are satisfied:*

(a) *$\mathrm{im}(\beta) = \mathrm{im}(\gamma)$ (in particular $p = m$).*

(b) *There exists an injective function $f : \{1,\ldots,m\} \to \{1,\ldots,k\}$ such that $a_{f(i)} \in B_i$ for all $i = 1,\ldots,m$.*

(c) *There exists an injective function $g : \{1,\ldots,p\} \to \{1,\ldots,k\}$ such that $a_{g(i)} \in C_i$ for all $i = 1,\ldots,p$.*

(iii) *$\beta \mathcal{H} \gamma$ if and only if $\beta = \gamma$ or all conditions (ia)–(ic) and (iia)–(iic) are satisfied.*

(iv) *$\beta \mathcal{D} \gamma$ if and only if one of the following mutually exclusive conditions is satisfied:*

(a) rank(β) = rank(γ) *and the conditions* (ib), (ic), (iib), (iic) *are satisfied.*

(b) *Conditions* (ia)–(ic) *are satisfied while at least one of the conditions* (iib) *or* (iic) *is not satisfied.*

(c) *Conditions* (iia)–(iic) *are satisfied while at least one of the conditions* (ib) *or* (ic) *is not satisfied.*

(d) $\beta = \gamma$, *at least one of the conditions* (iib) *or* (iic) *is not satisfied, and at least one of the conditions* (ib) *or* (ic) *is not satisfied.*

(v) $\mathcal{D} = \mathcal{J}$.

Proof. Assume that $\beta \neq \gamma$ and $\beta \mathcal{L} \gamma$. Then from Proposition 13.4.1(i) we have $\gamma \in S^\alpha \circ_\alpha \beta$, which yields the condition (ib). Similarly we get (ic). From Proposition 4.4.1 we have $\beta = \delta \alpha \gamma$ and $\gamma = \delta' \alpha \beta$ for some $\delta, \delta' \in S$. In particular, β and γ must be $\mathcal{L}$-connected in the original semigroup S. Hence the condition (ia) follows from Theorem 4.5.1.

Conversely, if all conditions (ia)-(ic) are satisfied, then $S^\alpha \circ_\alpha \beta = S^\alpha \circ_\alpha \gamma$ follows from Theorem 4.5.1. This proves (i). The proof of the statements (ii) is similar. The statement (v) follows from Theorem 5.4.1. The statements (iii) and (iv) follow from the statements (i) and (ii) and the definitions. $\square$

13.5 Addenda and Comments

13.5.1 The problem to study variants of semigroups seems to go back at least to Lyapin's monograph [Ly]. Various sandwich semigroups were studied by several authors, see for example some older paper [Sy1, Sy2, Hic1, Hic2, Ch1, Ch2, MS1, MS2, MMT] or some more recent papers [KL, Ji, Ts1, Ts2, Ts3, Ts4, Ku1, KT, MT].

The results of Sect. 13.2 are proved in [Sy1, Ts1, Ts3]. The results of Sect. 13.4 are partially proved in [Ts2]. Proposition 13.1.7 is taken from [Ku1]. The results of Sect. 13.3 are partially taken from [MT].

13.5.2 Theorem 13.2.1 roughly speaking says that the semigroups $\mathcal{T}_n$, $\mathcal{PT}_n$, and $\mathcal{IS}_n$ have lots of isomorphic variants. There exist semigroups, all variants of which are not isomorphic. For example, we think that the following statement is noteworthy:

Theorem 13.5.1 ([Ts1]) *Let* $\mathfrak{B}$ *denote the monoid given by the following presentation:* $\mathfrak{B} = \langle a, b | ab = 1 \rangle$. *Then for* $x, y \in \mathfrak{B}$ *we have* $\mathfrak{B}^x \cong \mathfrak{B}^y$ *if and only if* $x = y$.

The monoid $\mathfrak{B}$ defined above is called the *bicyclic semigroup*. The semigroup $\mathfrak{B}$ is an infinite inverse semigroup, hence the statement of Theorem 13.5.1 is rather remarkable, especially because of Exercise 13.6.8.

Proof. From the definition we have that every element of $\mathfrak{B}$ can be uniquely written in the form $\alpha_{i,j} = b^i a^j$, $i, j \geq 0$. We further have

$$\alpha_{i,j}\alpha_{k,l} = \begin{cases} \alpha_{i,l+j-k}, & j \geq k; \\ \alpha_{i+k-j,l}, & j < k, \end{cases}$$

in particular, $\alpha_{i,j}$ is an idempotent if and only if $i = j$.

Fix now $k, m \geq 0$, let $\alpha = \alpha_{k,m}$ and consider the semigroup $\mathfrak{B}^\alpha$.

Lemma 13.5.2 *(i)* $\mathcal{E}(\mathfrak{B}^\alpha) = \{\alpha_{m+l,k+l} : l \geq 0\}$.

(ii) If for $i \geq 0$ we set $e_i^{k,m} = \alpha_{m+i,k+i}$, then $e_i^{k,m} \circ_\alpha e_j^{k,m} = e_{\max(i,j)}^{k,m}$.

Proof. For $i, j \geq 0$ we have that $\alpha_{i,j} \in \mathcal{E}(\mathfrak{B}^\alpha)$ if and only if

$$b^i a^j \cdot b^k a^m \cdot b^i a^j = b^i a^j. \tag{13.7}$$

If $j < k$ and $i < m$, then a direct calculation reduces (13.7) to the equation $b^{i+k-j}a^{j+m-i} = b^i a^j$ and hence $k = j, m = i$, a contradiction.

If $j < k$ and $i \geq m$, then a direct calculation reduces (13.7) to the equation $b^{i+k-j+i-m}a^j = b^i a^j$. Thus, $0 > k - j = m - i \leq 0$, which is again a contradiction. Analogously, one gets a contradiction in the case $j \geq k$ and $i < m$.

In the remaining case $j \geq k$ and $i \geq m$ a direct calculation reduces (13.7) to the equation $i - m = j - k \geq 0$. The statement (i) follows.

The statement (ii) is now proved by a direct calculation. $\qquad\square$

For $i \geq 0$ consider the sets

$$P_i^{k,m} = \{\gamma \in \mathfrak{B}^\alpha : e_i^{k,m} \circ_\alpha \gamma \neq \gamma\},$$
$$Q_i^{k,m} = \{\gamma \in \mathfrak{B}^\alpha : \gamma \circ_\alpha e_i^{k,m} \neq \gamma\}.$$

Lemma 13.5.3 $|P_i^{k,m} \cap Q_j^{k,m}| = (k+j)(m+i)$ *for all $i, j \geq 0$.*

Proof. For $x, y \geq 0$ we have $\alpha_{x,y} \notin P_i^{k,m}$ if and only if $e_i^{k,m} \circ_\alpha \alpha_{x,y} = \alpha_{x,y}$, that is,

$$b^{m+i}a^{k+i} \cdot b^k a^m \cdot b^x a^y = b^{m+i}a^{m+i} \cdot b^x a^y = b^x a^y.$$

If $m + i > x$, then the latter equality is equivalent to $m + i = x$ and $m + i - x + y = y$. Hence $m + i = x$, a contradiction. If $m + i \leq x$, then the equality becomes an identity. This means $\alpha_{x,y} \notin P_i^{k,m}$ if and only if $m + i \leq x$, and hence $\alpha_{x,y} \in P_i^{k,m}$ if and only if $x < m + i$.

Analogously, one shows that $\alpha_{x,y} \in Q_j^{k,m}$ if and only if $y < k + j$. The statement of the lemma follows. $\qquad\square$

Assume now that $k, m, u, v \geq 0$, $\alpha = \alpha_{k,m}$, $\beta = \alpha_{u,v}$ and $\varphi : \mathfrak{B}^\alpha \to \mathfrak{B}^\beta$ is an isomorphism. Then φ maps idempotents to idempotents and preserves the multiplication of idempotents. Hence from Lemma 13.5.2 it follows that $\varphi(e_i^{k,m}) = e_i^{u,v}$ for all $i \geq 0$. This, in particular, yields that $\varphi(P_i^{k,m}) = P_i^{u,v}$ and $\varphi(Q_i^{k,m}) = Q_i^{u,v}$ for all $i \geq 0$. Thus we have

$$\varphi(P_1^{k,m} \cap Q_1^{k,m}) = P_1^{u,v} \cap Q_1^{u,v};$$
$$\varphi(P_1^{k,m} \cap Q_0^{k,m}) = P_1^{u,v} \cap Q_0^{u,v};$$
$$\varphi(P_0^{k,m} \cap Q_1^{k,m}) = P_0^{u,v} \cap Q_1^{u,v}.$$

However, from Lemma 13.5.3 we have

$$|P_1^{k,m} \cap Q_1^{k,m}| - |P_1^{k,m} \cap Q_0^{k,m}| = m + 1;$$
$$|P_1^{u,v} \cap Q_1^{u,v}| - |P_1^{u,v} \cap Q_0^{u,v}| = v + 1;$$
$$|P_1^{k,m} \cap Q_1^{k,m}| - |P_0^{k,m} \cap Q_1^{k,m}| = k + 1;$$
$$|P_1^{u,v} \cap Q_1^{u,v}| - |P_0^{u,v} \cap Q_1^{u,v}| = u + 1.$$

Thus $m = v$ and $k = u$. This completes the proof. $\qquad\square$

13.5.3 Proposition 13.1.7 says that a typical variant of a semigroup is an inflation of a certain subsemigroup. Automorphism groups of semigroup inflations have rather special structure, described in [Ku1]: Assume that we are in the situation of Proposition 13.1.7. Then $\mathrm{Aut}(S)$ is a semidirect product of two subgroups A_1 and A_2. The normal subgroup A_1 of $\mathrm{Aut}(S)$ is just the direct product of symmetric groups on equivalence classes of the relation $\mathfrak{h}$. The group A_1 acts on S in the natural way. The subgroup A_2 consists of all automorphisms of T, which are *extendable* to S, that is, which are obtained from an automorphism of S by restriction. For variants of $\mathcal{IS}_n$ this was proved in [KT]. The approach of [Ku1] unifies the results of [GTS, KT, Sy1, St2].

13.5.4 Variants of transformation semigroups are just a special case of the following more general and rather remarkable construction: If X and Y are two nonempty sets and $\alpha : Y \to X$ is a fixed mapping, then the set $F(X,Y)$ of all mappings from X to Y (note the direction!) becomes a semigroup with respect to the operation $\circ_\alpha$ defined as follows: For $f, g \in F(X,Y)$ we set $f \circ_\alpha g = f\alpha g$. The definition of this semigroup goes back at least to Lyapin, see [Ly]. Several properties of this semigroup were studied in [Sy1, Sy2].

13.6 Additional Exercises

13.6.1 Let S be a finite monoid. Show that all variants of S are isomorphic if and only if S is a group.

13.6.2 Let S be a finite semigroup with zero. Show that all variants of S are isomorphic if and only if $S^3 = 0$.

13.6.3 Let S be a semigroup. Assume that the group $\mathrm{Aut}(S)$ acts transitively on S (i.e., for any $x, y \in S$ there exists $\varphi \in \mathrm{Aut}(S)$ such that $\varphi(x) = y$). Show that all variants of S are isomorphic.

13.6.4 Show that for every S from the following list all variants of S are isomorphic:

(a) Semigroup with zero multiplication

(b) Left zero semigroup

(c) Right zero semigroup

(d) Rectangular band

(e) The semigroup of positive rational (or real) numbers with respect to the addition

13.6.5 For $a, b \in \mathbb{N}$ define $a \star b = \min(a, b)$. Show that any two variants of $(\mathbb{N}, \star)$ with respect to two different elements are not isomorphic.

13.6.6 For $a, b \in \mathbb{N}$ define $a \star b = \max(a, b)$. Show that any two variants of $(\mathbb{N}, \star)$ with respect to two different elements are not isomorphic.

13.6.7 Let $S = (\mathbb{N}, +)$. Show that $S^a \not\cong S^b$ if $a \neq b$.

13.6.8 Let S be a finite inverse semigroup, which is not a group. Show that there exist $a, b \in S$ such that $(S, \circ_a) \not\cong (S, \circ_b)$.

13.6.9 Let S be a semigroup and T be a variant of S. Show that every variant of T is isomorphic to some variant of S.

13.6.10 (a) For $a, b \in \mathbb{Z}$ define $a \star b = \min(a, b)$. Show that all variants of $(\mathbb{Z}, \star)$ are isomorphic.

(b) Let S be some variant of $(\mathbb{Z}, \star)$ from (a). Show that all variants of S are isomorphic to S.

13.6.11 An element x of a semigroup S is called a *mididentity* provided that $uxv = uv$ for any $u, v \in S$. Let S be a monoid and $a \in S$. Show that $x \in S^a$ is a mididentity if and only if $axa = a$.

13.6.12 ([Ts2]) Compute the number of one-element $\mathcal{R}$-classes and the number of one-element $\mathcal{L}$-classes in $\mathcal{IS}_n^\alpha$.

13.6.13 ([Ts2]) Compute the number of one-element $\mathcal{R}$-classes and the number of one-element $\mathcal{L}$-classes in $\mathcal{T}_n^\alpha$.

13.6.14 ([Ts2]) Show that $\mathcal{IS}_n^\alpha \cong \overleftarrow{\mathcal{IS}}_n^\alpha$ for any $\alpha \in \mathcal{IS}_n$.

13.6.15 ([Ts1]) Show that for $\alpha, \beta \in \mathcal{B}_n$ the semigroups $\mathcal{B}_n^\alpha$ and $\overleftarrow{\mathcal{B}}_n^\beta$ are isomorphic if and only if $\beta = \alpha^{-1}$.

13.6.16 ([Ts4]) Let $\alpha \in \mathcal{IS}_n$, and T_α denote the set of all $\beta \in \mathcal{IS}_n$ such that $\beta(\mathrm{im}(\alpha)) = \mathrm{dom}(\alpha)$.

(a) Show that T_α is a subsemigroup of $\mathcal{IS}_n^\alpha$.

(b) Prove that the only completely isolated subsemigroups of $\mathcal{IS}_n^\alpha$ are: $\mathcal{IS}_n^\alpha$, T_α and $\mathcal{IS}_n^\alpha \backslash T_\alpha$.

13.6.17 ([Ts4]) Let $\alpha \in \mathcal{IS}_n$, and T_α be as in Exercise 13.6.16. For $x \in \mathrm{im}(\alpha)$ let $G(x)$ denote the set of all $\beta \in \mathcal{IS}_n$ such that $\beta(x) \notin \mathrm{dom}(\alpha)$ and $\beta(y) \in \mathrm{dom}(\alpha) \backslash \alpha^{-1}(x)$ for all $y \in \mathrm{im}(\alpha) \backslash \{x\}$.

(a) Show that $G(x)$ is an isolated subsemigroup of $\mathcal{IS}_n^\alpha$.

(b) Prove that every isolated subsemigroup of $\mathcal{IS}_n^\alpha$ coincides with one of the following semigroups: $\mathcal{IS}_n^\alpha$, T_α, $\mathcal{IS}_n^\alpha \backslash T_\alpha$ or $G(x)$, $x \in \mathrm{im}(\alpha)$.

13.6.18 ([Ts4]) Let $\alpha \in \mathcal{IS}_n$. Classify all maximal nilpotent subsemigroups in $\mathcal{IS}_n^\alpha$.

13.6.19 ([Ch1]) Let S be a semigroup and $a \in S$. Show that for any Green's relation $\mathcal{X}$ the fact that elements $x, y \in S^a$ are $\mathcal{X}$-related in S^a implies that x, y are $\mathcal{X}$-related already in S.

Chapter 14

Order-Related Subsemigroups

14.1 Subsemigroups, Related to the Natural Order

In this chapter, we will study certain subsemigroups of $\mathcal{T}_n$, $\mathcal{PT}_n$, and $\mathcal{IS}_n$, related to the natural order $1 < 2 < \cdots < n$ on the set $\mathbf{N}$. One of these subsemigroups, namely, $\mathcal{IO}_n^{\leq}$ (which we simply denote by $\mathcal{IO}_n$ in the sequel), already appeared in Sect. 12.3 in the study of $\mathcal{H}$-cross-sections of $\mathcal{IS}_n$. In this chapter, we will study $\mathcal{IO}_n$ and some other similarly defined semigroups in more detail.

A partial transformation $\alpha \in \mathcal{PT}_n$ is called *order-preserving* provided that $x \leq y$ implies $\alpha(x) \leq \alpha(x)$ for all $x, y \in \mathrm{dom}(\alpha)$. The set of all order-preserving partial transformations from $\mathcal{PT}_n$ is denoted by $\mathcal{PO}_n$.

Exercise 14.1.1 Check that $\mathcal{PO}_n$ is a subsemigroup of $\mathcal{PT}_n$.

The semigroup $\mathcal{PO}_n$ is called the *semigroup of all order-preserving partial transformations* of the set $\mathbf{N}$. The subsemigroups $\mathcal{O}_n = \mathcal{PO}_n \cap \mathcal{T}_n$ and $\mathcal{IO}_n = \mathcal{PO}_n \cap \mathcal{IS}_n$ of $\mathcal{PO}_n$ are called the *subsemigroup of all order-preserving (total) transformations* and the *subsemigroup of all order-preserving partial permutations* of $\mathbf{N}$, respectively.

Example 14.1.2 The semigroup $\mathcal{PO}_2$ contains the following eight elements:

$$\begin{pmatrix} 1 & 2 \\ \varnothing & \varnothing \end{pmatrix}, \begin{pmatrix} 1 & 2 \\ 1 & \varnothing \end{pmatrix}, \begin{pmatrix} 1 & 2 \\ \varnothing & 1 \end{pmatrix}, \begin{pmatrix} 1 & 2 \\ 2 & \varnothing \end{pmatrix},$$

$$\begin{pmatrix} 1 & 2 \\ \varnothing & 2 \end{pmatrix}, \begin{pmatrix} 1 & 2 \\ 1 & 1 \end{pmatrix}, \begin{pmatrix} 1 & 2 \\ 1 & 2 \end{pmatrix}, \begin{pmatrix} 1 & 2 \\ 2 & 2 \end{pmatrix}.$$

The semigroup $\mathcal{O}_2$ contains the following three elements:

$$\begin{pmatrix} 1 & 2 \\ 1 & 1 \end{pmatrix}, \begin{pmatrix} 1 & 2 \\ 1 & 2 \end{pmatrix}, \begin{pmatrix} 1 & 2 \\ 2 & 2 \end{pmatrix}.$$

O. Ganyushkin, V. Mazorchuk, *Classical Finite Transformation Semigroups*, Algebra and Applications 9, DOI: 10.1007/978-1-84800-281-4_14, © Springer-Verlag London Limited 2009

The semigroup $\mathcal{IO}_2$ contains the following six elements:

$$\begin{pmatrix} 1 & 2 \\ \varnothing & \varnothing \end{pmatrix}, \begin{pmatrix} 1 & 2 \\ 1 & \varnothing \end{pmatrix}, \begin{pmatrix} 1 & 2 \\ \varnothing & 1 \end{pmatrix}, \begin{pmatrix} 1 & 2 \\ 2 & \varnothing \end{pmatrix}, \begin{pmatrix} 1 & 2 \\ \varnothing & 2 \end{pmatrix}, \begin{pmatrix} 1 & 2 \\ 1 & 2 \end{pmatrix}.$$

A partial transformation $\alpha \in \mathcal{PT}_n$ is called *order-decreasing* provided that $\alpha(x) \leq x$ for all $x \in \mathrm{dom}(\alpha)$. The set of all order-decreasing partial transformations from $\mathcal{PT}_n$ is denoted by $\mathcal{PF}_n$.

Exercise 14.1.3 Check that $\mathcal{PF}_n$ is a subsemigroup of $\mathcal{PT}_n$.

The semigroup $\mathcal{PF}_n$ is called the *semigroup of all order-decreasing partial transformations* of the set $\mathbf{N}$. The subsemigroups $\mathcal{F}_n = \mathcal{PF}_n \cap \mathcal{T}_n$ and $\mathcal{IF}_n = \mathcal{PF}_n \cap \mathcal{IS}_n$ of $\mathcal{PF}_n$ are called the *subsemigroup of all order-decreasing (total) transformations* and the *subsemigroup of all order-decreasing partial permutations* of $\mathbf{N}$, respectively.

Example 14.1.4 The semigroup $\mathcal{PF}_2$ contains the following six elements:

$$\begin{pmatrix} 1 & 2 \\ \varnothing & \varnothing \end{pmatrix}, \begin{pmatrix} 1 & 2 \\ 1 & \varnothing \end{pmatrix}, \begin{pmatrix} 1 & 2 \\ \varnothing & 1 \end{pmatrix}, \begin{pmatrix} 1 & 2 \\ \varnothing & 2 \end{pmatrix}, \begin{pmatrix} 1 & 2 \\ 1 & 1 \end{pmatrix}, \begin{pmatrix} 1 & 2 \\ 1 & 2 \end{pmatrix}.$$

The semigroup $\mathcal{F}_2$ contains the following two elements:

$$\begin{pmatrix} 1 & 2 \\ 1 & 1 \end{pmatrix}, \begin{pmatrix} 1 & 2 \\ 1 & 2 \end{pmatrix}.$$

The semigroup $\mathcal{IF}_2$ contains the following five elements:

$$\begin{pmatrix} 1 & 2 \\ \varnothing & \varnothing \end{pmatrix}, \begin{pmatrix} 1 & 2 \\ 1 & \varnothing \end{pmatrix}, \begin{pmatrix} 1 & 2 \\ \varnothing & 1 \end{pmatrix}, \begin{pmatrix} 1 & 2 \\ \varnothing & 2 \end{pmatrix}, \begin{pmatrix} 1 & 2 \\ 1 & 2 \end{pmatrix}.$$

We also define the following three subsemigroups, which consist of all transformations, which are both order-preserving and order-decreasing:

$$\mathcal{PC}_n = \mathcal{PO}_n \cap \mathcal{PF}_n, \quad \mathcal{C}_n = \mathcal{O}_n \cap \mathcal{F}_n, \quad \mathcal{IC}_n = \mathcal{IO}_n \cap \mathcal{IF}_n.$$

From Examples 14.1.2 and 14.1.4 we see that $\mathcal{PC}_2 = \mathcal{PF}_2$, $\mathcal{C}_2 = \mathcal{F}_2$, and $\mathcal{IC}_2 = \mathcal{IF}_2$. However, in the general case analogous equalities are no longer true.

Exercise 14.1.5 Prove that for $n \geq 3$ all nine semigroups $\mathcal{PC}_n, \mathcal{PO}_n, \mathcal{PF}_n$, $\mathcal{C}_n, \mathcal{O}_n, \mathcal{F}_n, \mathcal{IC}_n, \mathcal{IO}_n$, and $\mathcal{IF}_n$ are different.

We complete this section describing some properties of elements in the introduced semigroups. Every element $\alpha \in \mathcal{PO}_n$ can be written in the form

$$\alpha = \begin{pmatrix} A_1 & A_2 & \cdots & A_k \\ a_1 & a_2 & \cdots & a_k \end{pmatrix},$$

where $a_1 < a_2 < \cdots < a_k$. Then for every $i < j$ and every $a \in A_i$ and $b \in A_j$ we have $a < b$. In particular, in the case $\alpha \in \mathcal{O}_n$ this means that the sets $A_1, \ldots, A_k$ form a partition of $\mathcal{N}$ into *intervals* (i.e., if $x, y \in A_i$ and $x < y$, then A_i contains all z such that $x < z < y$).

For $\alpha \in \mathcal{IO}_n$ all sets A_i contain exactly one element, that is, $A_1 = \{b_1\}$, $A_2 = \{b_2\}, \ldots, A_k = \{b_k\}$. Hence $b_1 < b_2 < \cdots < b_k$, and α coincides with the unique increasing bijection from $\{b_1, \ldots, b_k\}$ to $\{a_1, \ldots, a_k\}$. In particular, α is uniquely determined by $\mathrm{dom}(\alpha)$ and $\mathrm{im}(\alpha)$ (see also Theorem 12.3.3 and its proof).

Proposition 14.1.6 *If $\alpha \in \mathcal{PT}_n$ is either order-preserving or order-decreasing, then the permutational part of α is the identity transformation.*

Proof. Assume that the permutational part of the element α contains some cycle $(a_1, a_2, \ldots, a_k)$. If α is order-decreasing, we have the inequalities $a_1 \geq a_2 \geq \cdots \geq a_k \geq a_1$ and hence $a_1 = a_2 = \cdots = a_k = a_1$ and $k = 1$.

Suppose now that α is order-preserving and $k > 1$. Consider first the case $a_1 < a_2$. Then $a_2 < a_3$, $a_3 < a_4, \ldots, a_{k-1} < a_k$, which yields $a_1 < a_k$. At the same time, as α is order-preserving, $a_{k-1} < a_k$ implies $a_k < a_1$, a contradiction. The case $a_1 > a_2$ is analogous. $\qquad\square$

Corollary 14.1.7 *Let α be a group element of any of the nine semigroups, defined in this section. Then α is an idempotent.*

Proof. It follows from Propositions 14.1.6 and 5.2.8. $\qquad\square$

A semigroup S is called *aperiodic* or *combinatorial* provided that all maximal subgroups of S have order one. From Corollary 14.1.7 it follows that all our nine semigroups are aperiodic.

14.2 Cardinalities

In this section, we compute the cardinalities of the semigroups defined in the previous section. The easiest computation is for the semigroup $\mathcal{F}_n$.

Proposition 14.2.1 $|\mathcal{F}_n| = n!$.

Proof. From the definition of $\mathcal{F}_n$ it follows that for an element $\alpha \in \mathcal{F}_n$ the image $\alpha(x)$ of some $x \in \mathbf{N}$ can be chosen in x different ways. Moreover, the images of different elements from $\mathcal{N}$ can be chosen independently. Hence $|\mathcal{F}_n| = 1 \cdot 2 \cdot \cdots \cdot n = n!$. $\qquad\square$

The computation of $|\mathcal{PF}_n|$ is completely similar. The only difference is that for every $x \in \mathcal{PF}_n$ there appears a new possibility for the image, namely, $\varnothing$.

Proposition 14.2.2 $|\mathcal{PF}_n| = 2 \cdot 3 \cdot \cdots \cdot (n+1) = (n+1)!$.

Proposition 14.2.3 $|\mathcal{O}_n| = \binom{2n-1}{n}$.

Proof. An element

$$\alpha = \left(\begin{array}{cccc} 1 & 2 & \cdots & n \\ a_1 & a_2 & \cdots & a_n \end{array} \right) \in \mathcal{O}_n$$

is uniquely determined by the second row $(a_1, \ldots, a_n)$. For $i = 1, \ldots, n$ set $b_i = a_i + i - 1$. Then the mapping

$$(a_1, \ldots, a_n) \mapsto (b_1, \ldots, b_n)$$

is an obvious bijection between the set of all vectors $(a_1, \ldots, a_n)$ such that $1 \leq a_1 \leq a_2 \leq \cdots \leq a_n \leq n$ and the set of all vectors $(b_1, \ldots, b_n)$ such that $1 \leq b_1 < b_2 < \cdots < b_n \leq 2n - 1$. The vector $(b_1, \ldots, b_n)$ uniquely determines the n-element subset $\{b_1, \ldots, b_n\}$ of $\{1, 2, \ldots, 2n - 1\}$. Hence $|\mathcal{O}_n| = \binom{2n-1}{n}$. $\square$

Proposition 14.2.4 $|\mathcal{IO}_n| = \binom{2n}{n}$.

Proof. As we have already seen in the previous section, an element $\alpha \in \mathcal{IO}_n$ is a unique increasing bijection from $\mathrm{dom}(\alpha)$ to $\mathrm{im}(\alpha)$, the latter being subsets of $\mathbf{N}$ of the same cardinality. Hence

$$|\mathcal{IO}_n| = \sum_{k=0}^{n} \binom{n}{k}^2 = \binom{2n}{n}.$$

$\square$

Corollary 14.2.5 $|\mathcal{IO}_n| = 2|\mathcal{O}_n|$.

Proposition 14.2.6 $|\mathcal{IF}_n| = B_{n+1}$.

Proof. The chain-cycle notation for an element from $\mathcal{IF}_n$ contains only cycles of length 1 and chains. With every partition ρ of the set $\mathbf{N}_* = \{*, 1, 2, \ldots, n\}$ we associate an element $\alpha_\rho \in \mathcal{IF}_n$ by the following rule: each $x \in \mathbf{N}$ such that $x\rho*$ corresponds to the cycle (x) of α; each block $A = \{a_1, \ldots, a_k\}$, $1 \leq a_1 \leq \cdots \leq a_k \leq n$, corresponds to the chain $[a_k, a_{k-1}, \ldots, a_1]$. Obviously, the mapping $\rho \mapsto \alpha_\rho$ is a bijection between the set of all partitions of $\mathbf{N}_*$ and $\mathcal{IF}_n$. Hence $|\mathcal{IF}_n| = B_{n+1}$. $\square$

The number $C_n = \frac{1}{n+1}\binom{2n}{n}$ is called the n-th *Catalan number*. Set also $C_0 = 1$.

Exercise 14.2.7 Show that Catalan numbers are uniquely determined by the following recursive relation:

$$C_0 = 1, \qquad C_{n+1} = \sum_{i=0}^{n} C_i C_{n-i}, n \geq 0.$$

Theorem 14.2.8   *(i)* $|\mathcal{C}_n| = \mathsf{c}_n$.

(ii) $|\mathcal{IC}_n| = \mathsf{c}_{n+1}$.

Proof. We start with the statement (i). Let $|\mathcal{C}_n| = t_n$ and set $t_0 = 1$. With every element

$$\alpha = \begin{pmatrix} 1 & 2 & \cdots & n \\ 1 & \alpha(2) & \cdots & \alpha(n) \end{pmatrix} \in \mathcal{C}_n$$

we associate the element

$$\hat{\alpha} = \begin{pmatrix} 1 & 2 & \cdots & n & n+1 \\ 1 & \alpha(2) & \cdots & \alpha(n) & n+1 \end{pmatrix} \in \mathcal{C}_{n+1}.$$

Now we partition $\mathcal{C}_n$ into n classes $C_n^{(2)}, \ldots, C_n^{(n+1)}$ depending on the minimal $k > 1$ such that $\hat{a}(k) = k$. Then for any element $\alpha \in C_n^{(k)}$ we have

$$\hat{\alpha} = \begin{pmatrix} 1 & 2 & \cdots & k-1 & k & k+1 & \cdots & n & n+1 \\ 1 & \alpha(2) & \cdots & \alpha(k-1) & k & \alpha(k+1) & \cdots & \alpha(n) & n+1 \end{pmatrix}$$

and for all i, $1 < i < k$, we have $\alpha(i) < i$. At the same time for all $j > k$ we have $\alpha(j) \geq k$. We separate the following two parts in $\hat{\alpha}$:

$$\hat{\alpha}_- = \begin{pmatrix} 2 & \cdots & k-1 \\ \alpha(2) & \cdots & \alpha(k-1) \end{pmatrix} \quad \hat{\alpha}_+ = \begin{pmatrix} k & k+1 & \cdots & n \\ k & \alpha(k+1) & \cdots & \alpha(n) \end{pmatrix}.$$

With $\hat{\alpha}_-$ we associate the following element:

$$\hat{\alpha}'_- = \begin{pmatrix} 1 & \cdots & k-2 \\ \alpha(2) & \cdots & \alpha(k-1) \end{pmatrix} \in \mathcal{C}_{k-2}$$

and with $\hat{\alpha}_-$ we associate the following element:

$$\hat{\alpha}'_+ = \begin{pmatrix} 1 & 2 & \cdots & n-k+1 \\ 1 & \alpha(k+1)-k+1 & \cdots & \alpha(n)-k+1 \end{pmatrix} \in \mathcal{C}_{n-k+1}.$$

It is easy to see that the correspondence

$$C_n^{(k)} \ni \alpha \leftrightarrow (\hat{\alpha}_-, \hat{\alpha}_+) \leftrightarrow (\hat{\alpha}'_-, \hat{\alpha}'_+) \in \mathcal{C}_{k-2} \times \mathcal{C}_{n-k+1}$$

is a bijection. Hence

$$t_n = |\mathcal{C}_n| = \sum_{k=2}^{n+1} |C_n^{(k)}| = \sum_{k=2}^{n+1} |\mathcal{C}_{k-2}| \cdot |\mathcal{C}_{n-k+1}| =$$

$$= \sum_{k=2}^{n+1} t_{k-2} \cdot t_{n-k+1} = \sum_{i=0}^{n-1} t_i \cdot t_{n-1-i}.$$

As $t_0 = 1 = \mathsf{c}_0$, the statement (i) follows now from Exercise 14.2.7.

The proof of the statement (ii) is quite similar. Let $|\mathcal{IC}_n| = f_{n+1}$ and set $f_0 = f_1 = 1$. With every element

$$\alpha = \begin{pmatrix} 1 & 2 & \cdots & n \\ \alpha(1) & \alpha(2) & \cdots & \alpha(n) \end{pmatrix} \in \mathcal{IC}_n$$

we associate the element

$$\hat{\alpha} = \begin{pmatrix} 1 & 2 & \cdots & n & n+1 \\ \alpha(1) & \alpha(2) & \cdots & \alpha(n) & n+1 \end{pmatrix} \in \mathcal{IC}_{n+1}.$$

Now we partition $\mathcal{IC}_n$ into $n+1$ classes $\mathcal{IC}_n^{(1)}, \ldots, \mathcal{IC}_n^{(n+1)}$ depending on the minimal k such that $\hat{\alpha}(k) = k$. Then for any element $\alpha \in \mathcal{IC}_n^{(k)}$ and for all $i \in \operatorname{dom}(\alpha)$, $1 < i < k$, we have $\alpha(i) < i$. At the same time for all $j \in \operatorname{dom}(\alpha)$ such that $j > k$ we have $\alpha(j) > k$. In particular, $\alpha(1) = \varnothing$ for all $k > 1$. Hence we can consider the mapping $\mathcal{IC}_n^{(k)} \to \mathcal{IC}_{k-2} \times \mathcal{IC}_{n-k}$ such that $\alpha \mapsto (\alpha_-, \alpha_+)$, where

$$\alpha_- = \begin{pmatrix} 1 & 2 & \cdots & k-2 \\ \alpha(2) & \alpha(3) & \cdots & \alpha(k-1) \end{pmatrix}$$

and

$$\alpha_+ = \begin{pmatrix} 1 & 2 & \cdots & n-k \\ \alpha(k+1)-k & \alpha(k+2)-k & \cdots & \alpha(n)-k \end{pmatrix}$$

(in the case $\alpha(j) = \varnothing$ we use the convention $\alpha(j) - k = \varnothing$). It is easy to see that the above correspondence is in fact a bijection (in the degenerate cases $k = 1, 2, n, n+1$ we obtain a bijection from $\mathcal{IC}_n^{(k)}$ to one of the factors in $\mathcal{IC}_{k-2} \times \mathcal{IC}_{n-k}$). Hence

$$f_{n+1} = |\mathcal{IC}_n| = \sum_{k=1}^{n+1} |\mathcal{IC}_n^{(k)}| = \sum_{k=1}^{n+1} |\mathcal{IC}_{k-2}| \cdot |\mathcal{IC}_{n-k}| = \sum_{k=0}^{n} f_k \cdot f_{n-k}.$$

The statement (ii) now also follows from Exercise 14.2.7. $\square$

Theorem 14.2.9 $|\mathcal{PO}_n| = \displaystyle\sum_{m=0}^{n} \binom{n}{m}\binom{n+m-1}{m}.$

Proof. We partition $\mathcal{PO}_n$ into blocks depending on $|\operatorname{dom}(\alpha)|$, $\alpha \in \mathcal{PO}_n$. If $|\operatorname{dom}(\alpha)| = m$, the set $\operatorname{dom}(\alpha)$ can be chosen in $\binom{n}{m}$ different ways. If $\operatorname{dom}(\alpha) = \{a_1, \ldots, a_m\}$ is fixed and $a_1 < a_2 < \cdots < a_m$, there is an obvious bijection between the elements

$$\alpha = \begin{pmatrix} a_1 & a_2 & \cdots & a_m \\ b_1 & b_2 & \cdots & b_m \end{pmatrix} \in \mathcal{PO}_n$$

and vectors $(b_1, \ldots, b_m)$ such that $b_1 \leq b_2 \leq \cdots \leq b_m$. At the same time, there is a bijection between the latter vectors and m-element subsets in

$\{1, \ldots, n + m - 1\}$ of the form $\{b_1, b_2 + 1, \ldots, b_m + m - 1\}$. Hence the number of elements $\alpha \in \mathcal{PO}_n$ such that $\mathrm{dom}(\alpha) = \{a_1, \ldots, a_m\}$ equals $\binom{n+m-1}{m}$. Therefore,

$$|\mathcal{PO}_n| = 1 + \sum_{m=1}^{n} \binom{n}{m}\binom{n+m-1}{m} = \sum_{m=0}^{n} \binom{n}{m}\binom{n+m-1}{m}.$$

$\square$

Exercise 14.2.10 Show that

$$(1 - x)^{-n} = \sum_{m \geq 0} \binom{n+m-1}{m} x^m.$$

Corollary 14.2.11 $|\mathcal{PO}_n|$ *coincides with the coefficient of x^n in the series* $(1 + x)^n (1 - x)^{-n}$.

Proof. This follows directly from Theorem 14.2.9, Exercise 14.2.10 and the binomial formula. $\square$

14.3 Idempotents

Recall from Sect. 2.7 that an element $\alpha \in \mathcal{IS}_n$ is an idempotent if and only if α is the identity transformation of some subset $A \subset \mathbf{N}$. In particular, every idempotent from $\mathcal{IS}_n$ is both order-preserving and order-decreasing, by definition. This means the following:

Proposition 14.3.1 *We have* $\mathcal{E}(\mathcal{IS}_n) = \mathcal{E}(\mathcal{IO}_n) = \mathcal{E}(\mathcal{IF}_n) = \mathcal{E}(\mathcal{IC}_n)$, *in particular, each of these sets contains 2^n elements.*

Proposition 14.3.2 *Every idempotent $\delta \in \mathcal{C}_n$ is uniquely determined by its image* $\mathrm{im}(\delta) = \{1, a_2, \ldots, a_k\}$, *moreover, the set $\{a_2, \ldots, a_k\}$ can be an arbitrary subset of $\{2, \ldots, n\}$. In particular, $|\mathcal{E}(\mathcal{C}_n)| = 2^{n-1}$.*

Proof. We have $\delta(1) = 1$ for all $\delta \in \mathcal{C}_n$. Let now $\delta \in \mathcal{E}(\mathcal{C}_n)$ and $\mathrm{im}(\delta) = \{1, a_2, \ldots, a_k\}$ be such that $1 = a_1 < a_2 < \cdots < a_k$. As any idempotent acts as the identity on its image, for any $x \in \{a_i, a_i + 1, \ldots, a_{i+1} - 1\}$ from $a_i \leq x < a_{i+1}$ we obtain

$$a_i = \delta(a_i) \leq \delta(x) \leq x < a_{i+1}.$$

This yields $\delta(x) = a_i$. Hence the element δ is uniquely determined by its image. Obviously, the set $\{a_2, \ldots, a_k\}$ can be an arbitrary subset of $\{2, \ldots, n\}$. The statement of the proposition follows. $\square$

Exercise 14.3.3 Prove that $\mathcal{C}_n$ contains $\binom{n-1}{k-1}$ idempotents of rank k.

Theorem 14.3.4 *The semigroup* $\mathcal{PC}_n$ *contains* $\frac{3^n+1}{2}$ *idempotents.*

Proof. We count the number of nonzero idempotents $\delta \in \mathcal{PC}_n$ for which the minimum element in $\mathrm{im}(\delta)$ equals k. Let $\mathrm{rank}(\delta) = i + 1$ and $\mathrm{im}(\delta) = \{k, b_1, \ldots, b_i\}$, where $k < b_1 < \cdots < b_i$. The subset $\{b_1, \ldots, b_i\}$ of the set $\{k+1, \ldots, n\}$ can be chosen in $\binom{n-k}{i}$ different ways.

Every $x < k$ does not belong to $\mathrm{dom}(\delta)$; every x, $k \leq x < b_1$, which belongs to $\mathrm{dom}(\delta)$ must be mapped to k; every x, $b_1 \leq x < b_2$, which belongs to $\mathrm{dom}(\delta)$ must be mapped to b_1 and so on. Hence there are 2^{n-k-i} ways to define the transformation δ on the set $\{k+1, \ldots, n\} \backslash \mathrm{im}(\delta)$. Therefore,

$$|\mathcal{E}(\mathcal{PC}_n)| = 1 + \sum_{k=1}^{n} \left(\sum_{i=0}^{n-k} \binom{n-k}{i} 2^{n-k-i} \right) =$$

$$= 1 + \sum_{k=1}^{n} (1+2)^{n-k} = 1 + \frac{3^n - 1}{2} = \frac{3^n + 1}{2}.$$

$\square$

Proposition 14.3.5 *(i)* $|\mathcal{E}(\mathcal{F}_n)| = \mathrm{B}_n$.

(ii) $|\mathcal{E}(\mathcal{PF}_n)| = \mathrm{B}_{n+1}$.

Proof. Let ρ be a partition of $\mathbf{N}$. Define the transformation α_ρ as follows: If X is some block of ρ, the transformation α_ρ maps all elements from X to the minimum element in X. Obviously, this transformation is idempotent. On the other hand, every idempotent $\delta \in \mathcal{E}(\mathcal{F}_n)$ has this form as, by definition, every block of ρ_δ must be mapped to the minimum element of this block. This gives us a bijection between $\mathcal{E}(\mathcal{F}_n)$ and partitions of $\mathbf{N}$. The statement (i) follows.

Let ρ be a partition of $\mathbf{N}_* = \mathbf{N} \cup \{*\}$. Define the transformation $\alpha_\rho \in \mathcal{PT}_n$ as follows: α is not defined on all elements which belong to the block of ρ containing $*$; if X is some block of ρ which does not contain $*$, the transformation α_ρ maps all elements from X to the minimum element in X. As in the previous paragraph one easily verifies that this correspondence defines a bijection between $\mathcal{E}(\mathcal{PF}_n)$ and partitions of $\mathbf{N}_*$. The statement (ii) follows and the proof is complete. $\square$

To count idempotents in the remaining semigroups we recall that *Fibonacci numbers* $\mathbf{f}_n$, $n \geq 1$, are defined recursively as follows: $\mathbf{f}_1 = \mathbf{f}_2 = 1$ and $\mathbf{f}_n = \mathbf{f}_{n-1} + \mathbf{f}_{n-2}$, $n \geq 3$.

Theorem 14.3.6 $|\mathcal{E}(\mathcal{O}_n)| = \mathbf{f}_{2n}$.

Proof. Let $q_n = |\mathcal{E}(\mathcal{O}_n)|$. Partition the set $\mathcal{E}(\mathcal{O}_n)$ into the following three disjoint subsets:

$$\begin{aligned}
\mathcal{E}_1(\mathcal{O}_n) &= \{\delta \in \mathcal{E}(\mathcal{O}_n) : \delta(n) < n\}; \\
\mathcal{E}_2(\mathcal{O}_n) &= \{\delta \in \mathcal{E}(\mathcal{O}_n) : \delta(n-1) < n, \delta(n) = n\}; \\
\mathcal{E}_3(\mathcal{O}_n) &= \{\delta \in \mathcal{E}(\mathcal{O}_n) : \delta(n-1) = n\}.
\end{aligned}$$

For every idempotent $\delta \in \mathcal{E}_1(\mathcal{O}_n)$ we have $\delta(n) = \delta(n-1)$ as δ is order-preserving, and hence $|\mathcal{E}_1(\mathcal{O}_n)| = q_{n-1}$. We also obviously have $|\mathcal{E}_2(\mathcal{O}_n)| = q_{n-1}$.

Let us now compute $|\mathcal{E}_3(\mathcal{O}_n)|$. If $\delta \in \mathcal{E}_3(\mathcal{O}_n)$, then the fact that δ is an idempotent implies $n - 1 \notin \operatorname{im}(\delta)$. Consider the transformation δ' : $\{1, \ldots, n-1\} \to \{1, \ldots, n-1\}$ defined as follows:

$$\delta'(x) = \begin{cases} \delta(x), & \delta(x) \neq n; \\ n-1, & \delta(x) = n. \end{cases}$$

Obviously, $\delta' \in \mathcal{E}(\mathcal{O}_{n-1})$. In this way we obtain exactly those idempotents from $\mathcal{E}(\mathcal{O}_{n-1})$, which fix the element $(n-1)$, that is, exactly the elements from $\mathcal{E}(\mathcal{O}_{n-1}) \backslash \mathcal{E}_1(\mathcal{O}_{n-1})$. This implies $|\mathcal{E}_3(\mathcal{O}_n)| = q_{n-1} - q_{n-2}$ and thus

$$q_n = q_{n-1} + q_{n-1} + (q_{n-1} - q_{n-2}) = 3q_{n-1} - q_{n-2}.$$

On the other hand, we obviously have $q_1 = 1 = \mathbf{f}_2$ and $q_2 = 3 = \mathbf{f}_4$ (see Example 14.1.2). For Fibonacci numbers we have

$$\mathbf{f}_{2n} = \mathbf{f}_{2n-1} + \mathbf{f}_{2(n-1)} = 2\mathbf{f}_{2(n-1)} + \mathbf{f}_{2n-3} = 3\mathbf{f}_{2(n-1)} - \mathbf{f}_{2(n-2)}.$$

Hence q_n satisfies the same recursive relation as the sequence $\mathbf{f}_{2n}$. The statement of the theorem follows. □

Theorem 14.3.7 $|\mathcal{E}(\mathcal{PO}_n)| = 1 + \sum_{k=1}^{n} \binom{n}{k} \mathbf{f}_{2k}.$

Proof. If δ is an idempotent, we have $\operatorname{im}(\delta) \subset \operatorname{dom}(\delta)$. Consider all idempotents from $\mathcal{PO}_n$ whose domain is the set $\{a_1, \ldots, a_k\}$, where $a_1 < \cdots < a_k$. Then each such idempotent has the form

$$\delta = \begin{pmatrix} a_1 & a_2 & \cdots & a_k \\ a_{i_1} & a_{i_2} & \cdots & a_{i_k} \end{pmatrix}.$$

With this element δ we associate the idempotent

$$\delta' = \begin{pmatrix} 1 & 2 & \cdots & k \\ i_1 & i_2 & \cdots & i_k \end{pmatrix} \in \mathcal{O}_k.$$

It is obvious that the mapping $\delta \mapsto \delta'$ is a bijection from the set of all idempotents from $\mathcal{PO}_n$ with domain $\{a_1, \ldots, a_k\}$ and $\mathcal{E}(\mathcal{O}_k)$. As the domain of a nonzero idempotent from $\mathcal{PO}_n$ may be an arbitrary nonempty subset of $\mathbf{N}$, using Theorem 14.3.6 we have

$$|\mathcal{E}(\mathcal{PO}_n)| = 1 + \sum_{k=1}^{n} \binom{n}{k} |\mathcal{E}(\mathcal{O}_k)| = 1 + \sum_{k=1}^{n} \binom{n}{k} \mathbf{f}_{2k}.$$

$\square$

14.4 Generating Systems

Each semigroup S from the nine semigroups defined in Sect. 14.1 contains the identity transformation ε. Since the set $S \backslash \{\varepsilon\}$ is a subsemigroup in S, every generating system of S contains ε. The main result of the present section is the following:

Theorem 14.4.1 *Each of the semigroups $\mathcal{O}_n$, $\mathcal{F}_n$, $\mathcal{C}_n$, $\mathcal{PO}_n$, $\mathcal{PF}_n$, $\mathcal{PC}_n$, $\mathcal{IO}_n$, $\mathcal{IF}_n$ and $\mathcal{IC}_n$ is generated by ε and elements of rank $(n-1)$.*

Proof. It is certainly enough to show that in each of these semigroups every element of rank r, $r < n-1$, can be decomposed into a product of elements of rank strictly greater than r.

Consider first the semigroup $\mathcal{PF}_n$. Let $\alpha \in \mathcal{PF}_n$ be such that $\text{rank}(\alpha) = r < n-1$. We consider the partition ρ_α of $\mathbf{N} \cup \{n+1\}$, associated to α, as in Sect. 4.3. For $x \in \mathbf{N} \cup \{n+1\}$ we denote by $\bar{x}$ the block of ρ_α, containing x. The block $\overline{n+1}$ will be called the *trivial* block.

Consider the case in which there exist nontrivial blocks of ρ_α, containing more than one element. Let $a_\alpha \in \mathbf{N}$ be the maximum element occurring in such blocks. Then there exists $b \in \overline{a_\alpha}$ such that $b < a_\alpha$, and as $\alpha(a_\alpha) = \alpha(b) \leq b$, we have $\alpha(a_\alpha) \neq a_\alpha$. Consider the following two transformations:

$$\beta_1(x) = \begin{cases} \alpha(x), & x < a_\alpha; \\ x, & x \geq a_\alpha \text{ and } x \in \text{dom}(\alpha); \\ \varnothing, & x \geq a_\alpha \text{ and } x \notin \text{dom}(\alpha); \end{cases}$$

and

$$\gamma_1(x) = \begin{cases} x, & x < a_\alpha; \\ \alpha(x), & x \geq a_\alpha \text{ and } x \in \text{dom}(\alpha); \\ x, & x \geq a_\alpha \text{ and } x \notin \text{dom}(\alpha). \end{cases}$$

Then $\alpha = \gamma_1 \beta_1$, moreover, both β_1 and γ_1 belong to $\mathcal{PF}_n$.

From the definition, we see that ρ_α and ρ_{β_1} share almost all blocks, the only exceptions the block $\overline{a_\alpha}$ of ρ_α, which in ρ_{β_1} decomposes into two blocks: $\{a_\alpha\}$ and $\overline{a_\alpha} \backslash \{a_\alpha\}$. As the rank of an element equals the number of nontrivial blocks, we obtain that $\text{rank}(\beta_1) = \text{rank}(\alpha) + 1 > \text{rank}(\alpha)$.

We have $\text{rank}(\gamma_1) \geq \text{rank}(\alpha)$ by Exercise 2.1.4(c). Consider the sets $A_1 = \{1, 2, \ldots, a_\alpha - 1\}$ and $A_2 = \mathbf{N} \backslash A_1$. Note that the restrictions $\gamma_1|_{A_1}$ and $\gamma_1|_{A_2}$ are injective. Hence nontrivial blocks of ρ_{γ_1} contain at most two elements, moreover, each two-element block contains one element from A_1 and one from A_2. Furthermore, $\text{dom}(\gamma_1) = \mathbf{N}$ by definition. Hence the equality $\text{rank}(\gamma_1) = r$ implies that γ_1 contains at least two two-element blocks. This yields $a_{\gamma_1} > a_\alpha$.

Now we repeat the same arguments for γ_1, which allows us to write $\gamma_1 = \gamma_2 \beta_2$, where $\text{rank}(\beta_2) > r$ and $a_{\gamma_2} > a_{\gamma_1}$ in the case $\text{rank}(\gamma_2) = r$. In the latter case, we apply the same arguments to γ_2 and so on. The sequence $a_\alpha < a_{\gamma_1} < a_{\gamma_2} < \cdots$ is finite, so in a finite number of steps, say k steps, we will get that $\gamma_{k-1} = \gamma_k \beta_k$, where both $\text{rank}(\beta_k)$ and $\text{rank}(\gamma_k)$ are strictly greater than r. Thus $\alpha = \gamma_k \beta_k \cdots \beta_1$ is a decomposition of α into a product of elements of rank strictly greater than r.

Now we consider the case for which all nontrivial blocks of ρ_α contain exactly one element. Then $\alpha \in \mathcal{IS}_n$ and we can write α in the following form:

$$\alpha = [a_1, \ldots, a_k][b_1, \cdots, b_l] \cdots [c_1, \cdots, c_m](d_1) \cdots (d_q),$$

where $a_1 > \cdots > a_k, \ldots, c_1 > \cdots > c_m$. As $\text{rank}(\alpha) < n - 1$, the element α contains at least two chains. Then $\alpha = \beta \gamma$, where

$$\begin{aligned} \beta &= [a_1, \cdots, a_k](b_1) \cdots (b_l) \cdots (c_1) \cdots (c_m)(d_1) \cdots (d_q), \\ \gamma &= (a_1) \cdots (a_k)[b_1, \ldots, b_l] \cdots [c_1, \ldots, c_m](d_1) \cdots (d_q). \end{aligned} \tag{14.1}$$

Obviously, $\text{rank}(\beta) = n - 1 > r$, $\text{rank}(\gamma) = r + 1$ and $\beta, \gamma \in \mathcal{PF}_n$. This completes the proof of our theorem for the semigroup $\mathcal{PF}_n$.

Note that the first part of our proof for $\mathcal{PF}_n$ (where we considered the case in which ρ_α has a nontrivial block consisting of at least two elements) proves the statement of the theorem for the semigroup $\mathcal{F}_n$. Moreover, the second part of the proof for $\mathcal{PF}_n$ proves the statement of the theorem for the semigroup $\mathcal{IF}_n$.

Let us now consider the case of the semigroup $\mathcal{PO}_n$. Let $\alpha \in \mathcal{PO}_n$ be such that

$$\alpha = \begin{pmatrix} A_1 & \cdots & A_r \\ b_1 & \cdots & b_r \end{pmatrix},$$

where $b_1 < \cdots < b_r$ and $r < n - 1$. Assume that there exists k such that $|A_k| > 1$ and let $A_k = \{a_1, \ldots, a_m\}$, where $a_1 < \cdots < a_m$. Consider the set $B' = \{b'_1, \ldots, b'_{r+2}\}$, $b'_1 < \cdots < b'_{r+2}$, which is obtained from the set $B = \{b_1, \ldots, b_r\}$ by adding two new elements b'_p and b'_q, $p < q$. Then $b'_1 = b_1, \ldots, b'_{p-1} = b_{p-1}, b'_{p+1} = b_p, \ldots, b'_{q-1} = b_{q-2}, b'_{q+1} = b_{q-1}, \ldots, b'_{r+2} = b_r$. If $q > k + 1$, then α decomposes into the product $\alpha = \gamma \beta$, where the element β equals

$$\begin{pmatrix} A_1 & \cdots & A_{k-1} & A_k \backslash \{a_m\} & a_m & A_{k+1} & \cdots & A_{q-2} & A_{q-1} & \cdots & A_r \\ b'_1 & \cdots & b'_{k-1} & b'_k & b'_{k+1} & b'_{k+2} & \cdots & b'_{q-1} & b'_{q+1} & \cdots & b'_{r+2} \end{pmatrix}$$

and the element γ equals

$$\left(\begin{array}{ccccccccccc} C_1 & C_2 & \cdots & C_{k-1} & C_k & C_{k+1} & \cdots & C_{q-2} & C_{q-1} & C_q & \cdots & C_{r+1} \\ b_1 & b_2 & \cdots & b_{k-1} & b_k & b_{k+1} & \cdots & b_{q-2} & b'_q & b_{q-1} & \cdots & b_r \end{array} \right),$$

where

$$\begin{aligned} C_1 &= \{1, \ldots, b'_1\}, \\ C_2 &= \{b'_1 + 1, \ldots b'_2\}, \end{aligned}$$

$$\cdots$$

$$\begin{aligned} C_{k-1} &= \{b'_{k-2} + 1, \ldots, b'_{k-1}\}, \\ C_k &= \{b'_{k-1} + 1, \ldots, b'_{k+1}\}, \\ C_{k+1} &= \{b'_{k+1} + 1, \ldots, b'_{k+2}\}, \end{aligned}$$

$$\cdots$$

$$C_{r+1} = \{b'_{r+1} + 1, \ldots, n\}.$$

If $q \leq k+1$, then $p < k+1$ and the elements β and γ are constructed playing a similar game with the element b'_p instead of the element b'_q. Obviously, both β and γ are elements of $\mathcal{PO}_n$ of rank $r+1$.

Assume now that all blocks $A_1, \ldots, A_r$ consist of oneelement and we have $A_i = \{a_i\}$ for all i. Extend the set $\{a_1, \ldots, a_r\}$ in any way to the set $A' = \{a'_1, \ldots, a'_{r+1}\}$, $a'_1 < \cdots < a'_{r+1}$ and the set $\{b_1, \ldots, b_r\}$ to the set $B' = \{b'_1, \ldots, b'_{r+1}\}$, $b'_1 < \cdots < b'_{r+1}$, by adding one more new element, a'_k and b'_m say, to each set, respectively. Consider also the set $C = \{1, 2, \ldots, r+2\}$. If $k \leq m$, we let β be the unique increasing bijection from A' to $C \backslash \{m+1\}$, and γ be the unique increasing bijection from $C \backslash \{k\}$ to B'. If $k > m$, we let β be the unique increasing bijection from A' to $C \backslash \{m\}$, and γ be the unique increasing bijection from $C \backslash \{k+1\}$ to B'. A direct calculation shows that $\alpha = \gamma\beta$. This proves our theorem for the semigroup $\mathcal{PO}_n$. At the same time, the first part of the above proof proves the statement of the theorem for the semigroup $\mathcal{O}_n$, and the second part of the proof proves the statement of our theorem for the semigroup $\mathcal{IO}_n$.

To proceed we will need the following statement:

Lemma 14.4.2 *Each transformation $\alpha \in \mathcal{IC}_n$ of rank k can be decomposed into a product $\alpha_m \cdots \alpha_1$ of transformations $\alpha_i \in \mathcal{IC}_n$ of rank k such that $x - 1 \leq \alpha_i(x) \leq x$ for all $x \in \mathrm{dom}(\alpha_i)$.*

Proof. Let

$$\alpha = \left(\begin{array}{ccc} a_1 & \cdots & a_k \\ b_1 & \cdots & b_k \end{array} \right) \in \mathcal{IC}_n,$$

where $a_1 < \cdots < a_k$. Then

$$\alpha = \left(\begin{array}{ccc} b'_1 & \cdots & b'_k \\ b_1 & \cdots & b_k \end{array} \right) \left(\begin{array}{ccc} a_1 & \cdots & a_k \\ b'_1 & \cdots & b'_k \end{array} \right), \qquad (14.2)$$

where

$$b'_j = \begin{cases} a_j, & a_j = b_j; \\ a_j - 1, & a_j > b_j. \end{cases}$$

We claim that both factors from (14.2) belong to $\mathcal{IC}_n$. Indeed, since $b_j \le b_j' \le a_j$, it is enough to check that $b_1' < \cdots < b_k'$. Assume that this is not the case and $b_j' \ge b_{j+1}'$ for some j. Then $b_j' = a_j$, $b_{j+1}' = a_{j+1} - 1$, which implies $b_j = a_j$ and

$$b_{j+1} \le b_{j+1}' = a_{j+1} - 1 \le b_j' = a_j = b_j,$$

a contradiction.

Both factors in (14.2) have rank k and the second factor obviously satisfies the inequality from the formulation. So, the second factor in (14.2) satisfies the condition from the formulation. For the first factor, which is an element of $\mathcal{IC}_n$, we obtain the following: For every j such that $a_j > b_j$ we have

$$b_j' - b_j = (a_j - b_j) - 1 < a_j - b_j.$$

Now we proceed by induction on $N = \sum_{j=1}^{k} (a_j - b_j)$. If the first factor in (14.2) does not satisfy the condition from the formulation, we apply to it the inductive assumption and find its decomposition into a product of elements from $\mathcal{IC}_n$, which satisfy the necessary condition. This completes the proof. □

Now the statement of our theorem is easy to prove for the semigroup $\mathcal{IC}_n$. Indeed, let $\alpha \in \mathcal{IC}_n$ be an element of rank $r < n - 1$. Consider some decomposition $\alpha = \alpha_m \cdots \alpha_1$ given by Lemma 14.4.2. Every element α_i has the form

$$\alpha_i = [u + v, u + v - 1, \dots, u] \cdots [t + s, t + s - 1, \dots, t](d_1) \cdots (d_q).$$

Now, taking β_i and γ_i as given by (14.1), we obtain the decomposition $\alpha_i = \beta_i \gamma_i$, where both β_i and γ_i belong to $\mathcal{IC}_n$ and have rank $\ge r + 1$.

Consider now the semigroup $\mathcal{PC}_n$. Let

$$\alpha = \begin{pmatrix} A_1 & \cdots & A_r \\ b_1 & \cdots & b_r \end{pmatrix} \in \mathcal{PC}_n,$$

where $b_1 < \cdots < b_r$ and $r < n - 1$. Let a_i denote the minimum element in A_i. From the definition of $\mathcal{PC}_n$ we have that $b_i \le a_i$ for all i, and $a < a_j$ for all $a \in A_i$ as soon as $i < j$. Hence for the elements

$$\alpha^* = \begin{pmatrix} A_1 & \cdots & A_r \\ a_1 & \cdots & a_r \end{pmatrix}, \qquad \hat{\alpha} = \begin{pmatrix} a_1 & \cdots & a_r \\ b_1 & \cdots & b_r \end{pmatrix}$$

we have $\alpha^* \in \mathcal{PC}_n$, $\hat{\alpha} \in \mathcal{IC}_n$, and $\alpha = \hat{\alpha}\alpha^*$.

Since we have already proved the theorem for the semigroup $\mathcal{IC}_n$, it is now enough to show that every α^* as above decomposes into a product of elements of rank $> r$. If $\alpha^* \in \mathcal{IC}_n$, this is already proved. If $\alpha^* \notin \mathcal{IC}_n$, we have the following three possibilities:

(i) There exists k such that $|A_k| \geq 3$.

(ii) There exists $i < j$ such that $|A_i| = |A_j| = 2$.

(iii) $|A_i| = 1$ for all $i \neq k$, $A_k = \{a_k, a'\}$ and $\mathrm{dom}(\alpha) = \mathrm{dom}(\alpha^*) \neq \mathbf{N}$.

In the case (i) let $A_k = A_k' \cup \{a', a''\}$, where $a_k < a' < a''$. Then we have

$$
\begin{pmatrix} A_k' & a' & a'' \\ a_k & a_k & a_k \end{pmatrix} = \begin{pmatrix} A_k' & a' & a'' \\ a_k & a_k & a'' \end{pmatrix} \begin{pmatrix} A_k' & a' & a'' \\ a_k & a' & a' \end{pmatrix},
$$

which implies the existence of the necessary decomposition for α^*.

In the case (ii) let $A_i = \{a_i, a'\}$ and $A_j = \{a_j, a''\}$. Then we have

$$
\begin{pmatrix} a_i & a' & a_j & a'' \\ a_i & a_i & a_j & a_j \end{pmatrix} = \begin{pmatrix} a_i & a' & a_j & a'' \\ a_i & a' & a_j & a_j \end{pmatrix} \begin{pmatrix} a_i & a' & a_j & a'' \\ a_i & a_i & a_j & a'' \end{pmatrix},
$$

which implies the existence of the necessary decomposition for α^*.

In the case (iii) let $A_k = \{a_k, a'\}$ and $a \in \mathbf{N} \backslash \mathrm{dom}(\alpha)$. If we have $a \notin \{a_k + 1, \ldots, a' - 1\}$, then the necessary decomposition for α^* follows from

$$
\begin{pmatrix} a_k & a' & a \\ a_k & a_k & \varnothing \end{pmatrix} = \begin{pmatrix} a_k & a' & a \\ a_k & a' & \varnothing \end{pmatrix} \begin{pmatrix} a_k & a' & a \\ a_k & a_k & a \end{pmatrix}.
$$

If $a \in \{a_k + 1, \ldots, a' - 1\}$, then the necessary decomposition for α^* follows from

$$
\begin{pmatrix} a_k & a & a' \\ a_k & \varnothing & a_k \end{pmatrix} = \begin{pmatrix} a_k & a & a' \\ a_k & a_k & a' \end{pmatrix} \begin{pmatrix} a_k & a & a' \\ a_k & \varnothing & a \end{pmatrix}.
$$

This completes the proof of our theorem for the semigroup $\mathcal{PC}_n$.

Finally, let us prove the statement of the theorem for the semigroup $\mathcal{C}_n$. Let $\alpha \in \mathcal{C}_n$. As $\mathcal{C}_n \subset \mathcal{PC}_n$, we again may write $\alpha = \hat{a}\alpha^*$ as above. Note that $\alpha^* \in \mathcal{C}_n$, and hence the constructed above decomposition for α^* will contain only factors from $\mathcal{C}_n$. Hence from the above arguments we have a decomposition

$$
\alpha = \gamma_k \cdots \gamma_1 \beta_m \cdots \beta_1,
$$

where all factors have rank $> r$, all γ_i's belong to $\mathcal{IC}_n$ and all β_j's belong to $\mathcal{C}_n$.

Every element

$$
\gamma = \begin{pmatrix} a_1 & \cdots & a_k \\ b_1 & \cdots & b_k \end{pmatrix} \in \mathcal{IC}_n,
$$

where $a_1 < \cdots < a_k$, can be extended to an element $\tilde{\gamma}$ from $\mathcal{C}_n$ as follows:

$$
\tilde{\gamma} = \begin{pmatrix} \{1, \ldots, a_1 - 1\} & \{a_1, \ldots, a_2 - 1\} & \cdots & \{a_k, \ldots, n\} \\ 1 & b_1 & \cdots & b_k \end{pmatrix}
$$

in the case $1 < b_1$, or

$$\tilde{\gamma} = \left(\begin{array}{cccc} \{1,\ldots,a_2-1\} & \{a_2,\ldots,a_3-1\} & \cdots & \{a_k,\ldots,n\} \\ b_1 & b_2 & \cdots & b_k \end{array} \right)$$

in the case $1 = b_1$. Note that $\mathrm{rank}(\tilde{\gamma}) \geq \mathrm{rank}(\gamma)$. A direct calculation shows that

$$\alpha = \gamma_k \cdots \gamma_1 \beta_m \cdots \beta_1 = \tilde{\gamma}_k \cdots \tilde{\gamma}_1 \beta_m \cdots \beta_1,$$

which completes the proof of the theorem. $\qquad\square$

For six of our semigroups, the statement of Theorem 14.4.1 can be strengthened.

Theorem 14.4.3 *The semigroups $\mathcal{F}_n$ and $\mathcal{PF}_n$ are generated by ε and all idempotents of rank $(n-1)$.*

Proof. Taking Theorem 14.4.1 into account, it is enough to show that every element of rank $(n-1)$ decomposes into a product of idempotents of rank $(n-1)$.

For the semigroup $\mathcal{F}_n$ we use induction on the number of fixed points. Obviously, an element $\alpha \in \mathcal{F}_n$ is an idempotent if and only if it has $(n-1)$ fixed points. Let $\alpha \in \mathcal{F}_n$ be an element of rank $(n-1)$. Then we can write

$$\alpha = \left(\begin{array}{ccccccccc} 1 & \cdots & k & \cdots & m-1 & m & m+1 & \cdots & n \\ a_1 & \cdots & a_k & \cdots & a_{m-1} & a_k & a_{m+1} & \cdots & a_n \end{array} \right)$$

for some $k < m$. Consider the element

$$\beta = \left(\begin{array}{ccccccccc} 1 & \cdots & k & \cdots & m-1 & m & m+1 & \cdots & n \\ a_1 & \cdots & a_k & \cdots & a_{m-1} & m & a_{m+1} & \cdots & a_n \end{array} \right).$$

Then $\alpha = \beta \varepsilon_{k,m}$. If $\mathrm{rank}(\beta) = n$, then $\beta = \varepsilon$ and $\alpha = \varepsilon_{k,m}$. If $\mathrm{rank}(\beta) = n-1$, then β has more fixed points than α and hence decomposes into a product of idempotents by the inductive assumption. This proves the statement of our theorem for the semigroup $\mathcal{F}_n$.

Let now α be an element of rank $(n-1)$ from $\mathcal{PF}_n \backslash \mathcal{F}_n$ and assume that $\mathbb{N}\backslash\mathrm{dom}(\alpha) = \{k\}$. If $k \notin \mathrm{im}(\alpha)$, then α acts on $\mathbb{N}\backslash\{k\}$ injectively and order-decreasing, and hence is the identity transformation. This implies that α is an idempotent. If $k \in \mathrm{im}(\alpha)$, the element α can be written in the following form:

$$\left(\begin{array}{cccccccccc} 1 & \cdots & k-1 & k & k+1 & \cdots & k+m-1 & k+m & k+m+1 & \cdots & n \\ 1 & \cdots & k-1 & \varnothing & k+1 & \cdots & k+m-1 & k & b_{k+m+1} & \cdots & b_n \end{array} \right)$$

for some $m > 0$. Consider the element β defined as follows:

$$\beta(x) = \begin{cases} x, & x < k+m; \\ \alpha(x), & x \geq k+m. \end{cases}$$

Then $\alpha = \beta \varepsilon_{(k)}$ and $\beta \in \mathcal{F}_n$ is an element of rank $(n-1)$. As we have already proved the statement of the theorem for the semigroup $\mathcal{F}_n$, β decomposes into a product of idempotents. Hence α decomposes into a product of idempotents as well. This completes the proof. $\qquad\square$

The next statement follows immediately from the definitions:

Lemma 14.4.4   (i) *Every element of rank $(n-1)$ from the semigroup C_n has the form*

$$\begin{pmatrix} 1 & \cdots & k & k+1 & \cdots & m & m+1 & \cdots & n \\ 1 & \cdots & k & k & \cdots & m-1 & m+1 & \cdots & n \end{pmatrix},$$

where $1 \leq k < m \leq n$.

(ii) *Every element of rank $(n-1)$ from the semigroup $\mathcal{O}_n$ either belongs to C_n or has the form*

$$\begin{pmatrix} 1 & \cdots & m & m+1 & \cdots & k & k+1 & \cdots & n \\ 1 & \cdots & m & m+2 & \cdots & k+1 & k+1 & \cdots & n \end{pmatrix},$$

where $1 \leq m < k \leq n$.

(iii) *Every element of rank $(n-1)$ from the semigroup $\mathcal{IO}_n$ has the form*

$$\begin{pmatrix} 1 & \cdots & & \cdots & & \cdots & k-1 & k+1 & \cdots & n \\ 1 & \cdots & m-1 & m+1 & \cdots & & \cdots & & \cdots & n \end{pmatrix},$$

where $1 \leq m \leq n$ and $1 \leq k \leq n$.

(iv) *Every element of rank $(n-1)$ from the semigroup $\mathcal{IC}_n$ has the form*

$$\begin{pmatrix} 1 & \cdots & k-1 & k+1 & \cdots & m & m+1 & \cdots & n \\ 1 & \cdots & k-1 & k & \cdots & m-1 & m+1 & \cdots & n \end{pmatrix},$$

where $1 \leq k \leq m \leq n$.

Theorem 14.4.5 *The semigroups $\mathcal{O}_n$, C_n, $\mathcal{PO}_n$, and $\mathcal{PC}_n$ are generated by ε and all idempotents of rank $(n-1)$.*

Proof. It is again enough to show that every element of rank $(n-1)$ decomposes into a product of idempotents. Let α be an element of rank $(n-1)$ from one of our semigroups. For the semigroup C_n the statement follows from the observation that the element α as in Lemma 14.4.4(i) has the following decomposition:

$$\alpha = \varepsilon_{m-1,m} \cdots \varepsilon_{k+1,k+2} \varepsilon_{k,k+1}.$$

For all elements from the semigroup $\mathcal{O}_n$ as in Lemma 14.4.4(ii) one constructs a decomposition similarly.

If $\alpha \in \mathcal{PO}_n \backslash \mathcal{O}_n$, then $\alpha \in \mathcal{IO}_n$ and has the form as in Lemma 14.4.4(iii). In the case $k \geq m$ we have

$$\alpha = \varepsilon_{m+1,m} \cdots \varepsilon_{k,k-1} \varepsilon_{(k)}.$$

The case $k \leq m$ is similar. This proves our statement for the semigroups $\mathcal{PO}_n$ and $\mathcal{PC}_n$ simultaneously. $\qquad \square$

Remark 14.4.6 In the semigroups $\mathcal{IO}_n$, $\mathcal{IC}_n$, and $\mathcal{IF}_n$ all idempotents form a commutative subsemigroup. In Examples 14.1.2 and 14.1.4 we saw that already $\mathcal{IO}_2$, $\mathcal{IC}_2$, and $\mathcal{IF}_2$ are not commutative. Hence none of the semigroups $\mathcal{IO}_n$, $\mathcal{IC}_n$, and $\mathcal{IF}_n$ is generated by idempotents for $n > 1$.

14.5 Addenda and Comments

14.5.1 The semigroup $\mathcal{O}_n$, disguised as the semigroup of all endomorphisms of a finite linearly ordered set, appears in the works [Ai2, Ai3, Ai4] of Aizenshtat. It was further studied by Howie in [Ho2] (some results of the latter paper overlap with [Ai2]). However, a really intensive study of order-preserving transformations started in the 1990s.

The semigroup $\mathcal{F}_n$ appears already in [Pi] in connection with the study of formal languages. In 1990–1992, Howie commented on the importance of the study of order-decreasing transformations in [Ho5]. A deeper study of $\mathcal{F}_n$ seems to start with the works [Um2, Um3] of Umar.

14.5.2 Instead of order-decreasing transformations one can of course study the dual notion of *order-increasing* transformation (that is, $\alpha(x) \geq x$ for all $x \in \mathbf{N}$). In this case instead of the semigroups $\mathcal{PF}_n$, $\mathcal{F}_n$, $\mathcal{IF}_n$, $\mathcal{PC}_n$, $\mathcal{C}_n$ and $\mathcal{IC}_n$ we obtain the corresponding semigroups $\mathcal{PF}_n^+$, $\mathcal{F}_n^+$, $\mathcal{IF}_n^+$, $\mathcal{PC}_n^+$, $\mathcal{C}_n^+$ and $\mathcal{IC}_n^+$. Obviously, the semigroups S and S^+ (where S is from the above list) are isomorphic.

14.5.3 The semigroups defined in Sect. 14.1 admit many variations. The most obvious one is to substitute the natural linear order on $\mathbf{N}$ by some other linear order. The obtained subsemigroups will be obviously isomorphic ($\mathcal{S}_n$-conjugate) to the corresponding original semigroups.

Another natural generalization is to take some partial order on $\mathbf{N}$ instead of a linear order. Additionally, one may assume that the partial order contains a minimal element. This leads to the definition of a huge variety of finite transformation semigroups. Not much is known about them.

Instead of $\mathcal{PO}_n$, $\mathcal{O}_n$, and $\mathcal{IO}_n$ one could consider slightly bigger "superversions" of these subsemigroups, consisting of all transformations, which are either order-preserving or *order-reversing* (i.e., $x \leq y$ implies $\alpha(x) \geq \alpha(y)$ for all $x, y \in \operatorname{dom}(\alpha)$). This idea appears, in particular, in [CH].

Another variation is connected with those transformations which preserve the cyclic orientation rather than the natural order (see for example [Cat, CH, Fe2]). A (partial) transformation α is called *orientation-preserving* provided that we have $\mathrm{dom}(\alpha) = \{a_1 < a_2 < \cdots < a_m\}$ and there exists k such that

$$\alpha(a_k) \leq \alpha(a_{k+1}) \leq \cdots \leq \alpha(a_m) \leq \alpha(a_1) \leq \alpha(a_2) \leq \cdots \leq \alpha(a_{k-1}).$$

14.5.4 Catalan numbers, their properties, and various algebraic and combinatorial interpretations are studied in many papers and even monographs, see for example [Sta1, Chap. 6] and [Sta2].

14.5.5 In the coordinate plane consider the piecewise linear paths from $(0,0)$ to (n,n), which satisfy the following two conditions:

- From the point (a,b) it is allowed to go directly either to the point $(a+1,b)$ or to the point $(a,b+1)$ or to the point $(a+1,b+1)$.

- It is not allowed to go over the diagonal $y = x$.

Such a path is called a *Schröder path* of order n. The number $\mathfrak{r}_n$ of such paths equals

$$\mathfrak{r}_n = \frac{1}{n+1} \sum_{k=0}^{n} \binom{n+1}{n-k}\binom{n+k}{k}.$$

The sequence $\{\mathfrak{r}_n : n \geq 0\}$ satisfies the following recursion:

$$\mathfrak{r}_0 = 1; \quad \mathfrak{r}_1 = 2; \quad (n+2)\mathfrak{r}_{n+1} = 3(2n+1)\mathfrak{r}_n - (n-1)\mathfrak{r}_{n-1}, \ n > 0.$$

Theorem 14.5.1 ([LU2]) $|\mathcal{PC}_n| = \mathfrak{r}_n$.

Proof. Each Schröder path contains equal numbers of horizontal and vertical steps. Let l be some Schröder path of order n and k be the number of horizontal (vertical) steps in l. For every vertical step $(i,j) \to (i,j+1)$ we write down the number $a = j+1$. The k numbers, which we obtain, are pairwise different and we write them in the natural order as follows: $a_1 < a_2 < \cdots < a_k$. After that for every horizontal step $(i,j) \to (i+1,j)$ we write down the number $b = j+1$. Some of the k numbers, which we obtain, may coincide; however, we write them (with multiplicities) in the natural order $b_1 \leq b_2 \leq \cdots \leq b_k$. Then

$$\alpha_l = \begin{pmatrix} a_1 & a_2 & \cdots & a_k \\ b_1 & b_2 & \cdots & b_k \end{pmatrix}$$

is an element from $\mathcal{PC}_n$. One checks that the mapping $l \mapsto \alpha_l$ is a bijection from the set of all Schröder paths to $\mathcal{PC}_n$. The statement follows. $\square$

14.5.6 The results about cardinalities of semigroups described in the present chapter are taken from the following papers: Propositions 14.2.1 and 14.2.2 are taken from [Ho6]; Proposition 14.2.3 is taken from [Ho2]; Proposition 14.2.4 is taken from [Ga1]; Proposition 14.2.6 is taken from [BRR]; Theorem 14.2.8(i) is taken from [Hi2]; Theorem 14.2.9 and Corollary 14.2.11 are taken from [GH2].

14.5.7 The paper [LU3] contains the following more compact formula for $|\mathcal{E}(\mathcal{OP}_n)|$:

$$|\mathcal{E}(\mathcal{OP}_n)| = 5^{\frac{n-1}{2}} \left(\left(\frac{\sqrt{5}+1}{2} \right)^n - \left(\frac{\sqrt{5}-1}{2} \right)^n \right) + 1.$$

14.5.8 The results about idempotents described in Sect. 14.3 are taken from the following papers: Proposition 14.3.2 is taken from [Hi2]; Theorem 14.3.4 is taken from [LU2]; Proposition 14.3.5(i) is taken from [Um2]; Theorem 14.3.6 is taken from [Ho6]; Theorem 14.3.7 is taken from [LU3].

14.5.9 Theorem 14.4.1 for the semigroup $\mathcal{IO}_n$ was proved in [Fe3]; Theorem 14.4.3 for the semigroup $\mathcal{F}_n$ was proved in [Um2]; Theorem 14.4.5 for the semigroup $\mathcal{O}_n$ was proved in [Ai2, Ho2]; Theorem 14.4.5 for the semigroup $\mathcal{PO}_n$ was proved in [GH2].

14.5.10 Let S be a semigroup. The minimum cardinality of a generating system of S is usually called the *rank* of S and denoted by rank(S). If one restricts attention to generating systems of special kind, for example consisting of idempotent or nilpotent elements, one speaks of the *idempotent rank* idrank(S) of S and the *nilpotent rank* nilrank(S) of S etc.

Several papers are dedicated to the study of various ranks for some of the semigroups considered in this chapter. For example, Aizenshtat proved in [Ai2] that $\mathcal{O}_n$ has a unique irreducible generating system, consisting of idempotents, namely, the set of all idempotents of rank n and $(n-1)$. In particular, this implies that idrank($\mathcal{O}_n$) $= 2n - 1$. Later on (and independently) this was rediscovered by Howie in [Ho2]. Analogous result about irreducible generating systems for $\mathcal{C}_n$ is obtained by Higgins in [Hi2]. In particular, idrank($\mathcal{C}_n$) $= n$. In [GH2] it is shown that rank($\mathcal{O}_n$) $= n + 1$, rank($\mathcal{PO}_n$) $= 2n$, idrank($\mathcal{PO}_n$) $= 3n - 1$. In [Fe3] it is shown that rank($\mathcal{IO}_n$) $= n + 1$.

We would like to note that for a monoid S one usually leaves the identity element out of all generating systems, which explains the difference between the above formulations and the original formulations, which can be found in the cited papers.

14.5.11 If M is a generating system of the semigroup S, then for every $a \in S$ there exists a decomposition of a into a product of generating element, which has the minimal possible length. The maximum of all such minimal lengths over all $a \in S$ is called the *depth* of M.

For the semigroups $\mathcal{O}_n$ and $\mathcal{C}_n$ the depths of certain generating systems were studied by Higgins in [Hi3]. In particular, Higgins showed that for both these semigroups the depth of the generating system consisting of all idempotents equals $(n-1)$; and the depth of the generating system consisting of all idempotents of rank at least $(n-1)$ equals $\lfloor n^2/4 \rfloor$.

14.5.12 Set

$$\mathcal{IO}_n(n-1) = \{\alpha \in \mathcal{IO}_n : \operatorname{rank}(\alpha) = n-1\}.$$

Let $\mathbf{K_N}$ denote the full directed graph with the set $\mathbf{N}$ of vertices (i.e., for every $x, y \in \mathbf{N}$ the graph $\mathbf{K_N}$ contains a unique oriented edge from x to y). With every $\alpha \in \mathcal{IO}_n(n-1)$ we associate the arrow $a \to b$ of the graph $\mathbf{K_N}$, where $a = \mathbf{N}\backslash\operatorname{dom}(\alpha)$ and $b = \mathbf{N}\backslash\operatorname{im}(\alpha)$. Then every subset $M \subset \mathcal{IO}_n(n-1)$ is uniquely determined by some subgraph Γ_M of $\mathbf{K_N}$ (on the same set of vertices) and vice versa. Recall that a directed graph Γ is called *strongly connected* provided that for any two vertices a, b in Γ there exists an oriented path in Γ going from a to b.

Theorem 14.5.2 ([GM2]) *A generating system of $\mathcal{IO}_n$ is irreducible if and only if it has the form $M \cup \{\varepsilon\}$, where $M \subset \mathcal{IO}_n(n-1)$ is such that the graph Γ_M is a minimal strongly connected directed subgraph of $\mathbf{K_N}$.*

Theorem 14.5.3 ([GM2]) *Every irreducible generating system of $\mathcal{IO}_n$ contains at least $n+1$ elements. In particular, $\operatorname{rank}(\mathcal{IO}_n) = n+1$ and $\mathcal{IO}_n$ contains exactly $(n-1)!$ irreducible generating systems of cardinality $n+1$.*

14.5.13 Let S denote one of the semigroups $\mathcal{F}_n, \mathcal{PF}_n, \mathcal{IF}_n, \mathcal{C}_n, \mathcal{PC}_n, \mathcal{IC}_n$. Each of these semigroups contains a zero element (for $\mathcal{F}_n$ and $\mathcal{C}_n$ this zero element is the transformation 0_1, for all other semigroups, this zero element is the transformation $\mathbf{0}$). The following statement follows directly from definitions:

Proposition 14.5.4 *(i) An element $\alpha \in S$ is nilpotent if and only if*

(a) $\alpha(x) < x$ for all $x \neq 1$ (for $S = \mathcal{F}_n, \mathcal{C}_n$), or
(b) $\alpha(x) < x$ for all $x \in \operatorname{dom}(\alpha)$ (for $S = \mathcal{PF}_n, \mathcal{IF}_n, \mathcal{PC}_n, \mathcal{IC}_n$).

(ii) The set $\operatorname{Nil}(S)$ of all nilpotent elements of S forms an ideal.

Proposition 14.5.5 ([Um2, LU5]) *(i) $|\operatorname{Nil}(\mathcal{F}_n)| = (n-1)!$.*

(ii) $|\operatorname{Nil}(\mathcal{C}_n)| = \mathsf{C}_{n-1}$.

Proof. Taking Proposition 14.5.4 into account, the statement (i) is obvious. To prove the statement (ii) we observe that there is a natural bijection between $\mathcal{C}_{n-1}$ and $\operatorname{Nil}(\mathcal{C}_n)$, constructed in the following way:

$$\mathcal{C}_{n-1} \ni \begin{pmatrix} 1 & 2 & \cdots & n-1 \\ 1 & a_2 & \cdots & a_{n-1} \end{pmatrix} \leftrightarrow \begin{pmatrix} 1 & 2 & 3 & \cdots & n-1 & n \\ 1 & 1 & a_2 & \cdots & a_{n-2} & a_{n-1} \end{pmatrix} \in \operatorname{Nil}(\mathcal{C}_n).$$

The statement (ii) now follows directly from Theorem 14.2.8(i). $\square$

14.5.14 Let S denote one of the semigroups $\mathcal{PO}_n$ or $\mathcal{IO}_n$. The semigroup S contains the zero element **0**. An element $\alpha \in S$ is nilpotent if and only if $\alpha(x) \neq x$ for all $x \in \mathrm{dom}(\alpha)$. However, in this case nilpotent elements do not form an ideal, they do not even form a subsemigroup. For each of these semigroups, the subsemigroup, generated by all nilpotent elements, was studied by Garba in [Ga1, Ga2]. In particular, every element of each of these semigroups is a product of at most three nilpotent elements.

14.5.15 Nilpotent subsemigroups of $\mathcal{IO}_n$ are studied in detail in [GM2].

Theorem 14.5.6 ([GM2]) *Let $\prec$ be an arbitrary linear order on **N**. Then the set*

$$T(\prec) = \{\alpha \in \mathcal{IO}_n \: : \: \alpha(x) \prec x \text{ for all } x \in \mathrm{dom}(\alpha)\}$$

*is a maximal nilpotent subsemigroup of $\mathcal{IO}_n$ and every maximal nilpotent subsemigroup of $\mathcal{IO}_n$ coincides with $T(\prec)$ for some linear order $\prec$ on **N**. In particular, $\mathcal{IO}_n$ contains exactly $n!$ maximal nilpotent subsemigroups.*

Theorem 14.5.7 ([GM2]) *If $<$ is the natural order on **N**, then we have $|T(<)| \geq |T(\prec)|$ for any linear order $\prec$ on **N**. Moreover, $|T(<)| = \mathsf{c}_n$.*

In connection with the latter theorem, we also refer the reader to Exercise 8.6.14.

14.5.16 In the following theorem another order-related subsemigroup appears:

Theorem 14.5.8 ([Ho4]) *(i) The semigroup $\mathcal{SPO}_n = \mathcal{PO}_n \backslash \mathcal{O}_n$ of all strictly partial order-preserving transformations of **N** is not generated by idempotents.*

(ii) $\mathrm{rank}(\mathcal{SPO}_n) = 2n - 2$.

14.5.17 A description of all maximal and maximal inverse subsemigroups of $\mathcal{IO}_n$ is given in [GM2]. The semigroup $\mathcal{IO}_n$ contains $2^n - 1$ maximal subsemigroups and 2^{n-1} maximal inverse subsemigroups. All maximal subsemigroups in $\mathcal{O}_n$ are described in [X.Ya2]. It turns out that $\mathcal{O}_n$ contains $n^2 - 2n + 2$ maximal subsemigroups. The paper [LY] contains a description of maximal subsemigroups of $\mathcal{O}_n$ among the semigroups with some additional properties, for example regular subsemigroups (there are $2n - 2$ maximal semigroups among all regular subsemigroups) or subsemigroups generated by idempotents (there are $2n - 3$ maximal subsemigroups among all subsemigroups which are generated by idempotents).

14.5.18 Presentations are known only for some of the semigroups which were studied in this chapter. A presentation for $\mathcal{O}_n$ with respect to the irreducible system of generators, consisting of idempotents of rank at least

$(n-1)$, was given already by Aizenshtat in [Ai2]. A presentation for $\mathcal{PO}_n$ was found by Solomon in [Sol]. In [Fe3] Fernandes constructs a system of defining relations for $\mathcal{IO}_n$ with respect to the $n+1$-element generating system, which corresponds to the oriented cycle $n \to n-1 \to \ldots \to 2 \to 1 \to n$ in terms of 14.5.12.

14.5.19 Not much seems to be known about congruences on the semigroups studied in this chapter. Recall that a semigroup S is said to be *semisimple* provided that all congruences on S are Rees congruences. In [Ai3] Aizenshtat shows that the semigroup $\mathcal{O}_n$ is semisimple. In [Fe3] Fernandes proves the same result for $\mathcal{IO}_n$. Thus there are exactly n congruences on $\mathcal{O}_n$ and $n+1$ congruences on $\mathcal{IO}_n$.

14.5.20 Description of both one-sided and two-sided ideals for the semigroups $\mathcal{O}_n$, $\mathcal{PO}_n$ and $\mathcal{IO}_n$ is the same as the corresponding description for the semigroups $\mathcal{T}_n$, $\mathcal{PT}_n$ and $\mathcal{IS}_n$, respectively (the latter one is given in Theorems 4.2.1, 4.2.4 and 4.2.8). In particular, all Green's relations on the semigroups $\mathcal{O}_n$, $\mathcal{PO}_n$ and $\mathcal{IO}_n$ are just restrictions of the corresponding Green's relations on the semigroups $\mathcal{T}_n$, $\mathcal{PT}_n$ and $\mathcal{IS}_n$, respectively. For details, we refer the reader to [Ai4] for the case $\mathcal{O}_n$, to [LU3] for the case $\mathcal{PO}_n$ and to [Fe1, GM2] for the case $\mathcal{IO}_n$.

The structure of Green's relations in $\mathcal{F}_n$ is more interesting:

Theorem 14.5.9 ([Pi, Um2]) *Let* $\alpha, \beta \in \mathcal{F}_n$.

(i) $\alpha \mathcal{L} \beta$ *if and only if* $\alpha = \beta$.

(ii) $\alpha \mathcal{R} \beta$ *if and only if* $\mathrm{im}(\alpha) = \mathrm{im}(\beta)$ *and for every* $a \in \mathbf{N}$ *we have* $\min\{x : \alpha(x) = a\} = \min\{y : \beta(y) = a\}$.

Corollary 14.5.10 ([Um2]) *In the semigroup* $\mathcal{F}_n$ *we have the equalities* $\mathcal{H} = \mathcal{L}$ *and* $\mathcal{R} = \mathcal{D} = \mathcal{J}$.

14.5.21 Finally, we present some asymptotic results.

Theorem 14.5.11 ([Hi2]) $\frac{|\mathcal{E}(\mathcal{C}_n)|}{|\mathcal{C}_n|} \sim \frac{(n+1)\sqrt{\pi n}}{2^{n+1}}$.

Theorem 14.5.12 ([LU6])   (i) $|\mathcal{E}(\mathcal{PO}_n)| \sim \frac{1}{\sqrt{5}} \left(\frac{5+\sqrt{5}}{2}\right)^n$.

(ii) $|\mathcal{PO}_n| = \frac{1}{\pi} \int_0^\pi (2+\sqrt{2}\cos t)(3+2\sqrt{2}\cos t)^{n-1} dt \sim \frac{(\sqrt{2}+1)^{2n}}{2^{3/4}\sqrt{\pi n}}$.

Theorem 14.5.13 ([LU6])

$$|\mathcal{PC}_n| = \frac{1}{\pi(n+1)} \int_0^\pi (4+3\sqrt{2}\cos t)(3+2\sqrt{2}\cos t)^{n-1} dt \sim \frac{(\sqrt{2}+1)^{2n+1}}{2^{3/4}n\sqrt{\pi n}}.$$

14.6 Additional Exercises

14.6.1 Prove the formula from 14.5.7 using Theorem 14.3.7.

14.6.2 Prove Lemma 14.4.4.

14.6.3 Compute the number of elements of rank $(n-1)$ in the semigroups

(a) $\mathcal{O}_n$

(b) $\mathcal{C}_n$

(c) $\mathcal{PO}_n$

(d) $\mathcal{PC}_n$

14.6.4 ([LU5]) Prove that the number of those elements from $\mathcal{O}_n$, the set of fixed points for which coincides with $\{1, n\}$, equals C_{n-1}.

14.6.5 ([LU5]) Let

$$f(n, r, k) = |\{\alpha \in \mathcal{O}_n : \alpha(n) = k \text{ and } \alpha \text{ has exactly } r \text{ fixed points}\}|.$$

Show that $f(n, r, k) = \binom{n+k-2}{k-r} - \binom{n+k-2}{k-r-1}$.

14.6.6 ([Hi2]) Let

$$F(n, r) = |\{\alpha \in \mathcal{O}_n : \alpha \text{ has exactly } r \text{ fixed points}\}|.$$

Show that $F(n, r) = \frac{r}{n}\binom{2n}{n+r}$.

14.6.7 ([LU5]) Prove that

$$|\{\alpha \in \mathcal{O}_n : \alpha(n) = k\}| = \binom{n+k-2}{k-1}.$$

14.6.8 ([LU5]) Prove that

$$|\{\alpha \in \mathcal{O}_n : \operatorname{rank}(\alpha) = r\}| = \binom{n}{r}\binom{n-1}{r-1}.$$

14.6.9 ([LU5]) Prove that

$$|\{\alpha \in \mathcal{O}_n : \operatorname{rank}(\alpha) = r \text{ and } \alpha(n) = k\}| = \binom{k-1}{r-1}\binom{n-1}{r-1}.$$

14.6.10 ([LU2]) Prove that the number of those idempotents $\delta \in \mathcal{PC}_n$, in which the maximum element from $\operatorname{im}(\delta)$ is k, equals $2^{n-k-1}(3^{k-1}+1)$.

14.6.11 ([LU3]) Prove that

$$|\{\alpha \in \mathcal{PO}_n : |\text{dom}(\alpha)| = r \text{ and } \max(\text{im}(\alpha)) = k\}| = \binom{n}{r}\binom{k+r-2}{k-1}.$$

14.6.12 ([LU3]) Prove that

$$|\{\alpha \in \mathcal{PO}_n : |\text{dom}(\alpha)| = r\}| = \binom{n}{r}\binom{n+r-1}{n-1}.$$

14.6.13 ([LU3]) Prove that the numbers $a_n = |\mathcal{E}(\mathcal{PF}_n)|$ satisfy the following recursion: $a_{n+1} = 1 + 5(a_n - a_{n-1})$.

14.6.14 ([Um1]) Let $F(\alpha)$ denote the set of all fixed points of an element $\alpha \in \mathcal{PF}_n$. Prove that $F(\alpha\beta) = F(\beta\alpha) = F(\alpha) \cap F(\beta)$.

14.6.15 For $1 < k \leq n$ set $I_k = \{\alpha \in \mathcal{PF}_n : \text{rank}(\alpha) \leq k\}$ and let $S_k = I_k/I_{k-1}$ be the corresponding Rees quotient. Prove that every element of S_k is either an idempotent or a nilpotent.

14.6.16 ([LU4]) Let

$$J(n, r, k) = |\{\alpha \in \mathcal{C}_n : \text{rank}(\alpha) = r \text{ and } \alpha(n) = k\}|.$$

Prove that

(a) $J(n, k, k) = \binom{n-1}{k-1}$.

(b) $J(n, r, k) = \frac{n-k+1}{n-r+1}\binom{n-1}{r-1}\binom{k-2}{r-2}$.

14.6.17 ([LU4]) Prove that

(a) $|\{\alpha \in \mathcal{C}_n : \text{rank}(\alpha) = r\}| = \frac{1}{n-r+1}\binom{n-1}{r-1}\binom{n}{r}$.

(b) $|\{\alpha \in \mathcal{C}_n : \alpha(n) = k\}| = \frac{n-k+1}{n}\binom{n+k-2}{n-1}$.

14.6.18 ([Hi2]) Let $N(n, k)$ denote the number of those elements from $\mathcal{C}_n$ which have exactly k fixed points. Prove that

(a) $N(n, k) = \frac{k}{2n-k}\binom{2n-k}{n}$.

(b) $N(n + 1, k) = N(n, k - 1) + 2N(n, k) + N(n, k + 1)$.

14.6.19 ([HMR]) Show that there exist two elements $\alpha, \beta \in \mathcal{T}_n$ such that $\mathcal{T}_n = \langle \mathcal{O}_n \cup \{\alpha, \beta\}\rangle$.

14.6.20 Prove that the semigroups $\mathcal{O}_n$, $\mathcal{PO}_n$ and $\mathcal{IO}_n$ are regular.

14.6.21 Prove that for $n > 2$ the semigroups $\mathcal{F}_n$, $\mathcal{PF}_n$, $\mathcal{IF}_n$, $\mathcal{C}_n$, $\mathcal{PC}_n$, and $\mathcal{IC}_n$ are not regular.

14.6.22 ([Ai2, GM2]) Prove that for $n > 1$ we have

$$|\mathrm{Aut}(\mathcal{O}_n)| = |\mathrm{Aut}(\mathcal{IO}_n)| = |\mathrm{Aut}(\mathcal{PO}_n)| = 2.$$

14.6.23 Prove Theorem 14.5.11.

14.6.24 Let $t(n,r) = |\{\alpha \in \mathcal{F}_n : \mathrm{rank}(\alpha) = r\}|$. Prove that

$$t(n,r) = r \cdot t(n-1,r) + (n-r+1)t(n-1,r-1).$$

14.6.25 ([Um1]) Prove that

$$|\{\alpha \in \mathcal{F}_n : \mathrm{rank}(\alpha) = r\}| = \sum_{k=0}^{r-1}(-1)^k \binom{n+1}{k}(r-k)^n.$$

14.6.26 ([Um1]) Recall that *Stirling numbers of the first kind* $\mathbf{s}(n,k)$ are defined as coefficients

$$x(x-1)(x-2)\cdots(x-n+1) = \sum_{k=1}^{n}\mathbf{s}(n,k)x^k.$$

Prove that

$$|\{\alpha \in \mathcal{F}_n : \alpha\text{ has exactly }k\text{ fixed points}\}| = (-1)^{n-k}\mathbf{s}(n,k).$$

14.6.27 ([LU3]) Let $\alpha \in \mathcal{PO}_n$ and $\mathrm{rank}(\alpha) = k$. Prove that in the semigroup $\mathcal{PO}_n$ we have:

(a) $|\mathcal{L}(\alpha)| = \binom{n}{k}$.

(b) $|\mathcal{R}(\alpha)| = \sum_{i=k}^{n}\binom{n}{i}\binom{i-1}{k-1}$.

(c) $|\mathcal{R}(\alpha)| = \sum_{i=1}^{n-k+1}\binom{n-i}{k-i}2^{n-k-i-1}$.

14.6.28 ([GH2]) Prove that the $\mathcal{D}$-class

$$D_{n-1} = \{\alpha \in \mathcal{O}_n : \mathrm{rank}(\alpha) = n-1\}$$

of the semigroup $\mathcal{O}_n$ contains exactly $(n-1)$ different $\mathcal{L}$-classes and exactly n different $\mathcal{R}$-classes.

14.6.29 Prove Theorem 14.5.9.

Answers and Hints
to Exercises

1.5.3 (a): 7, (b): 19, (c): 6, (d): 16, (e): 45.

1.5.4 8!.

1.5.6 (a): $(n-k)^n$ for $\mathcal{T}_n$ and $(n+1-k)^n$ for $\mathcal{PT}_n$,

(b): $\sum_{i=0}^{k}(-1)^i\binom{k}{i}(n-i)^n$ for $\mathcal{T}_n$ and $\sum_{i=0}^{k}(-1)^i\binom{k}{i}(n+1-i)^n$ for $\mathcal{PT}_n$,

(c): $\sum_{i=0}^{k}(-1)^i\binom{k}{i}(k-i)^n$ for $\mathcal{T}_n$ and $\sum_{i=0}^{k}(-1)^i\binom{k}{i}(k+1-i)^n$ for $\mathcal{PT}_n$.

Hint: For (b) and (c) use the inclusion–exclusion formula.

1.5.7 Hint: Use Cayley's theorem on the number of labeled trees.

1.5.8 Hint: See [Hi1, 6.1.1(b)]

1.5.9 (a): $n^n - (n-1)^n$ for at least one fixed element and $n \cdot (n-1)^{n-1}$ for exactly one fixed element,

(b): $(n+1)^n - n^n$ for at least one fixed element and n^n for exactly one fixed element.

2.10.3 Hint: To each $\alpha \in \mathcal{PT}_n$ associate $\overline{\alpha} \in \mathcal{T}_{n+1}$ as follows:

$$\overline{\alpha}(i) = \begin{cases} \alpha(i), & i \in \mathrm{dom}(\alpha); \\ n+1, & \text{otherwise.} \end{cases}$$

2.10.9 Hint: Use Corollary 2.7.4.

2.10.10 Hint: Use Corollary 2.7.4.

2.10.18 Hint: Rewrite the signless Lah number $L'(n, n-k)$ in the form $L'(n, n-k) = \binom{n}{k}\binom{n-1}{k}k!$ and use Theorem 2.5.1 and Corollary 2.8.6.

2.10.20 Hint: Use Theorem 2.5.1.

2.10.23 Hint: If $\alpha \neq \mathbf{0}$, the equation $\alpha \cdot x = \mathbf{0}$ may have at most n^n solutions, whereas the equation $x \cdot \alpha = \mathbf{0}$ may have at most $(n+1)^{n-1}$ solutions. But $n^n > (n+1)^{n-1}$.

3.3.2 (a) Hint: Use Cayley's theorem on the number of labeled trees.

3.3.3 (a) Hint: $\mathcal{S}_n$ is not commutative. (c) Hint: Show that one can write $(i, i+1)$, $i < n$, as a product of $(1,2)$ and powers of $(1, 2, \ldots, n)$.

4.4.10 Hint: For each $a \in S$ the condition of the exercise gives elements $e_a^{(l)}$ and $e_a^{(r)}$ such that $e_a^{(l)} a = a e_a^{(r)} = a$. Show first that $e_a^{(l)} b = b$ and $b e_a^{(r)} = b$ for all a, b and then that $e_a^{(l)} = e_a^{(r)} = e$ is the identity element of S.

4.8.1 Hint: To prove the first equality count how many different partitions of $\mathbf{N} \cup \{n+1\}$ generate the same partition of $\mathbf{N}$. To prove the second equality count the number of those partitions of $\mathbf{N} \cup \{n+1\}$ for which the block containing the element $n+1$ has cardinality $k+1$.

4.8.3 Hint: The number of anti-chains in $\mathcal{B}(\mathbf{N})$ is smaller than the cardinality of $\mathcal{B}(\mathbf{N})$. On the other hand, each collection of elements from $\mathcal{B}(\mathbf{N})$ having the same cardinality is an anti-chain.

4.8.4 (a): 5; (b): 19; (c): 167.

4.8.7 For $n > 4$ the function $f(x) = x^n / n^x$ satisfies the condition $f(2) > 1$ and $f(n) = 1$ and has on $[2, n]$ the unique local extremal point, namely, some local maximum.

4.8.8 (a): 1, 25, 200, 600, 600, 120; (b): 5, 300, 1, 500, 1, 200, 120; (c): 1, 155, 1, 800, 3, 900, 1, 800, 120.

4.8.9 Hint: If S is inverse, $e, f \in S$ are idempotents and $eS = fS$, then $fe = e$, $ef = f$ and thus $e = f$. If every principal one-sided ideal is generated by a unique idempotent and b, c are inverse to a, then for the idempotents ab, ac, ba and ca we have $abS = aS = acS$, $Sba = Sa = Sca$, which implies $ab = ac$, $ba = ca$ and $b = bac = c$.

4.8.10 $\mathcal{I}_n$ and $\mathcal{I}_{n-1}$.

4.8.11 $\mathcal{I}_n$ and $\mathcal{I}_{n-1}$.

4.8.12 $\mathcal{I}_n$.

4.8.13 (a): k^{n-k}; (b): $(k+1)^{n-k}$.

4.8.14 (a): $n_1 n_2 \cdots n_k$; (b): $n_1 n_2 \cdots n_k$.

4.8.18 (a): $(n+1)^{n-\mathrm{rank}(\alpha)}$; (b): $(|\overline{\mathrm{dom}}(\alpha)| + 1)^n$.

4.8.19 Hint: Each such ideal contains all idempotents of rank one. If $\epsilon \alpha = 0$ ($\alpha \epsilon = 0$) for every idempotent ϵ of rank one, then $\alpha = 0$.

5.1.5 Hint: Use the proof of Theorem 4.7.4.

5.5.1 Hint: Each maximal subgroup is an $\mathcal{H}$-class, but each $\mathcal{H}$-class of both $\mathcal{T}_n$ and $\mathcal{IS}_n$ is at the same time an $\mathcal{H}$-class of $\mathcal{PT}_n$.

5.5.2 (d) Hint: If b is an inverse to a and $ab = ba$, then $e = ab$ is an idempotent and a is invertible in eSe.

5.5.8 (a) Hint: Every generating system of $\langle a \rangle$ must contain a. (b) Hint: If $p_1, p_2, \ldots, p_k$ are pairwise different primes and $m = p_1 p_2 \cdots p_k$, then $\{m/p_1, m/p_2, \ldots, m/p_k\}$ is an irreducible generating system of $\mathbb{Z}_m$.

5.5.9 (a) Hint: An element of type (k, m) in $\mathcal{T}_n$ is uniquely determined by an ordered partition $\mathbf{N} = N_1 \cup N_2 \cup \cdots \cup N_{k+1}$, a collection of mappings $N_i \to N_{i+1}$, $i = 1, \ldots, k$, and a permutation of order m on N_{k+1}. (b) Hint: Additionally to (a) we have a (possibly empty) block $N_0 = \overline{\mathrm{dom}}(\alpha)$. (c) An element of type (k, m) in $\mathcal{IS}_n$ is uniquely determined by an ordered partition

$\mathbf{N} = N_1 \cup N_2 \cup \cdots \cup N_{k+1}$, a collection of injective mappings $N_i \to N_{i+1}$, $i = 1, \ldots, k$, and a permutation of order m on N_{k+1}.

5.5.11 Hint: $S = (\mathbb{Z}_3, +)$, $T = \{0\}$.

5.5.12 $n \cdot 2^{n-1} + \frac{n(n-1)}{2} + 3$.

5.5.13 (b) Hint: No. Consider the semigroup $(\mathbb{N}, \cdot)$ and its subsemigroups $p\mathbb{N}$ and $pq\mathbb{N}$, where p and q are different primes.

5.5.14 Hint: Consider an $\mathcal{R}$-class and an $\mathcal{L}$-class of a rectangular band.

6.6.3 These are all decompositions into left resp. right cosets with respect to some subgroup.

6.6.4 No.

6.6.8 Hint: We have to count the number of collections $k_1 < k_2 < \cdots < l_2 < l_1$ from Theorems 6.5.3 and 6.5.4. Use the fact that the number of ways to write n as a sum $n = n_1 + \cdots + n_t$ of positive integers (here t can vary), equals 2^{n-1}. For $\mathcal{IS}_n$ all necessary collections can be obtained in the following way: write $n + 1$ as the sum $n + 1 = n_1 + \cdots + n_t$, $t > 1$, if t is odd, delete the last summand, and then consider the collection $n_1 - 1$, $n_1 + n_2 - 1$, $n_1 + n_2 + n_3 - 1, \ldots$. For $\mathcal{PT}_n$ all necessary collections can be obtained in the following way: write $n + 1$ as the sum $n + 1 = n_1 + \cdots + n_t$, $t > 1$, and consider the collection $n_1 - 1$, $n_1 + n_2 - 1$, $n_1 + n_2 + n_3 - 1, \ldots$. If t is odd we can either delete the last summand or double the middle one. If t is even, we can either take this collection, or delete the last summand and double the middle one of those which are left. One has also to take into account that Δ coincides with its own transpose.

7.4.11 Hint: Follow the proof of Theorems 7.4.1 and 7.4.7.

7.5.7 Hint: Follow the proof of Corollaries 7.5.1 and 7.5.6.

7.7.2 $\mathbf{0}$ and 0_a, $a \in \mathbf{N}$.

7.7.4 14.

7.7.5 $\{\varepsilon, \mathbf{0}\}$ for $\mathcal{IS}_n$ and $\mathcal{PT}_n$ and $\{\varepsilon\}$ for $\mathcal{T}_n$.

7.7.6 Hint: Determine first all automorphism which induce the identity mapping on the set of all idempotents of $\mathcal{I}_1$.

7.7.7 Hint: Go through the lists given by Theorems 7.4.1, 7.4.7, and 7.4.10.

7.7.9 Hint: Each finite semigroup has an idempotent. At the same time mapping the whole semigroup to an idempotent is always an endomorphism.

7.7.11 Each endomorphism of $(\mathbf{N}, +)$ has the form $x \mapsto k \cdot x$ for some fixed $k \in \mathbf{N}$.

8.2.9 Hint: Show that N_{τ_m} can be mapped to N_{τ_n} using an inner automorphism of $\mathcal{PT}_n$. Analogously for $\mathcal{IS}_n$.

8.6.1 Hint: Consider the cyclic semigroup of type $(k, 1)$.

8.6.4 (b): For example, S is a nontrivial group, T consists of one element and φ is the unique mapping from S to T.

8.6.5 Hint: Take the product $a_1 \cdots a_{lk}$ and consider the factors $a_1 \cdots a_k$, $a_{k+1} \cdots a_{2k}, \ldots, a_{(l-1)k+1} \cdots a_{kl}$ of this product. Show that they all belong to X.

8.6.6 Hint: Consider all possible linear combinations of the elements of T and show that they form a nilpotent subalgebra of $\mathrm{Mat}_n(\mathbb{C})$.

8.6.7 Hint: Prove that each maximal subsemigroup of $\mathrm{Mat}_n(\mathbb{C})$ is a subalgebra and use flags of subspaces in $\mathbb{C}^n$ instead of partial orders.

8.6.9 Hint: Consider the subsemigroup $T = \mathcal{I}_n \cup \{[1,2,3][4] \ldots [n]\}$ of $\mathcal{IS}_n$. The maximal nilpotent subsemigroups of T, corresponding to the natural linear order and the linear order $3 < 2 < 1 < 4 < \cdots < n$, have different cardinalities.

8.6.10 $1 + \sum_{k=2}^{n}(m_1 + m_2 + \cdots + m_{i-1}) \cdot m_i$, where $m_i = |M_i|$ for all i.

8.6.11 Hint: Use Theorem 8.4.12.

9.6.2 $\{(a^t, a^{t+lm}) : t \geq k, l \geq 1\} \cup \{(a^n, a^n) : n \geq 1\}$.

9.6.9 Write α as a product $\alpha = \mu\nu$, where μ is a permutation and $\nu = \begin{pmatrix} \{i_1, \ldots, i_m\} & B_1 & \cdots & B_k \\ i_m & b_1 & \cdots & b_k \end{pmatrix}$ is an idempotent in $\mathcal{T}_n$.

10.1.1 Hint: For every $a \in M$ the set $\{\varphi(s)(a) : s \in S\}$ is invariant.

10.1.2 For example, the natural action of the subsemigroup S of $\mathcal{IS}_n$, generated by the element $[1, 2, \ldots, n]$, on $\mathbf{N}$.

10.7.1 This action is always faithful but never transitive. It is quasi-transitive unless $S = \mathcal{S}_n$, or $S = \mathcal{T}_n$.

10.7.2 Yes.

10.7.3 The action is faithful. It is neither transitive nor quasi-transitive unless $S = \mathcal{S}_1 = \mathcal{T}_1$.

10.7.4 The action is faithful. It is not transitive unless $S = \mathcal{S}_1 = \mathcal{T}_1$. It is not quasi-transitive unless $S = \mathcal{S}_1 = \mathcal{T}_1$, or $S = \mathcal{IS}_1 = \mathcal{PT}_1$.

10.7.5 The action is faithful and quasi-transitive but never transitive.

10.7.7 Only the trivial action.

10.7.12 Only the trivial action and the actions trivially induced from $\mathcal{S}_n$ (i.e., where only the action of invertible elements is different from $\mathbf{0}$). Hint: If $n > 1$ and $\varphi : \mathcal{T}_n \to \mathcal{IS}_m$ is a homomorphism, different from the ones listed above, then by Theorem 6.3.10 the homomorphism φ is injective on idempotents of rank $(n-1)$. However, such idempotents of $\mathcal{T}_n$ do not commute in general.

10.7.13 Only the trivial action and the actions trivially induced from $\mathcal{S}_n$ (i.e., where only the action of invertible elements is different from $\mathbf{0}$). Hint: See the hint to 10.7.12.

10.7.14 Hint: It is enough to consider the actions of the form $\xi_{e,H}$ and show that for any $\pi \in \mathcal{S}_n$ and $\alpha \in \mathcal{IS}_n$ such that $\pi|_{\mathrm{dom}(\alpha)} = \alpha$ we have $\alpha \cdot x = \pi \cdot x$.

10.7.15 (b) Hint: Consider the action $a \cdot x = ax$.

11.1.5 Hint: for $S = \mathcal{S}_n$ and $S = \mathcal{T}_n$ the subspace $\{(c_1, \ldots, c_n) \in \mathbb{C}^n : c_1 + \cdots + c_n = 0\}$ is a submodule of the natural module.

11.5.1 Hint: Use 11.4.4(ii).

11.7.3 If $A \subset \mathbf{N}$, then the simple module corresponding to A is $\mathbb{C}_A = \mathbb{C}$ with the action $B \cdot c = c$, $c \in \mathbb{C}$, $B \subset \mathbf{N}$, if $A \subset B$ and $B \cdot \mathbb{C} = 0$ otherwise. Every module is a direct sum of simple modules.

11.7.4 n.

11.7.5 For example, the module $V(M)$, where M is the trivial $\mathcal{H}(0_1)$-module.

11.7.9 (b) Hint: A finite semigroup is nilpotent if and only if it has a zero element and this zero element is the only idempotent.

11.7.11 Hint: All elements $s \in S \backslash G$ annihilate $M \otimes_{\mathbb{C}} V$.

11.7.17 Hint: Consider for example the semigroup $S = \langle x : x^2 = x^3 \rangle$.

11.7.18 A rectangular band has two simple modules: the trivial module, and the one-dimensional module with the zero multiplication. A rectangular band is a monoid if and only if it has only one element. Hence in the case, when a rectangular band has exactly one element, and we consider it as a monoid, there is only one simple module: the trivial one.

11.7.19 (b) Hint: Let X be a submodule of $V(M)$, which is not contained in $N(M)$. Show first that there exists $v \in X$ such that $e \cdot v \neq 0$. Show then that $e \cdot v \in X$ is a nonzero vector in the $\mathcal{H}(e)$-submodule M of $V(M)$. Finally, use the construction of $V(M)$ to show that such X contains $V(M)$.

11.7.20 (a) Hint: Let X be a simple S-module. Show first that there exists $e \in \mathcal{E}(S)$ such that $e \cdot X \neq 0$, while $f \cdot X = 0$ for all idempotents $f \in SeS$ such that $f \notin \mathcal{D}(e)$. Let $M = e \cdot X$ and M' be a simple submodule of M. Similarly to Lemma 11.4.3 show that there is a nonzero homomorphism from $V(M')$ to X. Finally, use Exercise 11.7.19 to conclude that $X \cong \overline{V}(M')$.

12.3.7 Hint: Show that any two $\mathcal{H}$-cross-sections can be transferred into each other by conjugation.

12.8.6 There are theree cross-sections of the form $\mathcal{IO}_3^{\prec}$, and an additional cross-section from Exercise 12.8.5.

12.8.8 Hint: Use the Miller-Clifford Lemma ([Hi1, Theorem 1.2.5]), which states that for any $a, b \in S$ we have $ab \in \mathcal{R}(a) \cap \mathcal{L}(b)$ if and only if $\mathcal{R}(b) \cap \mathcal{L}(a)$ contains an idempotent.

12.8.9 The $\mathcal{R}$-classes. Hint: Use Exercise 12.8.8.

13.6.3 Hint: Show that $\varphi : (S, \circ_a) \to (S, \circ_b)$ is an isomorphism.

13.6.4 Hint: Use Exercise 13.6.3.

13.6.5 Hint: Count the number of idempotents.

13.6.6 Hint: Count the number of those elements which cannot be written as a product of other elements.

13.6.10 Hint: Use Exercise 13.6.9.

13.6.12 Answer: $\displaystyle\sum_{p=0}^{n}\sum_{m=1}^{p}\binom{n-k}{m}\binom{k}{p-m}\binom{n}{p}p!.$

13.6.13 Answer: n^n if $k=1$, $n^n - \displaystyle\sum_{m=1}^{k}\binom{n}{m}\mathsf{S}(k,m)\sum_{j=1}^{m}\mathsf{S}(n-k,j)\binom{m}{j}j!$ if $k>1$.

13.6.15 Hint: Use the fact that $(\cdot)^{-1}:\mathfrak{B}\to\mathfrak{B}$ is an involution.

14.2.7 Reference: See for example [Grm, 10.5].

14.2.10 Hint: Use induction on n.

14.6.1 Hint: From the recursive relation $\mathbf{f}_{2k} = 3\mathbf{f}_{2(k-1)} - \mathbf{f}_{2(k-2)}$ deduce $\mathbf{f}_{2k} = \frac{1}{\sqrt{5}}\left(((3+\sqrt{5})/2)^k + ((3-\sqrt{5})/2)^k\right)$, insert this formula into the sum from Theorem 14.3.7 and use the binomial formula.

14.6.3 (a) $n(n-1)$. (b) $n(n-1)/2$. (c) $n(2n-1)$. (d) $n(n+1)/2$.

14.6.4 The graph of such transformation has two connected components with m and $n-m$ vertices, respectively. Such components are in bijection with nilpotent elements from $\mathcal{C}_n$ and $\mathcal{C}_{n-m}$, respectively. Using Proposition 14.5.5(ii), one obtains a recursive relation for the number of elements in question, and it remains to use Exercise 14.2.7.

14.6.6 Hint: Use Exercise 14.6.5.

14.6.7 Hint: Use Exercise 14.6.5.

14.6.9 Hint: The first factor gives the number of ways to choose $\mathrm{im}(\alpha)$, the second factor gives the number of partitions of $\mathbf{N}$ into r intervals.

14.6.15 If $\alpha\in\mathcal{PF}_n$ is not an idempotent (that is, not all $x\in\mathrm{im}(\alpha)$ are fixed points), then $\mathrm{rank}(\alpha^2) < \mathrm{rank}(\alpha)$.

14.6.16 (a) **Hint:** Choose maximal elements in $\alpha^{-1}(1),\dots,\alpha^{-1}(k-1)$.

14.6.23 Hint: Use Theorem 14.2.8(i), Proposition 14.3.1 and Stirling's formula for $n!$.

14.6.25 Hint: Use Exercise 14.6.24.

14.6.27 Hint: (b) $\mathcal{R}(\alpha) = \{\beta : \mathrm{im}(\beta)=\mathrm{im}(\alpha)\}$. Let $\mathrm{im}(\alpha)=\{a_1,\dots,a_k\}$, $a_1<\dots<a_k$. To determine all β we first choose the cardinality i such that $|\mathrm{dom}(\beta)|=i$, then $\mathrm{dom}(\beta)$, and, finally, the minimum elements from the inverse images $\beta^{-1}(a_1),\dots,\beta^{-1}(a_k)$. (c) To determine all β we first consequently choose the minimum elements in the inverse images $\beta^{-1}(a_1),\dots,$ $\beta^{-1}(a_k)$. Then every element $x\in\{i+1,\dots,n\}$, which is not yet chosen, is either included to $\mathrm{dom}(\beta)$ (in which case the value of β on it is uniquely determined) or not.

14.6.29 Hint: To prove (i) use Proposition 4.4.1. To prove (ii) use Proposition 4.4.2.

Bibliography

[APR]   R. Adin, A. Postnikov, Y. Roichman, Combinatorial Gelfand models. J. Algebra **320** (2008), no. 3, 1311–1325.

[Ai1]   A. Aizenshtat, Defining relations of finite symmetric semigroups. (Russian) Mat. Sb. N.S. **45(87)** (1958), 261–280.

[Ai2]   A. Aizenshtat, The defining relations of the endomorphism semigroup of a finite linearly ordered set. (Russian) Sibirsk. Mat. Zh. **3** (1962), 161–169.

[Ai3]   A. Aizenshtat, On the semi-simplicity of semigroups of endomorphisms of ordered sets. (Russian) Dokl. Akad. Nauk SSSR **142** (1962), 9–11.

[Ai4]   A. Aizenshtat, On homomorphisms of semigroups of endomorphisms of ordered sets. (Russian) Leningrad. Gos. Ped. Inst. Uchen. Zap. **238** (1962), 38–48.

[AAH]   G. Ayık, H. Ayık, J. Howie, On factorizations and generators in transformation semigroups. Semigroup Forum **70** (2005), no. 2, 225–237.

[BRR]   D. Borwein, S. Rankin, L. Renner, Enumeration of injective partial transformations. Discrete Math. **73** (1989), no. 3, 291–296.

[Ca]   R. Carmichael, Introduction to the theory of groups of finite order. Dover Publications, New York, 1956.

[Cat]   P. Catarino, Monoids of orientation-preserving transformations of a finite chain and their presentations. Semigroups and applications (St. Andrews, 1997), 39–46, World Sci. Publ., River Edge, NJ, 1998.

[CH]   P. Catarino, P. Higgins, The monoid of orientation-preserving mappings on a chain. Semigroup Forum **58(2)** (1999), 190–206.

[Ch1]   K. Chase, Sandwich semigroups of binary relations. Discrete Math. **28(3)** (1979), 231–236.

[Ch2]   K. Chase, Maximal groups in sandwich semigroups of binary re-
        lations. Pacific J. Math. **100(1)** (1982), 43–59.

[CP1]   A. Clifford, G. Preston, The algebraic theory of semigroups. Vol.
        I. Mathematical Surveys, No. **7**. American Mathematical Society,
        Providence, RI, 1961.

[CP2]   A. Clifford, G. Preston, The algebraic theory of semigroups. Vol.
        II. Mathematical Surveys, No. **7**. American Mathematical Society,
        Providence, RI, 1967.

[CR]   D. Cowan, N. Reilly, Partial cross-sections of symmetric inverse
        semigroups. Int. J. Algebra Comput. **5(3)** (1995), 259–287.

[C-O]   K. Cvetko-Vah, D. Kokol Bukovšek, T. Košir, G. Kudryavtseva,
        Y. Lavrenyuk, A. Oliynyk, Semitransitive and transitive sub-
        semigroups of the inverse symmetric semigroups, Preprint
        arXiv:0709.0743.

[DF]   M. Delgado, V. Fernandes, Abelian kernels of monoids of order-
        preserving maps and of some of its extensions. Semigroup Forum
        68(3) (2004), 335–356.

[Do]   C. Doss, Certain equivalence relations in transformation semi-
        groups, Ph.D. Thesis, Univ. of Tennessee, 1955.

[Fa]   C. Faith, Ring theory. Grundlehren der Mathematischen Wissen-
        schaften, No. **191**. Springer-Verlag, Berlin, 1976.

[Fe1]   V. Fernandes, Semigroups of order preserving mappings on a finite
        chain: a new class of divisors. Semigroup Forum **54(2)** (1997),
        230–236.

[Fe2]   V. Fernandes, A division theorem for the pseudovariety gener-
        ated by semigroups of orientation preserving transformations on
        a finite chain. Comm. Alg. **29(1)** (2001), 451–456.

[Fe3]   V. Fernandes, The monoid of all injective order preserving partial
        transformations on a finite chain. Semigroup Forum **62(2)** (2001),
        178–204.

[FP]   D. FitzGerald, G. Preston, Divisibility of binary relations. Bull.
        Aust. Math. Soc. **5** (1971), 75–86.

[GK1]   O. Ganyushkin, T. Kormysheva, The chain decomposition of
        partial permutations and classes of conjugate elements of the
        semigroup $\mathcal{IS}_n$. (Ukrainian) Visnyk of Kyiv University, 1993,
        no. 2, 10–18.

[GK2]   O. Ganyushkin, T. Kormysheva, Isolated and nilpotent subsemi-
        groups of a finite inverse symmetric semigroup. Dopov./Dokl.
        Akad. Nauk Ukr. (9) (1993), 5–9.

[GK3]   O. Ganyushkin, T. Kormysheva, On nilpotent subsemigroups
        of a finite symmetric inverse semigroup. Mat. Zametki 56(3)
        (1994), 29–35, 157; translation in Math. Notes 56(3–4) (1994),
        1023–1029 (1995).

[GK4]   O. Ganyushkin, T. Kormysheva, The structure of nilpotent sub-
        semigroups of a finite inverse symmetric semigroup. Dopov. Nats.
        Akad. Nauk Ukr. 1995, (1) 8–10.

[GM1]   O. Ganyushkin, V. Mazorchuk, The structure of subsemigroups
        of factor powers of finite symmetric groups. Mat. Zametki 58(3)
        (1995), 341–354, 478; translation in Math. Notes 58(3–4) (1995),
        910–920 (1996).

[GM2]   O. Ganyushkin, V. Mazorchuk, On the structure of $\mathcal{IO}_n$. Semi-
        group Forum 66(3) (2003), 455–483.

[GM3]   O. Ganyushkin, V. Mazorchuk, $\mathcal{L}$- and $\mathcal{R}$-cross-sections in $\mathcal{IS}_n$.
        Comm. Alg. 31(9) (2003), 4507–4523.

[GM4]   O. Ganyushkin, V. Mazorchuk, Combinatorics of nilpotents
        in symmetric inverse semigroups. Ann. Combin. 8(2) (2004),
        161–175.

[GM5]   O. Ganyushkin, V. Mazorchuk, Combinatorics and distributions
        of partial injections. Australas. J. Combin. 34 (2006), 161–186.

[GM6]   O. Ganyushkin, V. Mazorchuk, On classification of maximal
        nilpotent subsemigroups. J. Algebra 320 (2008), no. 8, 3081–3103.

[GMS]   O. Ganyushkin, V. Mazorchuk, B. Steinberg, On the irreducible
        representations of a finite semigroup, Preprint arXiv:0712.2076.

[GP]   O. Ganyushkin, M. Pavlov, On the cardinalities of a class of nilpo-
        tent semigroups and their automorphism groups, In: Algebraic
        structures and their applications, Proceedings of the Ukrainian
        Mathematical Congress 2001, Kyiv, Institute of Mathematics of
        the National Academy of Sciences of Ukraine, 2002, 17–21.

[GTS]   O. Ganyushkin, O. Temnikov, G. Shafranova (Kudryavtseva),
        Groups of automorphisms for maximal nilpotent subsemigroups
        of the semigroup $\mathcal{IS}(M)$. (Ukrainian) Mat. Stud. 13(1) (2000),
        11–22.

[Ga1]   G. Garba, Nilpotents in semigroups of partial one-to-one order-preserving mappings. Semigroup Forum **48(1)** (1994), 37–49.

[Ga2]   G. Garba, Nilpotents in semigroups of partial order-preserving transformations. Proc. Edinburgh Math. Soc. (2) **37(3)** (1994), 361–377.

[GH1]   G. Gomes, J. Howie, Nilpotents in finite symmetric inverse semigroups. Proc. Edinburgh Math. Soc. (2) **30(3)** (1987), 383–395.

[GH2]   G. Gomes, J. Howie, On the ranks of certain semigroups of order-preserving transformations. Semigroup Forum **45(3)** (1992), 272–282.

[Gr]   J. Green, On the structure of semigroups. Ann. Math. **54** (1951), 163–172.

[Gri]   P. Grillet, Semigroups. An introduction to the structure theory. Monographs and Textbooks in Pure and Applied Mathematics, **193**. Marcel Dekker, New York, 1995.

[Grm]   R. Grimaldi, Discrete and combinatorial mathematics, an applied introduction, 4th ed. Addison Wesley Longman, 2000.

[Grd]   C. Grood, A Specht module analog for the rook monoid. Electron. J. Combin. **9(1)** (2002), Research Paper 2, 10 pp.

[HS]   B. Harris, L. Schoenfeld, The number of idempotent elements in symmetric semigroups. J. Combin. Theory **3** (1967), 122–135.

[HZ1]   E. Hewitt, H. Zuckerman, The l_1-algebra of a commutative semigroup. Trans. Amer. Math. Soc. **83** (1956), 70–97.

[HZ2]   E. Hewitt, H. Zuckerman, The irreducible representations of a semi-group related to the symmetric group. Illinois J. Math. **1** (1957), 188–213.

[Hic1]   J. Hickey, Semigroups under a sandwich operation. Proc. Edinburgh Math. Soc. (2) **26(3)** (1983), 371–382.

[Hic2]   J. Hickey, On variants of a semigroup. Bull. Aust. Math. Soc. **34(3)** (1986), 447–459.

[Hi1]   P. Higgins, Techniques of semigroup theory. Oxford Science Publications. The Clarendon Press, Oxford University Press, New York, 1992.

[Hi2]   P. Higgins, Combinatorial results for semigroups of order-preserving mappings. Math. Proc. Cambridge Philos. Soc. **113(2)** (1993), 281–296.

[Hi3]   P. Higgins, Idempotent depth in semigroups of order-preserving mappings. Proc. Roy. Soc. Edinburgh Sect. A **124(5)** (1994), 1045–1058.

[HMR]   P. Higgins, J. Mitchell, N. Ruškuc, Generating the full transformation semigroup using order preserving mappings. Glasgow Math. J. **45(3)** (2003), 557–566.

[Hoe]   O. Hölder, Bildung zusammengesetzter Gruppen. Math. Ann. **46** (1895), 321–422.

[Ho1]   J. Howie, The subsemigroup generated by the idempotents of a full transformation semigroup. J. London Math. Soc. **41** (1966), 707–716.

[Ho2]   J. Howie, Products of idempotents in certain semigroups of transformations. Proc. Edinburgh Math. Soc. (2) **17** (1970/71), 223–236.

[Ho3]   J. Howie, An introduction to semigroup theory. L.M.S. Monographs, No. **7**. Academic Press, London, 1976.

[Ho4]   J. Howie, Semigroups and combinatorics. Monash Conference on Semigroup Theory (Melbourne, 1990), 135–139. World Sci. Publ., River Edge, NJ, 1991.

[Ho5]   J. Howie, Combinatorial and arithmetical aspects of the theory of transformation semigroups. Seminario do Centro de Algebra, University of Lisbon (1992), 1–14.

[Ho6]   J. Howie, Combinatorial and probabilistic results in transformation semigroups. Words, languages and combinatorics, II (Kyoto, 1992), 200–206. World Sci. Publ., River Edge, NJ, 1994.

[Ho7]   J. Howie, Fundamentals of semigroup theory. London Mathematical Society Monographs. New Series, **12**. Oxford University Press, London, 1995.

[HM]   J. Howie, R. McFadden, Ideals are greater on the left. Semigroup Forum **40(2)** (1990), 247–248.

[II]   Na. Iwahori, No. Iwahori, On a set of generating relations of the full transformation semigroups. J. Combin. Theory Ser. A **16** (1974), 147–158.

[JM]   S. Janson, V. Mazorchuk, Some remarks on the combinatorics of $\mathcal{IS}_n$. Semigroup Forum **70(3)** (2005), 391–405.

[Ji]   C. Jianmiao, On sandwich semigroups in Boolean group algebras.
            J. Pure Appl. Alg. **169(2–3)** (2002), 249–265.

[KM]   M. Kargapolov, J. Merzljakov, Fundamentals of the theory of
            groups. Graduate Texts in Mathematics, **62**. Springer-Verlag,
            Berlin, 1979.

[KS]   M. Katsura, T. Saito, Maximal inverse subsemigroups of the full
            transformation semigroup. Semigroups with applications, 101–
            113. World Sci. Publ., River Edge, NJ, 1992.

[Ka]   L. Katz, Probability of indecomposability of a random mapping
            function. Ann. Math. Stat. **26** (1955), 512–517.

[KL]   T. Khan, M. Lawson, Variants of regular semigroups. Semigroup
            Forum **62(3)** (2001), 358–374.

[KRS]   D. Kleitman, B. Rothschild, J. Spencer, The number of semi-
            groups of order n. Proc. Amer. Math. Soc. **55(1)** (1976), 227–232.

[Ko]   G. Köthe, Über maximale nilpotente Unterringe und Nilringe.
            Math. Ann. **103(1)** (1930), 359–363.

[Ku1]   G. Kudryavtseva, The structure of automorphism groups of semi-
            group inflations. Alg. Discrete Math. **6(1)** (2007), 62–67.

[Ku2]   G. Kudryavtseva, On conjugacy in regular epigroups. Preprint
            arXiv:math/0605698.

[KuMa1]   G. Kudryavtseva, V. Mazorchuk, On conjugation in some trans-
            formation and Brauer-type semigroups. Publ. Math. Debrecen
            70(1–2) (2007), 19–43.

[KuMa2]   G. Kudryavtseva, V. Mazorchuk, On presentations of Brauer-type
            monoids. Cent. Eur. J. Math. **4(3)** (2006), 413–434.

[KuMa3]   G. Kudryavtseva, V. Mazorchuk, On the semigroup of square ma-
            trices, Alg. Colloq. **15(1)** (2008), 33–52.

[KuMa4]   G. Kudryavtseva, V. Mazorchuk, On three approaches to conju-
            gacy in semigroups, Preprint arXiv:0709.4341, to appear in Semi-
            group Forum.

[KuMa5]   G. Kudryavtseva, V. Mazorchuk, Schur-Weyl dualities for
            symmetric inverse semigroups. J. Pure Appl. Algebra **212** (2008),
            no. 8, 1987–1995.

[KuMa6]   G. Kudryavtseva, V. Mazorchuk, Combinatorial Gelfand mod-
            els for some semigroups and q-rook monoid algebras, Preprint
            arXiv:0710.1972, to appear in Proc. Edinburgh Math. Soc.

[KT]   G. Kudryavtseva, G. Tsyaputa, The automorphism group of the
          sandwich inverse symmetric semigroup. (Ukrainian) Bulletin of
          the University of Kiev, Series: Mechanics and Mathematics 2005,
          no. 13–14, 101–105.

[KMM]   G. Kudryavtseva, V. Maltcev, V. Mazorchuk, $\mathcal{L}$- and $\mathcal{R}$-cross-
          sections in the Brauer semigroup. Semigroup Forum **72(2)** (2006),
          223–248.

[La]   E. Landau, Über die Maximalordnung der Permutation gegebenen
          Grades. Arch. Math. Phys. Ser. 3. **5** (1903), 92–103.

[LU1]   A. Laradji, A. Umar, On the number of nilpotents in the partial
          symmetric semigroup. Comm. Alg. **32(8)** (2004), 3017–3023.

[LU2]   A. Laradji, A. Umar, Combinatorial results for semigroups of
          order-decreasing partial transformations. J. Integer Seq. **7(3)**
          (2004), Article 04.3.8, 14 pp. (electronic).

[LU3]   A. Laradji, A. Umar, Combinatorial results for semigroups of
          order-preserving partial transformations. J. Alg. **278(1)** (2004),
          342–359.

[LU4]   A. Laradji, A. Umar, On certain finite semigroups of order-
          decreasing transformations. I. Semigroup Forum **69(2)** (2004),
          184–200.

[LU5]   A. Laradji, A. Umar, Combinatorial results for semigroups of
          order-preserving full transformations. Semigroup Forum **72(1)**
          (2006), 51–62.

[LU6]   A. Laradji, A. Umar, Asymptotic results for semigroups of order-
          preserving partial transformations. Comm. Alg. **34(3)** (2006),
          1071–1075.

[Law]   M. Lawson, Inverse semigroups. The theory of partial symmetries.
          World Sci. Publ., River Edge, NJ, 1998.

[Le]   I. Levi, Congruences on normal transformation semigroups. Math.
          Japon. **52(2)** (2000), 247–261.

[Lev]   J. Levitzki, Über nilpotente Unterringe. Math. Ann. **105(1)**
          (1931), 620–627.

[Lib]   A. Liber, On symmetric generalized groups. Mat. Sb. N.S. **33(75)**,
          (1953), 531–544.

[LPS]   M. Liebeck, C. Praeger, J. Saxl, A classification of the maximal subgroups of the finite alternating and symmetric groups. J. Alg. **111(2)** (1987), 365–383.

[Li]   S. Lipscomb, Symmetric inverse semigroups. Mathematical Surveys and Monographs, **46**. American Mathematical Society, Providence, RI, 1996.

[LY]   C. Lu, X. Yang, Maximal properties of some subsemigroups in finite order-preserving transformation semigroups. Comm. Alg. **28(7)** (2000), 3125–3135.

[Ly]   E. Lyapin, Semigroups. (Russian) Gosudarstv. Izdat. Fiz.-Mat. Lit., Moscow, 1960.

[MS1]   K. Magill, Jr., S. Subbiah, Green's relations for regular elements of sandwich semigroups. I. General results. Proc. London Math. Soc. (3) **31(2)** (1975), 194–210.

[MS2]   K. Magill, Jr., S. Subbiah, Green's relations for regular elements of sandwich semigroups. II. Semigroups of continuous functions. J. Aust. Math. Soc. Ser. A **25(1)** (1978), 45–65.

[MMT]   K. Magill, Jr., P. Misra, U. Tewari, Structure spaces for sandwich semigroups. Pacific J. Math. **99(2)** (1982), 399–412.

[MR]   M. Malandro, D. Rockmore, Fast Fourier transforms for the rook monoid, Preprint arXiv:0709.4175.

[Ma1]   A. Mal'cev, Symmetric groupoids. Mat. Sb. N.S. **31(73)**, (1952), 136–151.

[Ma2]   A. Mal'cev, Nilpotent semigroups. Ivanov. Gos. Ped. Inst. Uchen. Zap. Fiz.-Mat. Nauki **4** (1953), 107–111.

[MT]   V. Mazorchuk, G. Tsyaputa, Isolated subsemigroups in the variants of $\mathcal{T}_n$. Acta Math. Univ. Com., Vol. LXXVII, **1** (2008), 63–84.

[Mi]   W. Miller, The maximum order of an element of a finite symmetric group. Amer. Math. Monthly **94(6)** (1987), 497–506.

[Mo1]   M. Mogilevskiĭ, Order relations on symmetric semigroups of transformations and on their homomorphic images. Semigroup Forum **19(4)** (1980), 283–305.

[Mo2]   M. Mogilevskiĭ, A remark on the orderings of the semigroup of total transformations. Ordered sets and lattices, No. **2**, p. 59. Izdat. Saratov Univ., Saratov, 1974.

[Moo] E. Moore, Concerning the abstract groups of order $k!$ and $\frac{1}{2}k!$ holohedrically isomorphic with the symmetric and alternating substitution groups on k letters, Proc. London Math. Soc. **28** (1897), 357–366.

[Mu1] W. Munn, On semigroup algebras. Proc. Cambridge Philos. Soc. **51** (1955), 1–15.

[Mu2] W. Munn, Matrix representations of semigroups. Proc. Cambridge Philos. Soc. **53** (1957), 5–12.

[Mu3] W. Munn, The characters of the symmetric inverse semigroup. Proc. Cambridge Philos. Soc. **53** (1957), 13–18.

[Mu4] W. Munn, Irreducible matrix representations of semigroups. Quart. J. Math. Oxford Ser. (2) **11** (1960), 295–309.

[Ne] P. Neumann, A lemma that is not Burnside's. Math. Sci. **4(2)** (1979), 133–141.

[Ni] J. Nichols, A class of maximal inverse subsemigroups of T_X. Semigroup Forum **13(2)** (1976/77), 187–188.

[Pek1] V. Pekhterev, Retracts of the semigroup T_n. (Ukrainian) Visnyk of Kyiv University, 2003, no. 10, 127–129.

[Pek2] V. Pekhterev, $\mathcal{H}$- and $\mathcal{R}$-cross-sections of the full finite semigroup T_n. Alg. Discrete Math. **2(3)** (2003), 82–88.

[Pek3] V. Pekhterev, $\mathcal{R}$-cross-sections of the semigroup T_X. (Ukrainian) Mat. Stud. **21(2)** (2004), 133–139.

[Pek4] V. Pekhterev, $\mathcal{H}$-, $\mathcal{R}$- and $\mathcal{L}$-cross-sections of the infinite symmetric inverse semigroup $\mathcal{IS}_X$. Alg. Discrete Math. **4(1)** (2005), 92–104.

[Pe] M. Petrich, Introduction to semigroups. Merrill Research and Lecture Series. Charles E. Merrill Publishing Co., Columbus, OH, 1973.

[Pi] J. Pin, Variétés de langages formels. (French) With a preface by M. P. Schützenberger. Études et Recherches en Informatique. Masson, Paris, 1984. 160 pp.

[Po1] I. Ponizovskiy, Transitive representations by transformations of semi-groups of a certain class. (Russian) Sibirsk. Mat. Zh. **5** (1964), 896–903.

[Po2]   I. Ponizovskiy, Representations of inverse semigroups by partial one-to-one transformations. (Russian) Izv. Akad. Nauk SSSR Ser. Mat. **28** (1964), 989–1002.

[Po3]   I. Ponizovskiy, On matrix representations of associative systems. (Russian) Mat. Sb. N.S. **38 (80)** (1956), 241–260.

[Po4]   I. Ponizovskiy, Some examples of semigroup algebras of finite representation type. (Russian) Zap. Nauchn. Sem. Leningrad. Otdel. Mat. Inst. Steklov. (LOMI) **160** (1987), Anal. Teor. Chisel i Teor. Funktsii. 8, 229–238, 302; translation in J. Sov. Math. **52(3)** (1990), 3170–3178.

[Pp]   L. Popova, Defining relations of certain semigroups of partial transformations of a finite set. (Russian) Leningrad. Gos. Ped. Inst. Uchen. Zap. **218** (1961), 191–212.

[Pu1]   M. Putcha, Complex representations of finite monoids. Proc. London Math. Soc. (3) **73(3)** (1996), 623–641.

[Pu2]   M. Putcha, Complex representations of finite monoids. II. Highest weight categories and quivers. J. Alg. **205(1)** (1998), 53–76.

[Pu3]   M. Putcha, Reciprocity in character theory of finite semigroups. J. Pure Appl. Alg. **163(3)** (2001), 339–351.

[Re1]   N. Reilly, Embedding inverse semigroups in bisimple inverse semigroups. Quart. J. Math. Oxford Ser. (2) **16** (1965), 183–187.

[Re2]   N. Reilly, Maximal inverse subsemigroups of $T_X^{(1)}$. Semigroup Forum **15(4)** (1977/78), 319–326.

[Ri]   C. Ringel, The representation type of the full transformation semigroup T_4. Semigroup Forum **61(3)** (2000), 429–434.

[Sa]   B. Sagan, The symmetric group. Representations, combinatorial algorithms, and symmetric functions. Second edition. Graduate Texts in Mathematics, **203**. Springer-Verlag, New York, 2001.

[SYT]   S. Satoh, K. Yama, M. Tokizawa, Semigroups of order 8. Semigroup Forum **49(1)** (1994), 7–29.

[Sc1]   B. Schein, Representations of generalized groups. (Russian) Izv. Vyssh. Uchebn. Zaved. Mat. **28(3)** (1962), 164–176.

[Sc2]   B. Schein, Representation of semigroups by binary relations. (Russian) Dokl. Akad. Nauk SSSR **142** (1962), 808–811.

[Sc3] B. Schein, Transitive representations of semigroups. (Russian) Usp. Mat. Nauk **18(3)** (1963), 215–222.

[Sc4] B. Schein, Lectures on transformation semigroups. (Russian) Special course. Izdat. Saratov Univ., Saratov, 1970.

[Sc5] B. Schein, A symmetric semigroup of transformations is covered by its inverse subsemigroups. Acta Math. Acad. Sci. Hung. **22** (1971/72), 163–171.

[ST1] B. Schein, B. Teclezghi, Endomorphisms of finite symmetric inverse semigroups. J. Alg. **198(1)** (1997), 300–310.

[ST2] B. Schein, B. Teclezghi, Endomorphisms of finite full transformation semigroups. Proc. Amer. Math. Soc. **126(9)** (1998), 2579–2587.

[ST3] B. Schein, B. Teclezghi, Endomorphisms of symmetric semigroups of functions on a finite set. Comm. Alg. **26(12)** (1998), 3921–3938.

[Sch] J. Schreier, Über Abbildungen einer abstrakten Menge auf ihre Teilmengen. Fundam. Math. **28** (1936), 261–264.

[Sw1] Š. Schwarz, The theory of characters of finite commutative semigroups. (Russian) Czech. Math. J. **4(79)** (1954), 219–247.

[Sw2] Š. Schwarz, Characters of commutative semigroups as class functions. (Russian) Czech. Math. J. **4(79)** (1954), 291–295.

[Sw3] Š. Schwarz, On a Galois connexion in the theory of characters of commutative semigroups. (Russian) Czech. Math. J. **4(79)** (1954), 296–313.

[Sh1] G. Shafranova, Maximal nilpotent subsemigroups of the semigroup $\mathrm{PAut}_q(V_n)$. Probl. Alg. **13** (1998), 69–83.

[Sh2] G. Shafranova, Nilpotent subsemigroups of transformation semigroups. (Ukrainian) Ph.D. Thesis, Kyiv University, Kyiv, Ukraine, 2000.

[She] L. Shevrin, On nilsemigroups. (Russian) Usp. Mat. Nauk **14(5)** (1959), 216–217.

[Sol] A. Solomon, Monoids of order-preserving transformations of a finite chain, Res. report **94-8**, School of Math. and Stat., University of Sydney, 1994.

[So] L. Solomon, Representations of the rook monoid. J. Alg. **256(2)** (2002), 309–342.

[Sta1]   R. Stanley, Enumerative combinatorics. Vol. 2. Cambridge Stud-
           ies in Advanced Mathematics, **62**. Cambridge University Press,
           Cambridge, 1999.

[Sta2]   R. Stanley, Catalan addendum, manuscript, available at:
           http://www-math.mit.edu/~rstan/ec/catadd.pdf

[Ste1]   B. Steinberg, Möbius functions and semigroup representation the-
           ory. J. Combin. Theory Ser. A **113(5)** (2006), 866–881.

[Ste2]   B. Steinberg, Möbius functions and semigroup representa-
           tion theory II: Character formulas and multiplicities, Preprint
           arXiv:math/0607564, to appear in Adv. Math.

[St1]   G. Stronska, Nilpotent subsemigroups of the semigroup of order-
           decreasing transformations, Mat. Stud. **22(2)** (2004), 184–197.

[St2]   G. Stronska, On the automorphisms for the nilpotent subsemi-
           group of the order-decreasing transformation semigroup. Proceed-
           ings of Gomel University, 2007.

[Su1]   A. Suschkewitsch, Untersuchung über verallgemeinerte Substitu-
           tionen. Atti del Congresso Internazionale dei Matematici Bologna,
           1 (1928), 147–157.

[Su2]   A. Suschkewitsch, Theory of generalized groups. (Russian)
           Kharkov-Kyiv, GNTI, 1937

[Sut1]   E. Sutov, Defining relations in finite semigroups of partial trans-
           formations. Dokl. Akad. Nauk SSSR **132** 1280–1282 (Russian);
           translated as Sov. Math. Dokl. **1** (1960), 784–786.

[Sut2]   E. Sutov, Homomorphisms of the semigroup of all partial transfor-
           mations. Izv. Vyssh. Uchebn. Zaved. Mat. **22(3)** (1961), 177–184.

[Sy1]   J. Symons, On a generalization of the transformation semigroup.
           J. Aust. Math. Soc. **19** (1975), 47–61.

[Sy2]   J. Symons, Some results concerning a transformation semigroup.
           J. Aust. Math. Soc. **19(4)** (1975), 413–425.

[Sz]   M. Szalay, On the maximal order in S_n and S_n^*. Acta Arith. **37**
           (1980), 321–331.

[Ts1]   G. Tsyaputa, Transformation semigroups with the deformed mul-
           tiplication. (Ukrainian) Bulletin of the University of Kiev, Series:
           Physics and Mathematics, 2003, no. 4, 82–88.

[Ts2]   G. Tsyaputa, Green's relations on the deformed transformation semigroups. Alg. Discrete Math. **3(1)** (2004), 121–131.

[Ts3]   G. Tsyaputa, Deformed multiplication in the semigroup $\mathcal{PT}_n$. (Ukrainian) Bulletin of the University of Kiev, Series: Mechanics and Mathematics, 2004, no. 11–12, 35–38.

[Ts4]   G. Tsyaputa, Isolated and nilpotent subsemigroups in the variants of $\mathcal{IS}_n$. Alg. Discrete Math. **5(1)** (2006), 89–97.

[Um1]   A. Umar, Semigroups of order-decreasing transformations, Ph.D. Thesis, University of St. Andrews, 1992.

[Um2]   A. Umar, On the semigroups of order-decreasing finite full transformations. Proc. Roy. Soc. Edinburgh Sect. A **120(1–2)** (1992), 129–142.

[Um3]   A. Umar, On the semigroups of partial one-to-one order-decreasing finite transformations. Proc. Roy. Soc. Edinburgh Sect. A **123(2)** (1993), 355–363.

[Vl1]   T. Voloshyna, Effective transitive representations of the finite symmetric inverse semigroup. (Ukrainian) Visn. Kyiv Univ. **1998(2)** (1998), 16–21.

[Vl2]   T. Voloshyna, Closed inverse subsemigroups of monogenic inverse semigroups. (Ukrainian) Visn. Kyiv Univ. **2000(3)**, 16–23.

[Vl3]   T. Voloshyna, Effective transitive representations of monogenic inverse semigroups. (Ukrainian) Mat. Stud. **16(1)** (2001), 25–36.

[Vo1]   N. Vorob'ev, On ideals of associative systems. Dokl. Akad. Nauk SSSR (N.S.) **83** (1952), 641–644.

[Vo2]   N. Vorob'ev, On symmetric associative systems. Leningrad. Gos. Ped. Inst. Uchen. Zap. **89** (1953), 161–166.

[H.Ya]   H. Yang, Two notes of maximal inverse subsemigroups of full transformation semigroups. J. Math. (Wuhan) **24(4)** (2004), 375–380.

[X.Ya1] X. Yang, A classification of maximal inverse subsemigroups of the finite symmetric inverse semigroups. Comm. Alg. **27(8)** (1999), 4089–4096.

[X.Ya2] X. Yang, A classification of maximal subsemigroups of finite order-preserving transformation semigroups. Comm. Alg. **28(3)** (2000), 1503–1513.

[YY]   H.Yang, X. Yang, $\mathcal{L}$ (or $\mathcal{R}$)-cross-sections of finite symmetric inverse semigroups, Preprint 2002.

List of Notation

$(\tilde{\Gamma}_1, \tilde{\Gamma}_2, \ldots, \tilde{\Gamma}_k)$ Linear notation for a connected element with a cycle

$(x_0, x_1, \ldots, x_{k-1})$ Oriented cycle

$[\tilde{\Gamma}_1, \tilde{\Gamma}_2, \ldots, \tilde{\Gamma}_k; a]$ Linear notation for a tree with a sink

$[x]_m$ The polynomial $x(x-1)(x-2)\ldots(x-m+1)$

α A transformation

$\alpha(x)$ The value of α at x

$\alpha(x) = \varnothing$ α is not defined at x

$\alpha : M \to M$ α is a transformation of M

$\alpha|_B$ The restriction of α to B

$\alpha^{(K)}$ Restriction of α to the orbit K

α_j^i An element of the form $[u_1^i, u_2^i, \ldots, u_j^i](a)(b)\cdots(c)$

α_k The element $[1, 2, \ldots, k](k+1)\cdots(n)$

β_j^i An element of the form $[u_j^i, u_{j+1}^i, \ldots, u_{m_i}^i](a)(b)\cdots(c)$

η_i The canonical identification of M and $M^{(i)}$

Γ_α The graph of α

ι_S The identity congruence on S

$\lambda \vdash n$ λ is a partition of n

λ_u The mapping $x \mapsto xu$

$\langle A \rangle$ The set of all elements generated by A

$\langle a \rangle$ A cyclic semigroup, generated by a

$\langle A | \Sigma \rangle$ Semigroup generated by A with defining relations Σ

$\mathbb{C}[S]$ The semigroup algebra of S over $\mathbb{C}$

$\mathbb{C}_{\text{triv}}$ The trivial S-module

$\mathbb{N}$ The set of all positive integers

$\mathbb{Z}_y$ The set of all residue classes modulo y

$\mathbf{0}$ The nowhere defined partial transformation

$\mathbf{0}_n$ The nowhere defined partial transformation of $\mathbf{N}$

$\mathbf{GL}_n$ The group of invertible $n \times n$ matrices

$\mathbf{K_N}$ The full directed graph on $\mathbf{N}$

$\mathbf{m}$ A partition of $\mathbf{N}$ into a disjoint union of nonempty subsets

$\mathbf{N}$ The set $\{1, 2, \ldots, n\}$

$\mathbf{N}_n$ The set $\{1, 2, \ldots, n\}$

$\mathbf{Z}(S)$ The center of the semigroup S

$\mathcal{A}_k$ The alternating group

$\mathcal{B}(X)$ The Boolean of X

$\mathcal{C}_{\mathcal{S}_n}(\alpha)$ The centralizer of α in $\mathcal{S}_n$

$\mathcal{D}$ Green's relation $\mathcal{D}$

$\mathcal{D}(a)$ The $\mathcal{D}$-class of a

$\mathcal{D}_k$ The set of all elements of rank k in $\mathcal{T}_n$, $\mathcal{PT}_n$, or $\mathcal{IS}_n$

$\mathcal{E}(S)$ The set of idempotents of S

$\mathcal{E}_k$ The trivial subgroup of $\mathcal{S}_k$

$\mathcal{F}_n$ The set of all order-decreasing total transformations

$\mathcal{H}$ Green's relation $\mathcal{H}$

$\mathcal{H}(a)$ The $\mathcal{H}$-class of a

$\mathcal{IF}_n$ The set of all order-decreasing partial permutations

$\mathcal{IO}_n$ The set of all order-preserving partial permutations

$\mathcal{IO}_n$ The set of all order-preserving partial permutations

$\mathcal{IO}_n^{\prec}$ The set of all $\prec$-order-preserving partial permutations

$\mathcal{IS}_n$ The symmetric inverse semigroup on **N**

$\mathcal{I}_k$ The set of all elements of rank at most k

$\mathcal{I}_k^*$ The dual symmetric inverse semigroup

$\mathcal{I}_\rho$ The unique ideal of the congruence ρ

$\mathcal{J}$ Green's relation $\mathcal{J}$

$\mathcal{J}(a)$ The $\mathcal{J}$-class of a

$\mathcal{L}$ Green's relation $\mathcal{L}$

$\mathcal{L}(a)$ The $\mathcal{L}$-class of a

$\mathcal{M}(e, H, \mathfrak{a})$ The set used to construct transitive actions by partial transformations

$\mathcal{N}_n$ The set of all nilpotent elements of $\mathcal{IS}_n$

$\mathcal{O}_n$ The set of all order-preserving total transformations

$\mathcal{PF}_n$ The set of all order-decreasing partial transformations

$\mathcal{PO}_n$ The set of all order-preserving partial transformations

$\mathcal{PT}(M)$ The set of all partial transformations of M

$\mathcal{PT}_n$ The set of all partial transformations of **N**

$\mathcal{R}$ Green's relation $\mathcal{R}$

$\mathcal{R}(a)$ The $\mathcal{R}$-class of a

$\mathcal{SPO}_n$ Semigroup of strictly partial order-preserving transformations of **N**

$\mathcal{S}_n$ The symmetric group on **N**

$\mathcal{T}(M)$ The set of all total transformations of M

$\mathcal{T}_n$ The set of all total transformations of **N**

$\mathcal{V}_4$ The Klein 4-group as a subgroup of $\mathcal{S}_4$

$\mathfrak{B}$ The bicyclic semigroup

$\mathfrak{h}$ Congruence associated with variants

i_n The canonical inclusion of $\mathcal{PT}_n$ to $\mathcal{T}_{n+1}$

$\mathfrak{O}_n$ The set of all partial orders on **N**

$o_\alpha(x)$ The orbit of x in Γ_α

$\mathfrak{p}(n)$ The partition function

$\mathfrak{r}_n$ The number of Schröder paths of order n

$\mathfrak{t}(\alpha)$ The type of α

$\mathfrak{t}_k(\alpha)$ The number of $x \in \mathbf{N}$ for which $|\{y \in \mathbf{N} : \alpha(y) = x\}| = k$

0_a The constant transformation with the image $\{a\}$

$\mathrm{Ann}_l(A)$ The left annihilator of the set A

$\mathrm{Ann}_l(a)$ The left annihilator of the element a

$\mathrm{Ann}_r(a)$ The right annihilator of the element a

$\mathrm{Ann}_r(A)$ The right annihilator of the set A

$\mathrm{Aut}(S)$ The set of all automorphisms of S

$\mathrm{ct}(\pi)$ The cyclic type of the permutation π

$\mathrm{def}(\alpha)$ The defect of α

$\mathrm{dom}(\alpha)$ The domain of α

$\mathrm{End}(S)$ The set of all endomorphisms of S

$\mathrm{fi}(g)$ The number of fixed points of g

$\mathrm{Hom}_S(V, W)$ The set of all S-homomorphisms from V to W

$\mathrm{idrank}(S)$ Idempotent rank of the semigroup S

$\mathrm{im}(\alpha)$ The image of α

$\mathrm{Inn}(S)$ The set of all inner automorphisms of S

inv_σ The number of inversions for σ

$\mathrm{Ker}(\varphi)$ The kernel of the homomorphism φ

lcm The least common multiple

$\mathrm{Mat}_n(\mathbb{C})$ The algebra of all complex $n \times n$ matrices

$\min(A, \prec)$ The minimal element of A with respect to $\prec$

$\mathrm{nd}(a)$ The nilpotency degree of a

$\mathrm{nd}(S)$ The nilpotency degree of a nilpotent semigroup S

nilrank(S) Nilpotent rank of the semigroup S

Nil(S) The set of all nilpotent subsemigroups of S

Nil$_k(S)$ The set of all nilpotent subsemigroups of S of nilpotency degree k

rank(α) The rank of α

rank(S) Rank of the semigroup S

stim(α) The stable image of α

strank(α) The stable rank of α

St$_G(m)$ stabilizer of the point m with respect to the action of G

triv$_S$ trivial action of a semigroup S

tr$_\alpha(x)$ The trajectory of x in Γ_α

B$_n$ The nth Bell number

C$_n$ The n-th Catalan number

C$_n$ The n-th Catalan number

I$_n$ The cardinality of $\mathcal{IS}_n$

L$_n$ The total number of chains in the chain decompositions of all elements of $\mathcal{IS}_n$

M$_n$ The total number of chains in the chain decompositions of all nilpotent elements of $\mathcal{IS}_n$

n($\mathcal{PT}_n$) The total number of nilpotent elements in $\mathcal{PT}_n$

N$_n$ The total number of nilpotent elements in $\mathcal{IS}_n$

N$_n$ The total number of nilpotent elements in $\mathcal{IS}_n$

O$_G$ The number of G-orbits

P$_n$ The total number of fixed points of all elements of $\mathcal{IS}_n$

S(n,k) Stirling numbers of the second kind

s(n,k) Stirling numbers of the first kind

ω_S The uniform congruence on S

ω_α Binary relation describing orbits of α

$\dotplus$ The addition of residue classes

$\overline{\operatorname{dom}}(\alpha)$ The codomain of α

$\overline{a}$ The equivalence class of a

$\overline{a}_\rho$ The equivalence class of a with respect to the equivalence relation ρ

$\overline{K}_n$ The semigroup, generated by $\alpha_1, \ldots, \alpha_n$ in Section 12.5

$\overline{x}$ The residue class of x

$\overleftarrow{\prec}$ The linear order opposite to $\prec$

$\overleftarrow{S}$ The dual of S

π'_α A special binary relation on $\mathbf{N} \cup \{n+1\}$

π_k The transposition $(1, k)$

π_α A special binary relation on $\mathbf{N}$

π_ρ The canonical projection $S \twoheadrightarrow S/\rho$

$\preceq$ Partial order on the set of all partitions of $\mathbf{N}$

$\rho(S, A)$ Kernel of the natural epimorphism $A^+ \twoheadrightarrow S$

ρ_I The Rees congruence with respect to the ideal I

ρ_α A partition of $\mathbf{N}$, associated to α

ρ_Σ The minimal congruence containing all relations from Σ

Σ_A^B System of relations for B induced from that for A

$\sim_{pS}$ Relation of primary S-conjugation

$\sim_S$ Relation of S-conjugation

$\sqrt{e}$ The set of all x such that $x^m = e$ for some m

τ_T A binary relation on $\mathbf{N}$ associated with a nilpotent subsemigroup T of $\mathcal{PT}_n$

Υ_i exceptional endomorphisms

Δ The equality relation

ε The identity transformation

ε_A The unique idempotent of $\mathcal{IS}_n$ with domain A

ε_n The identity transformation on $\mathbf{N}$

$\varepsilon_{(k)}$ The idempotent $\varepsilon_{\{1,\ldots,k-1,k,\ldots,n\}}$

$\varepsilon_{i,J}$ The product $\varepsilon_{i,j_1}\varepsilon_{i,j_2}\cdot\varepsilon_{i,j_k}$, where $J = \{j_1,\ldots,j_k\}$

$\varepsilon_{m,k}$ The idempotent of rank $(n-1)$ satisfying $\varepsilon_{m,k}(m) = \varepsilon_{m,k}(k) = m$

Λ_a The inner automorphisms corresponding to $a \in S^*$

Ω_φ An exceptional endomorphism

Φ_α The binary relation associated with α

Φ_ϵ An endomorphism of rank one

$\Psi_{\epsilon,\delta}$ An endomorphism of rank two

$\Theta^\tau_{\epsilon,\delta}$ An endomorphism of rank three

$|a|$ The order of the element a

$|S|$ The cardinality of S

Ξ An exceptional endomorphism

$\xi \circ \eta$ Product of binary relations ξ and η

$a\,\rho\,b$ The pair (a,b) belongs to the binary relation ρ

$a \cdot b$ The product of a and b

$a \equiv b$ a is equivalent to b

$a \equiv b(\rho)$ a is equivalent to b with respect to the equivalence relation ρ

$a\mathcal{D}b$ a and b are $\mathcal{D}$-equivalent

$a\mathcal{H}b$ a and b are $\mathcal{H}$-equivalent

$a\mathcal{J}b$ a and b are $\mathcal{J}$-equivalent

$a\mathcal{L}b$ a and b are $\mathcal{L}$-equivalent

$a\mathcal{R}b$ a and b are $\mathcal{R}$-equivalent

$a \sim_{pS} b$ a and b are primarily S-conjugate

$a \sim_S b$ a and b are S-conjugate

A^+ The set of all (finite) words over the alphabet A

a^{-1} The inverse of a in an inverse semigroup

a^{-1} The inverse of a

AB The product of the sets A and B

ab The product of a and b

E_α Arrows of Γ_α

$F(X,Y)$ The set of all mappings from X to Y

$f : X \to Y$ A mapping from X to Y

G/H The set of all left cosets of G modulo H

$g \cdot m$ The element $\varphi(g)(m)$ for the action φ of G on M

G_e The maximal subgroup corresponding to the idempotent e

$H \lhd G$ H is a normal subgroup of G

K_n The semigroup $\overline{K}_n \cup \{\varepsilon\}$

$L(\prec_1, \ldots, \prec_k)$ An $\mathcal{L}$-cross-section of $\mathcal{IS}_n$

n A nonnegative integer

N'_τ The nilpotent subsemigroup of $\mathcal{IS}_n$ associated to the partial order τ on $\mathbf{N}$

N_τ The nilpotent subsemigroup of $\mathcal{PT}_n$ associated to the partial order τ on $\mathbf{N}$

$R(\prec_1, \ldots, \prec_k)$ An $\mathcal{R}$-cross-section of $\mathcal{IS}_n$

S/ρ The quotient of the semigroup S modulo the congruence ρ

S/I The quotient of the semigroup S modulo the Rees congruence ρ_I

$S \cong T$ S is isomorphic to T

S^* The group of units of the semigroup S

S^0 S if $0 \in S$, and $S \cup \{0\}$ otherwise

S^1 S if $1 \in S$, and $S \cup \{1\}$ otherwise

S^a The variant of S with respect to the sandwich element a

S^k $\{a_1 a_2 \cdots a_k : a_i \in S, \quad i = 1, \ldots, k\}$, $k > 1$

$T < S$ T is a subsemigroup of S

$V(M)$ S-module $\mathcal{L}$-induced from M

$V \cong W$ V is isomorphic to W

$V_S(a)$ The set of elements, inverse to $a \in S$

V_α Vertices of Γ_α

$x \bmod y$ The residue of x modulo y

Index